# Current Topics in Microbiology and Immunology

260

# Antibodies in Viral Infection

Edited by Dennis R. Burton

With 51 Figures and 7 Tables

Springer

Professor Dennis R. Burton, Ph.D.
Departments of Immunology and Molecular Biology
The Scripps Research Institute
10550 North Torrey Pines Road
La Jolla, CA 92037, USA
*e-mail*: burton@scripps.edu

*Cover Illustration: Picornavirus neutralization. One model suggests viruses are neutralized when the coating density of neutralizing antibodies bound to the viral surface reaches a critical density as discussed by Burton, Ollmann Saphire and Parren in this issue. The figure depicts a picornavirus particle with 6 antibodies attached; binding of this number of antibody molecules results in neutralization, and for human rhinovirus has been shown to inhibit binding to its target cells.*

ISSN 0070-217X
ISBN 3-540-41611-0 Springer-Verlag Berlin Heidelberg New York



Springer-Verlag Berlin Heidelberg New York
a member of BertelsmannSpringer Science + Business Media GmbH

http://www.springer.de


Library of Congress Catalog Card Number 15-12910
Printed in Germany



Cover Design: *design & production GmbH*, Heidelberg
Typesetting: Scientific Publishing Services (P) Ltd, Madras
Printed on acid-free paper SPIN: 10718388 27/3020M - 5 4 3 2 1 0

# Preface

At the end of the 1960s, the US Surgeon General declared that it was time to close the book on infectious diseases. We now know this was very premature. The AIDS pandemic has ravaged the world and has become a catastrophe of immense proportions in sub-Saharan Africa. Novel viral pathogens such as Ebola virus, Hantavirus, and West Nile virus have begun to attract attention. Nevertheless, there has also been good news. Smallpox has been eradicated and the power of vaccination in protecting against a range of viral pathogens has been established. However, it has also become increasingly clear that the largely empirical strategies employed in the past are unlikely to be successful against all viral pathogens. More understanding of viral pathogenesis and the interaction of viruses with the immune system is required. Antibodies form a critical part of that system.

In this monograph, a number of experts describe different aspects of the interaction of antibodies and viruses at the molecular level. Taken together, these contributions show that considerable advances are being made that will lead us, I believe, to a new era in vaccine design. The first four chapters have a strong structural theme. Tom Smith describes structural studies on the interaction of antibodies and human rhinovirus and lessons for mechanisms of neutralization and viral evolution in the face of an antibody response. Liz Hewat and Dieter Blaas describe structural studies on rhinovirus and foot-and-mouth disease virus in particular and again draw lessons for mechanisms of neutralization. Vidana Epa and Peter Colman look at shape and electrostatic complementarity of antibody and viral antigen and the influence of these parameters on features such as viral escape. Thierry Bizebard, John Skehel, Marcel Knossow, and colleagues present structural studies of antibody interacting with influenza virus in particular and show how these studies help us to understand viral escape. Neil Greenspan provides an overview of the fundamental properties of antibody–antigen recognition that govern the interaction between antibodies and viruses. P.J. Klasse and Quentin Sattentau discuss mechanisms of antibody neutralization with a focus on HIV-1. Paul Parren, Erica Ollmann-Saphire and I provide another overview of antibody neutralization with an emphasis on a model in which antibody coating of virus is crucial for neutralization. Nancy Sullivan discusses the mechanisms by which antibody can enhance viral infection. Walter Gerhard summarizes his unparalleled experience on the role of the antibody response in the prevention and control of influenza virus infection. Similarly, Diane Griffin and colleagues report on their detailed understanding of the novel mechanisms by which antibody controls alphavirus infection of neurons. Many viruses infect via the mucosal surfaces and Jerry McGhee and his colleagues discuss how antibodies can act anti-virally at these surfaces. Antibody can trigger the complement cascade and Greg Spear and colleagues show how this can have important consequences in antibody activity against viruses. Finally, two chapters deal with novel man-made molecules based on antibodies that have anti-viral activity. One is by Wayne Marasco describing intrabodies – antibody molecules expressed intracellularly. The other is by Guy Gorochov and Zelig Eshhar and colleagues describing T bodies, which are antibody variable domains grafted onto T cell receptor signaling molecules.

I should like to express my great appreciation to all the contributors to this volume for their time and effort. Their generosity in sharing their insights is a real service to the virology community. I think they have done a magnificent job.

March 2001, La Jolla DENNIS R. BURTON

# List of Contents

# List of Contributors

(Their addresses can be found at the beginning of their respective chapters.)

# Antibody Interactions with Rhinovirus: Lessons for Mechanisms of Neutralization and the Role of Immunity in Viral Evolution

T.J. SMITH



## 1 Introduction

Over the past two decades, advances in molecular and structural biology have made a tremendous impact on our understanding of many aspects of the viral life cycle. Advances in molecular biology techniques have allowed us to sequence virus genomes, isolate mutations that allow viruses to circumvent the host antibody response and antiviral agents, and to dissect the function of the various gene products. Crystallography has developed to the point where the structures of

Department of Biological Sciences, Purdue University, West Lafayette, IN 47907, USA extremely large viruses and virus complexes have been realized. Finally, cryo-electron microscopy and image reconstruction methods have allowed us to examine viral complexes, assembly/uncoating intermediates, and macro–macromolecular assemblages that are beyond the scope of any other structural methodology. Recently, these fields have not only continued to develop independently, but have also begun to merge. These hybrid technologies have allowed us to examine biological processes that any one field cannot address. This has especially proven to be the case in the studies on human rhinovirus. Using a combination of these disciplines, we have been able to examine the process of antibody-mediated neutralization of human rhinovirus. These results have led us to reevaluate the proposed interplay of antibody response and the evolution of viral capsid structure.

## 2 Background

Picornaviruses are among the largest of the animal virus families and include polio-, rhino-, foot-and-mouth disease, coxsackie, and hepatitis A viruses. The rhinoviruses, of which there are more than 100 serotypes, are major causative agents of the common cold in humans (Rueckert 1996). The virus is non-enveloped and has a ~300Å-diameter protein shell that encapsidates a single-stranded, plus-sense, RNA genome of about 7,200 bases. The human rhinovirus 14 (HRV14) capsid exhibits a pseudo T=3 (P=3) icosahedral symmetry and consists of 60 copies each of four viral proteins, VP1, VP2, VP3, and VP4. VP1–3 each has an eight-stranded anti-parallel β-barrel motif and these proteins comprise most of the capsid structure (Fig. 1). VP4 is smaller, has an extended structure, and lies at the RNA/capsid interface (Rossmann et al. 1985). A ~20Å-deep canyon lies roughly at the junction of VP1 (forming the "north" rim) with VP2 and VP3 (forming the "south" rim), and surrounds each of the 12 icosahedral fivefold vertices. The canyon regions of HRV14 and HRV16, both major receptor group rhinoviruses, were shown to contain the binding site of the cellular receptor intercellular adhesion molecule 1 (ICAM-1) (Colonno et al. 1988; Kolatkar et al. 1999; Olson et al. 1993). Four major neutralizing immunogenic (NIm) sites, NIm-IA, NIm-IB, NIm-II, and NIm-III, were identified by studies of neutralization-escape mutants with monoclonal antibodies (Sherry et al. 1986; Sherry and Rueckert 1985) and then mapped to four protruding regions on the viral surface (Rossmann et al. 1985).

Antibodies are a major defense against picornavirus infection, and several mechanisms of antibody-mediated neutralization have been proposed (Fig. 2). Perhaps the simplest is based on aggregation of virions (Brioen et al. 1985; Thomas et al. 1985, 1986), which generally occurs over a narrow range of antibody:virus ratios. This limited range of effective concentrations has raised questions as to how important of a role aggregation plays in vivo. Large changes in pI have been shown to occur when antibodies bind to virions (Dulbecco et al. 1956; Emini

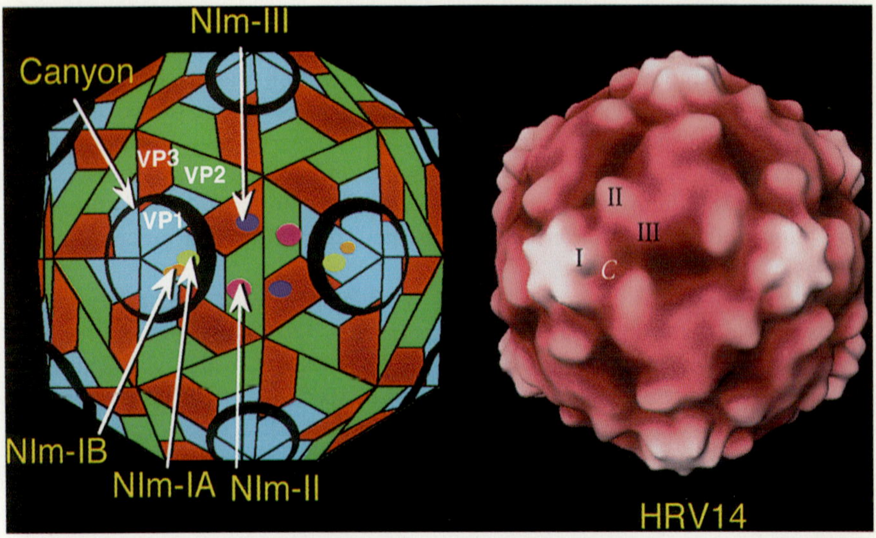


**Fig. 1.** Architecture of human rhinovirus (HRV) 14. A schematic of HRV14 is shown on the *left*, and a surface rendering of the atomic structure is shown on the *right*, with the various NIm sites and canyon region labeled

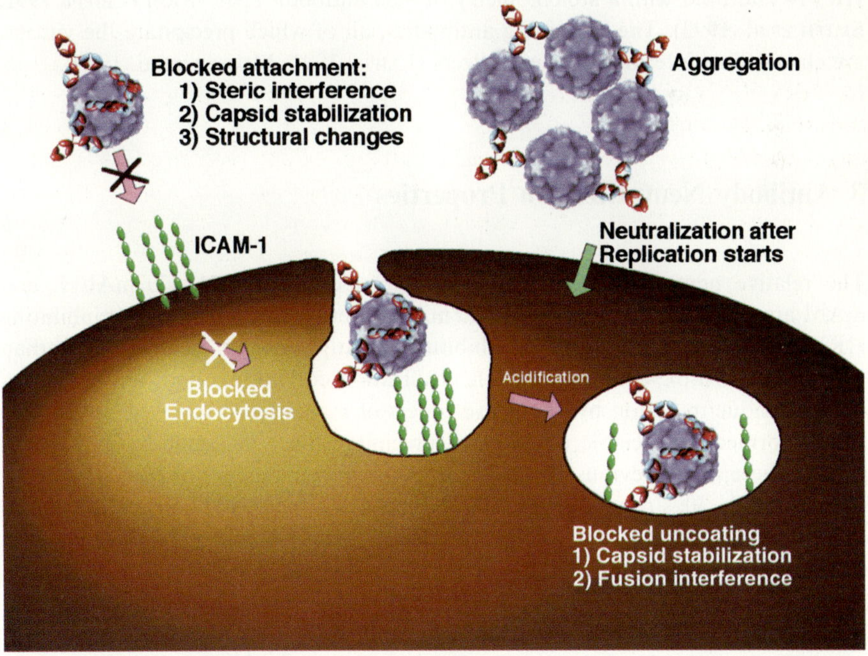


**Fig. 2.** Proposed neutralization mechanisms for picornaviruses. The antibody heavy and light chains are colored *red* and *blue*, respectively, with each immunoglobulin domain being represented by an ellipsoid. ICAM-1, composed of five immunologlobulin domains, is colored *green*. HRV14 is shown as a surface representation and colored *lavender*

et al. 1983b; MANDEL 1976). This led to a popular hypothesis that antibody-mediated neutralization of picornaviruses is due to the induction of large conformational changes upon antibody binding. This contention was later supported by studies that suggested neutralization might occur when only one to two antibodies bind to the virion surface (Wetz et al. 1986), abrogate virus attachment to the host cell (COLONNO et al. 1989; EMINI et al. 1983a), or prevent uncoating (VRIJSEN et al. 1993). There is no universal acceptance of a single neutralization mechanism, and the various monoclonal antibodies (mAbs) may neutralize with different combinations of these mechanisms.

Neutralizing mAbs against HRV14 have been divided into three groups: strong, intermediate, and weak neutralizers (LEIPPE 1991; MOSSER et al. 1989). All strongly neutralizing antibodies bind to the NIm-IA site, which was defined by natural escape mutations at residues D1091 and E1095 of VP1 on the loop between the β-B and β-C strands of the VP1 β-barrel (the letter designates the amino acid, the first digit identifies the viral protein, and the remaining three digits specify the sequence number). Because strongly neutralizing antibodies form stable, monomeric virus/antibody complexes with a maximum stoichiometry of 30 antibodies per virion, it was concluded that they bind bivalently to the virions (LEIPPE 1991; MOSSER et al. 1989).

Weakly neutralizing antibodies form unstable, monomeric complexes with HRV14 and bind with a stoichiometry of ~60 antibodies per virion (LEIPPE 1991; SMITH et al. 1993). The remaining antibodies, all of which precipitate the virions, are classified as intermediate neutralizers (LEIPPE 1991; MOSSER et al. 1989).

## 3 Antibody Neutralization Properties

The relative neutralization and aggregation efficacies of mAb17, mAb12, and mAb1 are shown in Fig. 3. mAb17 and mAb12 were highly efficacious at inhibiting HRV14 infection, with mAb12 exhibiting slightly stronger neutralization than mAb17 (Fig. 3a,b). Under optimal neutralization conditions, both antibodies inhibited plaque initiation by about five orders of magnitude. However, neither antibody formed appreciable amounts of precipitate over the same antibody:virus ratios. These, and previous results demonstrating a maximal antibody:virus stoichiometry of ~30, strongly suggest that both mAb12 and mAb17 bind bivalently to the virion surface and that virion aggregation cannot account for neutralization (MOSSER et al. 1989).

In contrast, mAb1 only weakly inhibited plaque formation, with a maximum inhibition of ~2 orders of magnitude and with a slight enhancement in neutralization at intermediate antibody:virus ratios (Fig. 3c). Unlike mAb12 and mAb17, mAb1 strongly aggregated the virions over the same range of ratios in which neutralization enhancement was observed. At very high antibody:virus ratios, there was little precipitation but yet significant (~1.5log) neutralization.

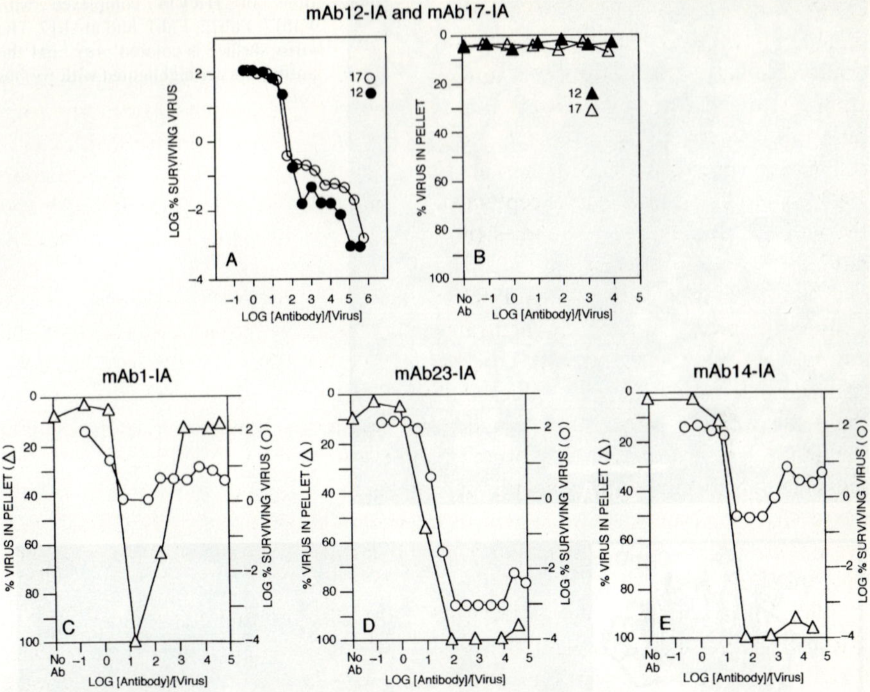


**Fig. 3.** Aggregation and neutralization profiles for several NIm-IA antibodies

Therefore, mAb1 does not neutralize HRV14 solely by precipitating it, but neutralization is enhanced by aggregation at intermediate antibody:virus ratios. Since mAb1 strongly aggregates HRV14, cannot form stable, monomeric virus/antibody species, and binds with a maximal stoichiometry of ~60 antibodies/virion, this antibody binds monovalently to the virion surface.

These results are in contrast with those from mAb23-IA and mAb14-IA (Fig. 3c,d) which also bind to the NIm-IA site. Both of these antibodies precipitate HRV14 over a wider range of antibody concentrations compared to mAb1. However, unlike mAb1, mAb23 is a strongly neutralizing antibody. This implies that bivalent binding is not a prerequisite for efficient neutralization. Instead, it suggests that the intrinsic binding of the Fab arm to HRV14 is higher in mAb23 than in mAb1, and it is the binding strength itself (avidity or affinity) that determines neutralization efficacy. Finally, mAb14 neutralizes slightly better than mAb17 but aggregates over a wider range of antibody:virus ratios than mAb1. In this case, it may be that the binding orientation of mAb14 is better suited for interparticle cross-linking than mAb1.

It is important to note, however, that neutralization efficacy is very dependent upon how the assay is performed. In the experiments discussed above, samples of HRV14 were incubated with antibody overnight and added to monolayers of HeLa cells. After incubation for 1h at room temperature to allow for viral attachment to

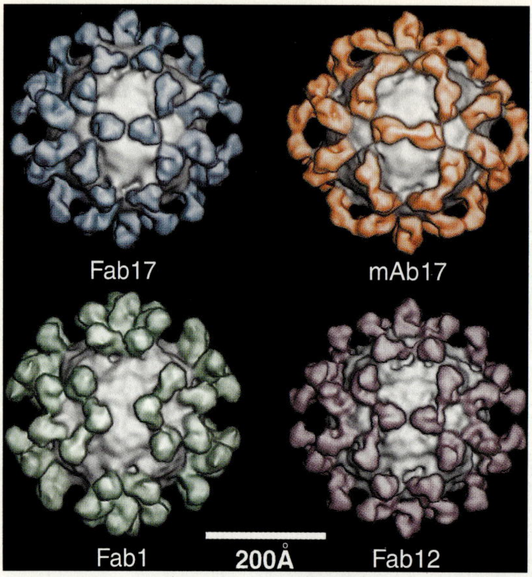


**Fig. 4.** Cryo-transmission electron microscopy (TEM) image reconstructions of HRV14 complexed with Fab17, Fab12, Fab1, and mAb17. The virus surface is colored *gray*, and the antibodies are highlighted with *various colors*

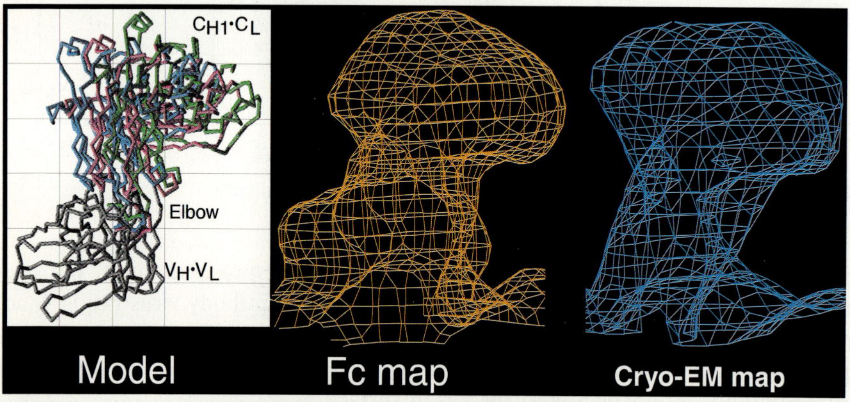


**Fig. 5.** (Caption opposite)

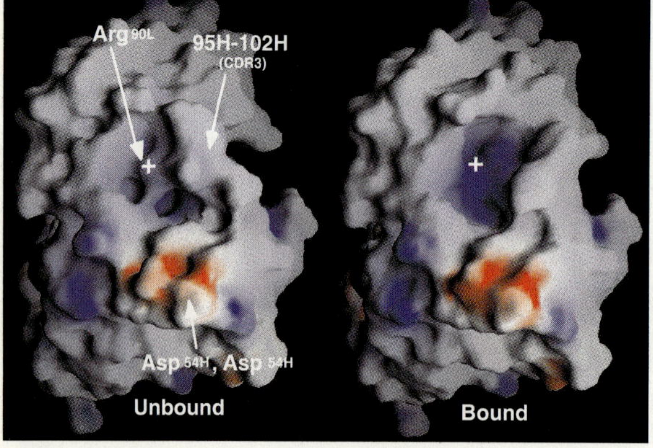


**Fig. 6.** (Caption opposite)

the cells, unbound virus/antibody complexes were washed away. In contrast, when antibody was kept in the plaque assay overlays, all three antibodies neutralized HRV14 infectivity with comparable efficacy. This difference is due to the fact that, in the latter assays, the antibody was around to inhibit secondary infections which are needed for plaque development. This demonstrates that one must be careful when assessing the efficacy of an antibody based solely on in vitro assays. While the latter assay may more accurately represent in vivo conditions, in which antibodies are not removed after the first round of viral attachment, the assay shown in Fig. 3 is more sensitive to differences in antibody affinities.

## 4 Cryo-Transmission Electron Microscopy Structures of Antibody-Virus Complexes

We have examined the structures of several antibodies bound to HRV14. For these experiments, we chose to examine antibodies that react to the NIm-IA site. While antibodies that bind to the NIm-IA site exhibit both strongly and weakly neutralizing properties, the other three neutralizing sites only produce weakly neutralizing, aggregating antibodies. It was thought that comparison of strongly vs weakly neutralizing antibodies would allow us to directly compare these potentially different neutralizing mechanisms without having to compensate for differing binding regions.

Using a combination of molecular biology, cryo-transmission electron microscopy (cryo-TEM), and crystallographic studies, we hoped to find structural evidence for some of these mechanisms of neutralization. For example, using these methods we expected to observe bivalent binding of an antibody to HRV14 and to see large antibody-induced conformational changes in the virion. However, we were also aware that we might not be able to observe structural changes in the virion if the antibodies neutralized by other mechanisms, such as capsid stabilization.

**Fig. 5.** Simulated motion about the elbow axis in the bound Fab17 cryo-TEM structure. The heavy- and light-chain variable domains are represented by *gray* C-α backbone models, and three different positions for the constant domains are shown in *green*, *mauve*, and *cyan*. Using an occupancy of 0.33 for each of the constant domain models, an electron density was calculated (*orange cage*). This calculated electron density compares well with the actual cryo-TEM map (*blue cage*). This suggests that domain motions on this scale may be undetectable at the resolution of these cryo-TEM studies (20–25Å)

**Fig. 6.** Movement in the paratope that accommodates the epitope. On the *left* is the Van der Waal surface of the Fab17 paratope in the unliganded structure. The negative and positive electrostatic fields are colored *red* and *blue*, respectively. On the *right*, is the paratope surface as observed in the Fab17/HRV14 crystal structure. The majority of the structural changes are due to motion in the heavy-chain CDR3 loop. This conformational change results in the exposure of the highly basic cleft between the heavy and light chains

We have now examined the structures of three different Fab/HRV14 (Fab17, Fab12, and Fab1) complexes and of one mAb/HRV14 (mAb17) complex (Fig. 4). mAb17 and mAb12 are both strongly neutralizing antibodies, whereas mAb1 is a weakly neutralizing antibody. What was immediately apparent was that these different antibodies had different binding orientations. Fab17 and Fab12 both bound to the NIm-IA site at a somewhat tangential orientation that placed the constant domains ($C_{H1} \cdot C_L$) of twofold-related Fabs in close proximity to each other. This suggested that the intrinsic paratope/epitope interactions of these antibodies places the Fab arms in an orientation that facilitates bidentate binding to twofold-related NIm-IA sites. In contrast, Fab1 binds almost vertically to the virion surface with a "twist" that made it seem unlikely that these antibodies could bind bivalently.

These studies also demonstrated that the Fab fragments did not induce gross conformational changes upon binding to the virion. However, it was possible that conformational changes are only induced by mAb binding – perhaps by having the Fab arms "pry-up" portions of the capsid. To this end, we used cryo-TEM to determine the structure of the mAb17/HRV14 complex. In this complex, there was a clear connection between the Fab arms. However, the Fc region was only observed as a weak island of density above the connection. This is more than likely due to the fact that the hinge region is very flexible, causing the Fc region to be averaged away during the reconstruction procedure. These results supported the contention that strongly neutralizing antibodies bind bivalently, while weakly antibodies bind monovalently. The fact that antibodies to the other NIm sites appeared to only bind monovalently suggests that the distance and orientation of the other icosahedrally related antigenic sites make it difficult for antibodies to bind with both Fab arms. This is in contrast to some antibodies to human rhinovirus 2 and calicivirus that have been suggested to undergo a great deal of contortion about the hinge region to bind bivalently (Hewat and Blaas 1996; Thouvenin et al. 1997). Nevertheless, there was no evidence of changes in the capsid structure in this mAb17/HRV14 complex. Therefore, it seemed unlikely that bivalently bound antibodies are more efficacious due to induced structural changes in the capsid.

This result, therefore, suggests that strongly neutralizing, poorly aggregating antibodies are more efficacious due to the fact that bivalent binding increases the apparent affinity of the antibodies for the virion. This effect of valency on apparent affinity (avidity) has been eloquently modeled and tested using hapten-labeled phage particles (Hornick and Karush 1972). One way to explain this phenomenon is to note that, when one antibody arm binds to an antigenic surface, the location of the second arm is confined to a small area of space immediately adjacent to the antigenic surface. This increases the apparent antigenic concentration of antigen for the second antibody arm. Alternatively, one can consider that, once the two arms are bound to the surface, both must release simultaneously in order for the antibody to disassociate. As was shown in the case of the hapten/phage conjugate, this process increases the apparent affinity $10^2$- to $10^3$-fold, going from a monovalent Fab to a bivalent IgG, and by $\sim 10^5$-fold when going to a decavalent IgM. In reality, these apparent increases can be demonstrably smaller due to limitations in

Fab arm flexibility. This is especially true in the case of an IgM in which the arms have far less hinge flexibility than found in an IgG. It is important to note that increased avidity can come from intra-particle cross-linking or inter-particle cross-linking at the equivalence zone of the immuno-precipitation curve.

Finally, it seems likely that bivalent binding is not essential for neutralization since Fabs also neutralize the virus, although less effectively than the intact mAbs. If bivalent binding was required to induce conformational changes, and these changes are essential to neutralization, then one would expect Fabs to be incapable of neutralizing the virus. Instead, the Fab activity was diminished in accordance with the loss in avidity (Colonno et al. 1989). Therefore, it seemed likely that, if antibodies neutralize via induction of conformational changes, then these changes must occur with Fabs. Such conformational changes were not observed with any of the studied Fabs or mAbs used for these cryo-TEM studies.

## 5 Atomic Models of the Fab/Virus Complexes

While the cryo-TEM studies are only at relatively low resolution (20–25Å), we could use the previously determined structure of HRV14 (Rossmann et al. 1985) and our structures of Fab17 (Liu et al. 1994) and Fab1 (Che et al. 1998) to glean further details from these cryo-TEM structures. Modeling and site-directed mutagenesis studies demonstrated that electrostatic interactions play a key role in the binding of Fab17 to HRV14 (Smith et al. 1993). In the Fab17/HRV14 complex, the loop of the NIm-IA site on HRV14 sits clamped in the cleft between the heavy and light chain hypervariable regions and forms complementary electrostatic interactions with, Lys-58$^{H}$ (on heavy chain) and Arg-91$^{L}$ (on light chain) of Fab17. In addition, a cluster of lysines on HRV14 (K1236, K1097, K1085) interact with two acidic residues, Asp-45$^{H}$ and Asp-54$^{H}$, in the CDR2 region (CDR: complementarity determining region) of the Fab heavy chain (Smith et al. 1996). These modeling results were tested using site-directed mutagenesis. Indeed, it was found that even though K1236, K1097, and K1085 were not identified as sites of naturally occurring escape mutations, they do affect antibody binding (Table 1). This is likely due to

**Table 1.** Residual infectivity (% of residual plaques) of HRV14 wildtype and mutant viruses after treatment with NIm-IA antibodies

| Antibody | LP1(WT) | K1085E | K1236E | K1097Q | K1097E |
|---|---|---|---|---|---|
| mAb1 | 0.0021 | 0.032 | 0.0088 | 0.0054 | 0.27 |
| mAb3 | 0.0064 | 0.060 | 0.014 | 0.011 | 83 |
| mAb4 | 0.0031 | 0.045 | 0.0082 | 0.0059 | 0.44 |
| mAb6 | 0.0045 | 0.050 | 0.014 | 0.012 | 71 |
| mAb7 | 0.0064 | 0.063 | 0.014 | 0.015 | 106 |
| mAb14 | 0.0 | 0.0 | 0.0 | 0.0 | 0.19 |
| mAb17 | 0.0190 | 0.052 | 0.016 | 3.1 | 87 |
| mAb20 | 0.0000 | 0.003 | 0.0013 | 0.00066 | 0.018 |

the fact that these mutations yielded poorly replicating viruses (Che et al. 1998). Therefore, these results demonstrate two major points: Firstly, electrostatic interactions can dominate paratope/epitope interactions. Secondly, the naturally occurring escape mutations are clearly only a small subset of residues crucial for antibody binding. Indeed, mutating K1097, which is only halfway down the canyon, is very deleterious to the virus. Therefore, one cannot use the location of epitopes as unequivocal evidence that the convolutions in the virion surface hide crucial residues.

The 3D reconstruction of the Fab1/HRV14 complex clearly differs from the other two virus-Fab reconstructions (Fig. 4). The Fab arms of the weakly neutralizing, strongly aggregating mAb1 bind in a more radial orientation on the capsid, and they are rotated ~25° about their long axes compared to Fab12 and Fab17. In this orientation, the constant domains of the symmetry-related, bound Fab1 fragments point away from each other. Unlike mAb17 (Smith et al. 1993a), a model for bivalently bound mAb1 can only be generated if the structure is forced to assume unacceptable conformations about the elbow axis region. The large separation of neighboring Fab1 molecules and the more radial orientation of the constant region in Fab1 presumably favors monovalent binding. Though the hinge region of antibodies is known to be highly flexible (Thouvenin et al. 1997; Wade et al. 1989), the distance between icosahedral, twofold-related NIm-IA sites is sufficiently large so the relatively small difference between Fab17 and Fab1 binding orientations produces a profound effect on binding valency. Notably, an antibody/FMDV virus complex has a similar distance constraint on bivalent binding (Hewat et al. 1997).

It is interesting to note that, while Fab17 and Fab1 have disparate binding orientations, they share some key residues that maintain electrostatic interactions. The cluster of basic groups in the cleft between Fab17 heavy and light chains that interact with D1091 and E1095 are also found in Fab1. In addition, the D55 and D57 in Fab17 that form important interactions with a cluster of basic groups on the HRV14 surface (e.g., K1097 and R1094) are also found in the CDR2 loop of Fab1. The important difference is that the two homologous Asp residues in Fab1 are three residues upstream from where they are located in Fab17. This places these acidic groups in a counter clockwise position on the paratope relative to Fab17. What we may be observing is that the charged groups on the paratope are dictating the binding orientation of the Fab. This can be best explained by the fact that the coulombic interactions between the paratope and epitope will be felt at a much longer distance than any other chemical interactions. In this way, the docking of the Fab is directed and optimized by the charged residues.

## 6 Crystal Structure of the Fab17/HRV14 Complex

The above studies consistently demonstrated that antibodies do not induce conformational changes in the virion upon binding. This has been also shown in the reconstructions of virus/Fab complexes of cowpea mosaic virus (PORTA et al. 1994; WANG et al. 1992), canine parvovirus (WIKOFF et al. 1994), FMDV (HEWAT et al. 1997), Sindbis virus (SMITH et al. 1995), Ross River virus (SMITH et al. 1995), and HRV-2 (HEWAT and BLAAS 1996). However all of these cryo-TEM studies are limited to 20–30Å resolution, where relatively small conformational changes may not observable. To address this problem, we crystallized and determined the crystal structure of the Fab17/HRV14 complex. Again, as noted above, previous results clearly demonstrate that, if conformational changes are associated with neutralization, they should be observed in this structure.

Using the atomic models for Fab17, HRV14, and the pseudo-atomic model of the complex, we determined the structure of the Fab17/HRV14 complex to 4Å resolution. The most notable property of the bound Fab17 in this structure was that the entire constant domain ($C_{H1} \cdot C_L$) was disordered starting at the elbow region. This domain was, however, observable in the cryo-TEM studies. One interpretation of this is that the hinge region is so flexible that, even at the low resolution of the cryo-TEM studies, the electron density of the Fc region is averaged away. Accordingly, the elbow region is less flexible than the hinge region and the constant domains are observable at cryo-TEM resolutions. However, at 4Å resolution, this motion causes the electron density for the constant regions to be uninterpretable. To test for this, the constant domains were placed in several different positions according to the known flexibility of the elbow region. Each of these models was given partial occupancy and a 20Å electron density map was calculated (Fig. 5). Indeed, the resulting map very closely correlated to the observed cryo-TEM electron density. This demonstrates that there is a great deal of motion about the elbow region, but this motion is apparently less than that at the hinge region.

The second major observation was that the pseudo-atomic model created using the cryo-TEM electron density was a fairly accurate representation of the actual structure. When the Fab17 model was placed into the cryo-TEM density, the limited resolution was insufficient to justify adjustments to the model to improve the paratope/epitope interactions. Therefore, in spite of a number of crystallographic studies that demonstrated marked conformational changes (especially in the heavy chain CDR3 loop), we only performed a rigid-body alignment of the Fab17 model into the cryo-TEM envelope. This was clearly the root of most of the error in the cryo-TEM fitting process. Indeed, the CDR3 of Fab17 moves by ~5Å to better accommodate the epitope (Fig. 6), which places the Fab closer to the surface and further down into the canyon than previously modeled. Interestingly, Fab17 binds poorly to peptides representing the NIm-IA loop. Our recent Fab17/peptide crystallographic studies suggest that this may be, in part, due to the fact that the peptide is insufficient to induce conformational changes in the paratope cleft that are necessary to accommodate the βB-βC loop (J. Wu and T. Smith, unpublished results).

This structure also clearly showed that there were no significant changes in the virion due to Fab binding. The only observable changes were localized in a few side chains on the β-βC loop on which the naturally occurring escape mutations lie.

D1091 and E1095 rotate slightly to form salt bridges with the basic residues in the paratope cleft, and K1097 and R1094 rotate to better interact with D55 and D57. There were not, however, any significant changes in the NIm-IA loop itself. If antibodies were capable of inducing conformational changes, it seems likely that such changes would start in the antigenic loop itself. This result is not entirely surprising since none of the crystallographic structures of antigen/Fab complexes determined to date have demonstrated large antibody-induced conformational changes. This structure demonstrates that antibodies do not deform intact virus capsids.

The resolution of this Fab17/HRV14 structure permits us to more quantitatively analyze the antibody/virus interactions. Firstly, it is quite apparent that, in contradiction to the "canyon hypothesis", Fab17 penetrates into the canyon region. It is able to do this by binding somewhat on its side with the $V_H$ domain, making extensive interactions with the north and south walls of the canyon. The hypervariable residues contact the entire north wall, the bottom of the canyon, and part of the lower south wall (Table 2). The framework residues contact the upper south wall only. The vast majority of these interactions are via either charged or polar residues. This clearly demonstrates that antibodies are much more facile at recognizing crevasses in viral surfaces than originally thought.

Interestingly, there is a great deal of contact between the framework region of Fab17 and the upper, south wall of the canyon (Fig. 7). This kind of extensive framework interaction has never been previously observed. However, one might argue that these interactions are merely a consequence of the fact that these antibodies bind bivalently, and such interactions do not significantly contribute to the overall binding energies. The results of sequencing Fab12 suggest that these contacts may indeed contribute to antibody binding. Fab12 was chosen as a second example of a neutralizing antibody, since it appeared to be a better neutralizer than Fab17, and the Fab12 appeared to have different physiochemical properties. However, when Fab12 was sequenced, there were relatively few differences in the nucleic acids sequences between it and Fab17. From the small number of basic changes, and because these two antibodies came from the same mouse, it seemed

**Table 2.** Fab17-HRV14 contacts

| Residue type | Area ($Å^2$) | Total (%) |
|---|---|---|
| Acidic (D, E) | 96 | 11 |
| Base (H, K, R) | 142 | 17 |
| Polar (N, Q, S, T) | 187 | 22 |
| Small (A, G) | 61 | 7 |
| Hydrophobic (C, I, L, M, P, V) | 23 | 3 |
| Aromatic polar (W, Y) | 336 | 38 |
| Total | 845 | |
| North wall contacts (CDR) | 532 | 63 |
| South wall contacts | 313 | 37 |
| Hypervariable (CDR) contacts | 103 | 12 |
| Framework contacts | 207 | 24 |

likely that they came from the same progenitor cell and their sequences now differed due to somatic hypermutation. This infers that the sequence differences between these antibodies led to selectable changes in binding affinity during clonal expansion. While some changes in framework residues have been suggested to indirectly affect antibody binding, it seems rather fortuitous that three amino acid differences cluster about the framework/south-wall contact region. These potential interactions are currently under examination using an ScFv version of Fab17.

The area of the HRV14 surface, to which Fab17 binds, directly overlaps the contact surface of the receptor, ICAM-1 (Fig. 8). From the cryo-TEM structure of the ICAM-1/HRV14 complex, it has been suggested that ICAM-1 interacts with the south canyon wall. This area of contact is immediately adjacent to the NIm-II site, approximately two thirds of which is encompassed by the Fab17 contact area. Therefore, it appears that the canyon does not, in fact, adequately hide the ICAM-1 binding surface from antibody recognition. This might lead to the conclusion that Fab17 neutralizes by directly interfering with ICAM-1 binding. However, it has been shown that antibodies to all four antigenic sites can abrogate cell attachment (Colonno et al. 1989) even though some sites (e.g., NIm-III) are quite distal to the canyon region. The simplest explanation for this is that the antibodies are large relative to the virion itself. Since the length of an antibody (~150Å) is approximately equal to the radius of the particles, it only takes a few antibodies bound to the surface to potentially interfere with how the virus interacts with the host cell membrane. In addition, the aggregating antibodies would further exacerbate binding by clumping the virions together.

## 7 Mechanism of Antibody-Mediated Neutralization of HRV14

Using the results from these structural studies, we can better evaluate the previously proposed neutralization mechanisms: aggregation, virion stabilization, induction of conformational changes, and abrogation of cellular attachment.

### 7.1 Aggregation

It has been suggested that aggregation and neutralization occur concomitantly and that virus:antibody ratios in vivo favor aggregation (Brioen et al. 1983, 1985; Thomas et al. 1985). However, our studies strongly suggest that aggregation is not a major factor in neutralization of HRV14. Firstly, antibodies that bind bivalently to virions do not induce aggregation over antibody:virus ratios that range several orders of magnitude, yet such antibodies are strong neutralizers (Che et al. 1998). Secondly, antibodies that are strong aggregators neutralize virus even at antibody:virus concentrations that do not favor aggregation. Neutralization for aggregating antibodies is often optimal at ratios at which immunoprecipitation

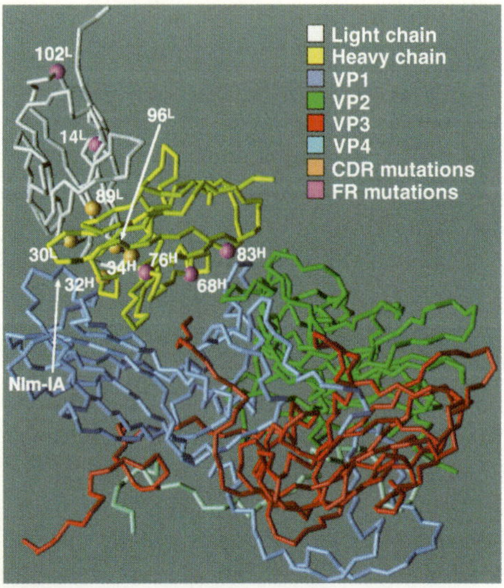


**Fig. 7.** Crystal structure of the Fab17/HRV14 complex. The $V_H$, $V_L$, VP1, VP2, VP3, and VP4 C-α backbones are represented by *yellow*, *white*, *blue*, *green*, *red*, and *cyan*, respectively. Comparing the Fab12 and Fab17 sequences, the framework differences are represented by *mauve balls*, and the CDR mutations are represented by *yellow balls*. The view here is parallel to the canyon, with the nearest fivefold axis running vertically along the *left side* of the diagram and the RNA interior lies toward the *bottom*. Note the penetration of the Fab into the canyon and the extensive contact along the south wall (the *right side* of the central depression)

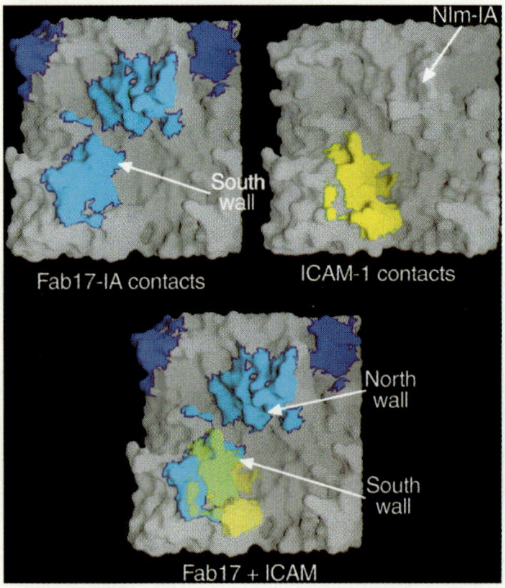


**Fig. 8.** Contact areas of Fab17 and ICAM-1 on the HRV14 surface. The central depression is the canyon and the fivefold axis it at the *top, center*. The Fab17 contact area is highlighted in *cyan*, the ICAM-1 area in *yellow*, and the overlapping area in *green*. The *dark blue* surface represents icosahedrally related, Fab17 contact areas

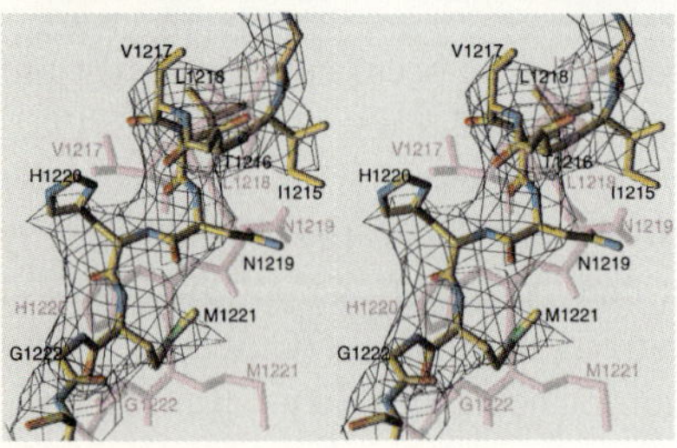


**Fig. 9.** (Caption opposite)

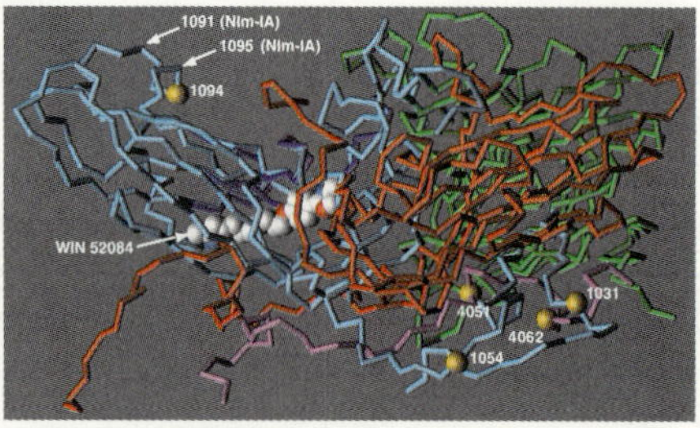


**Fig. 10.** (Caption opposite)

is greatest. In this circumstance, neutralization may be enhanced within a narrow range of antibody:virus ratios that favors precipitation. This enhancement may result from a decrease of independent infectious particles or from avidity effects (apparent increase in affinity due to multivalent binding of antibodies) caused by antibodies bound bivalently to neighboring particles in the large immunocomplexes. Though aggregation probably does not play a significant role in vitro, it may facilitate innate immunological responses such as opsonization in vivo.

## 7.2 Virion Stabilization

Antibodies have also been thought to neutralize virions by stabilizing the capsid, a process that might then prevent uncoating (MOSSER et al. 1989). All NIm-IA antibodies to HRV14, including both aggregating and non-aggregating ones, stabilize virions to varying extents against low pH (CHE et al. 1998). None of the non-NIm-IA antibodies we tested caused such stabilization. However, some of the NIm-IA antibodies are weakly neutralizing whereas some of the antibodies that bind to other antigenic sites are quite efficacious neutralizers. These stabilization effects thus do not correlate well with neutralization efficacy or binding valency. Furthermore, antibodies to all four NIm sites prevent cellular attachment (COLONNO et al. 1989), thereby blocking infectivity prior to possible stabilization effects. Notably, no escape mutation has yet been observed that prevents neutralization without affecting antibody binding. If capsid stabilization/destabilization was a major determinant of neutralization, it seems likely that some escape mutations would to arise that could abrogate these effects. As evidence for the capacity of these viruses to overcome stabilizing ligands, analogous distal-site drug resistance mutants do occur (HEINZ et al. 1989; MOSSER and RUECKERT 1993; MOSSER et al. 1994).

## 7.3 Induction of Conformational Changes

Antibodies and Fab fragments cause an apparent decrease in the pI of the viral capsid concomitant to neutralization (COLONNO et al. 1989; MANDEL 1976). This

◄

**Fig. 9.** Stereo image of the electron density of the VP1 strand immediately above the drug pocket that moves upon binding WIN drugs. The *black cage* represents the electron density of this region in the Fab17/HRV14 crystal structure. The transparent *mauve model* is the native HRV14 structure and the *stick model* (colored by atom type) is taken from the WIN/HRV14 structure. There is a clear density beneath this conformational change in the same position as the WIN compound (not shown) and is likely due to a PEG contaminant

**Fig. 10.** Location of trypsin digestion sites in HRV14, as determined by mass spectroscopy analysis. VP1–4 C-α backbones are colored *blue*, *green*, *red*, and *cyan*, respectively. The digestion sites are represented by *yellow balls*, and the WIN compound is represented by the CPK model immediately under the canyon floor. The native, unliganded structure around the bound drug is represented by *dark blue* strands fact has been cited as evidence that antibodies neutralize by distorting the capsid structure. The crystal structure of the Fab17/HRV14 complex clearly demonstrated that efficacious neutralization does not require large conformational changes in the capsid (Smith et al. 1996). Instead, large conformational changes in the paratope region of Fab17 permit the Fab to better accommodate the epitope without inducing structural changes in the virion (Smith et al. 1996). All antibodies to the four different antigenic sites that were tested (mAbs 13, 17, 21, 28, 29, 33, 34, 35) did cause apparent changes in the pI of the capsid (Colonno et al. 1989), but it seems unlikely that such dissimilar antibodies, which bind to disparate epitopes, would all cause the same conformational change in the capsid. Antibodies might cause conformational changes in protein structure upon binding, but such changes would more likely be expected to occur on flexible portions of the viral structure. Indeed, binding affinity is proportional to the sum of all chemical interactions at the paratope/epitope interface. Subtracted from this sum would be the energy required to induce conformational changes upon binding. Therefore, antibody-induced conformational changes on less flexible regions would require a greater cost in binding energy, thereby potentially diminishing antibody affinity. For these reasons, we believe that induced conformational changes play little or no significant role in the mechanism of antibody neutralization.

## 7.4 Abrogation of Cellular Attachment

Antibodies to all four HRV14 antigenic sites clearly block cellular attachment (Colonno et al. 1989). The three NIm-IA antibodies we have studied clearly bind in a manner that overlaps the binding site footprint of ICAM-1 as determined by cryo-TEM (Olson et al. 1993). Competition between receptor and Nim-IA antibody binding might simply be explained as a steric hindrance effect. NIm-II is immediately adjacent to the ICAM binding region, and the $\sim$600–900Å$^2$ contact region of NIm-II antibodies probably overlaps the ICAM site or at least hinders attachment to ICAM. However, NIm-III is quite distal ($\sim$40Å away) to the receptor binding region, yet NIm-III antibodies also compete with receptor binding (Colonno et al. 1989). Hence, antibody competition with receptor may merely be a consequence of the bulkiness of an antibody molecule and does not require direct overlap with the ICAM binding region.

## 7.5 Significance of In Vitro Neutralization Mechanisms In Vivo

Taken together, results of many studies with HRV14 suggest that neutralization in vitro may occur by a much simpler mechanism than previously envisioned: the presence of antibodies bound to the surface of HRV14 is sufficient to block attachment of virus receptors. For poliovirus and rhinovirus, interactions with their receptors appear to be essential for the proper release of the genomic RNA into the cytoplasm of the host cell. When antibody/poliovirus complexes enter cells, the viral

RNA is quickly digested (Mandel 1967). In contrast, FMDV/antibody complexes are infectious if allowed to enter cells via artificially expressed Fc receptors (Mason et al. 1993). Therefore, these two viruses have different roles for their receptor/virus interactions; rhinovirus and poliovirus require the receptor to initiate uncoating, whereas FMDV just needs its receptor to deliver the virus to cytoplasm. In both cases, antibodies can block normal receptor entry, and therefore infection, quite effectively. Therefore, a simple steric model (antibody binding blocks receptor binding) is sufficient to explain the neutralization behavior of many antibodies to these viruses. Binding of some antibodies might cause secondary effects to viral capsids (Delaet and Boeye 1993; Dulbecco et al. 1956; Li et al. 1994; Wien et al. 1995), but these effects seem to be dispensable for neutralization, and, perhaps more importantly, antibodies that induce such effects would not be exclusively selected for during B cell clonal expansion. This is because B cell proliferation is simply driven by the strength of the interaction between the antigen and surface immunoglobulins and is independent of potential secondary effects of antibody binding.

It is possible, however, that some antibodies might induce conformational changes, but they may be the exception rather than the rule. For example, both receptor and antibody binding in polioviruses occurs at the top of the canyon region (Harber et al. 1995; Wien et al. 1997). Here, it is possible that some antibodies might bind to this region and mimic receptor binding by causing conformational changes in the viral capsid, or antibodies could block changes in this region. Also, antibodies to the influenza virus hemagglutinin (HA) can prevent attachment and replication, but antibodies to the neuraminidase (NA) only interfere with virus release (Kilbourne et al. 1968).

It is important to note, however, that these in vitro mechanisms we have described may not represent the primary mode by which antibodies protect animals from viral infections. For example, antibodies against Sindbis virus (Schmaljohn et al. 1982) and FMDV (McCullough et al. 1992) that are not efficacious in vitro still protect animals from viral challenge. Also, non-neutralizing antibodies against the NA of influenza do affect disease progression in vivo (Schulman 1975). Therefore, the primary role of antibodies in vivo may be to act synergistically with other components of the immune system. Vaccine design strategies might therefore benefit by focusing on the production of high affinity antibodies rather than on a particular in vitro neutralization property. This has been shown to be true for HIV, where antibody neutralization efficacy is directly proportional to binding affinity and does not depend upon which epitope is recognized (Parren et al. 1998).

## 8 Canyon Hypothesis

When the high-resolution structure of influenza virus N9 neuraminidase was solved, it was noted that the conserved residues involved in sialic acid binding were located in a crevasse. The authors suggested that such architecture might have several functional implications (Colman et al. 1983). Analogous to most enzymes, a cavity or pocket-like structural feature may have evolved to facilitate contact with receptor. It was further noted that a concave morphological feature like that which occurs on the NA spike would also offer some protection against binding of host antibodies to functionally important residues. However, even if some of the conserved residues were accessible to antibodies, the quaternary structure of the NA spike required antibody contacts with some of the variable loops at the top of the sialic acid binding site. Influenza virus could therefore escape antibody binding without any need to alter crucial portions of its NA spike.

The X-ray structure of HRV14 exhibited some of the same crevasse-like features as NA (Rossmann et al. 1985). In HRV14, a 20-Å deep and 20-Å wide "canyon" encircles each of the icosahedral fivefold axes (Fig. 1). These dimensions, and the knowledge that the sites of escape mutations map to the canyon rim, led to the hypothesis that the canyon region is involved in receptor binding (Rossmann 1985, 1989). It was also suggested that HRV evolved this quaternary structure as a means to protect conserved residues necessary for receptor recognition (Rossmann et al. 1985). However, the X-ray structure of the HRV14/Fab complex demonstrated that antibodies do bind into deep recesses of virion surfaces and do contact conserved residues (Smith et al. 1996).

These results make it important to fully reevaluate the canyon hypothesis and its validity in describing the evolution of viral capsid structure. In the original hypothesis, it was predicted that the receptor binding regions of all viruses that replicate in mammals would be hidden in deep convolutions of the capsid surface. This is a reasonable and necessary thesis if antibody response is a major factor in viral evolution. Again, the two major points of this thesis are: (1) all viruses will have receptor binding regions in convoluted portions of the capsid, and (2) conserved residues necessary for receptor binding will be hidden from antibodies. It has been clearly demonstrated in several viruses that these contentions are not correct.

## 8.1 Foot-and-Mouth Disease Virus

The Arg-Gly-Asp (RGD) sequence in FMDV, which probably interacts with the integrin receptor, is highly exposed on the βG-βH loop in VP1 (Acharya et al. 1989). The residues flanking the RGD motif residues vary with serotypes. This led these investigators to suggest that FMDV can use "camouflage" to hide crucial residues "in plain sight". According to this hypothesis, crucial residues might be exposed to antibodies but simply do not change in response to antibodies, since doing so would be lethal to the virus. On the other hand, residues adjacent to the RGD sequence can change and thwart antibody binding. The exposed nature of these residues has been further demonstrated by structural studies showing that these conserved residues make extensive interactions with a neutralizing antibody (Verdaguer et al. 1995). Furthermore, the functional overlap between antibody and receptor binding is made evident by the fact that adaptation of FMDV to different growth conditions often leads to changes in antigenicity (Diez et al. 1990; Rieder et al. 1994). Perhaps even more importantly, these results demonstrate that antigenic drift can be independent of host immunity.

It could be argued that a hidden, secondary receptor that binds to a different region might compensate for this immunologically exposed RGD region. This possibility was recently eliminated by resolution of the structure of a FMDV complexed with a heparin sulfate (Fry et al. 1999). Heparin sulfate has been suggested to be involved in a two-step attachment process in which low affinity interactions with heparin sulfate at one site is followed by high affinity binding to an integrin receptor via the RGD sequence (Jackson et al. 1996). Like the RGD sequence, this oligosaccharide binding site is not only exposed but is also part of one of the antigenic sites (Fry et al. 1999). Therefore, it is very clear that FMDV is exactly contrary to what was predicted by the canyon hypothesis.

## 8.2 Alphaviruses

Anti-idiotypic antibodies to a neutralizing antibody were found to compete with virus for cell receptors (Wang et al. 1991). This suggests that, not only is the crucial receptor binding region on the virus exposed to antibodies, but it can also be mimicked by anti-idiotypic antibodies. In our previous cryo-TEM studies, this region was shown to lie on the outermost tip of the trimeric viral spikes (Smith et al. 1995). In these studies, a similar virus, Ross River virus, was also examined. Here an antibody was used that binds to an epitope immediately adjacent to residues that change upon adaptation to different hosts. Therefore, it seems likely that these antibodies also recognize a portion of the spikes involved in receptor recognition. Indeed, while these antibodies bound in a different orientation than that observed in the Fab/Sindbis complex, the contact area was identical and highly exposed.

## 8.3 Influenzavirus

As described above, the sialic binding site in the NA spike of influenza lies in a depression on its spike. While the residues within this deep depression are conserved, the residues about the rim vary with serotype. This suggests that conserved residues are hidden from antibody recognition. However, subsequent studies demonstrated that approximately one third of the conserved binding region in this depression is contacted by a neutralizing antibody (Bizebard et al. 1995). To explain how viruses might evade antibody attack while leaving conserved residues immunologically exposed, Colman (1997) has proposed that this capability may reflect the potential for different proteins to recognize identical protein surfaces. In this way, receptors and antibodies can bind to overlapping areas of the viral surface, but can exhibit differing sensitivities to mutations at these contact surfaces.

## 8.4 Poliovirus

Similar to human rhinovirus, poliovirus has a deep canyon about the icosahedral fivefold axes, and escape mutations are located on top of the canyon. However, in contradiction to the canyon hypothesis, there is a direct link between the receptor binding site and the serotypic determinants in poliovirus. It has been shown that upper, exposed regions of the poliovirus canyon are crucial for receptor interactions, but residues at the bottom of the poliovirus canyon are not. In fact, changes at the top of the canyon that affect antibody-neutralizing sites also alter receptor-virus interactions (Harber et al. 1995). Yielding similar conclusions, other studies showed that mutations at the north and south walls of the canyon overcome deleterious defects in the poliovirus receptor. These mutations, which are quite distal to the canyon floor, lie very close to the antigenic sites and appear to represent destabilizing mutations (Wien et al. 1997). These results are further evidence that the canyon plays a pivotal role in capsid structural homeostasis, and conservation of canyon residues is probably due to viability issues rather than accessibility to antibodies. Indeed, it has been noted that, if the canyon exists solely to assist in hiding the receptor-binding site, then one would expect to see far more than the three poliovirus serotypes (Harber et al. 1995). It was also noted that some RNA viruses have been shown to diverge antigenically in the absence of the immune system, therefore questioning the idea that antibodies are the only selective pressure driving antigenic diversification. Finally, the cryo-TEM structure of the poliovirus/receptor complex (Belnap et al. 2000b; Yongning et al. 2000) clearly shows that some of the mutation sites that affect serotype-specific binding to mutant receptors lie within the receptor contact area (Harber et al. 1995).

## 8.5 The Validity of the Canyon Hypothesis

To summarize, the following results allow us to assess the validity of the canyon hypothesis: (1) Key, conserved residues are not always found in deep crevasses on viral surfaces. (2) Antibodies have been shown to recognize conserved regions in all of these viruses that are involved in receptor attachment. (3) There is not only contact, but also functional overlap between antibody and receptor binding in these viruses. (4) Antigenic drift can occur in the absence of antibodies. (5) Naturally occurring mutations are only a subset of residues crucial for antibody binding and do not necessarily denote the full antibody contact area.

We are left with the observation that, while the canyon is indeed the receptor-binding region, it is not clear whether the canyon hypothesis correctly assigns the selective pressure that created it. As with many evolutionary arguments, this is difficult to either prove or disprove, but the current data seem to shed doubt on the canyon hypothesis. Firstly, it was not predictive as to the location of receptor binding regions in viruses that developed in animals with immune systems. This hypothesis also failed to predict the fact that antibodies could, in fact, bind to critical, conserved residues. While one could argue about the importance of a particular residue bound by the receptor and not the antibody, the functional overlap elsewhere makes this discussion moot. Secondly, for the canyon to have evolved to hide important residues implies that such is the most important function of the canyon. This is indeed not the case with poliovirus and rhinovirus, where receptor-induced uncoating is absolutely essential for infection and is therefore most likely to be the driving force behind capsid evolution. Indeed, we would argue that the fact that poliovirus and rhinovirus require receptor/canyon interactions whereas FMDV does not (Mason et al. 1993) may be due to differences in capsid stability (Oliveira et al. 1999) rather than differences in how host immunity influenced viral evolution.

A more basic question is whether these viruses truly evade the immune system. For poliovirus, there are only three serotypes, and no additional serotypes have emerged during world-wide vaccination efforts. Indeed, many viruses do not propagate by evading the infected host's immune system, but do so by spreading to an immunologically naive animal. Therefore, the necessity for immune evasion and the genetic mechanisms to do so are not as developed as that observed, for example, in trypanosomes. Here, the parasitic life cycle requires continual immune evasion. To facilitate a persistent infection, trypanosomes utilize a thousand different genes, each of which encodes a different surface glycoprotein (Watson et al. 1983). These genes, or their promoters, are constantly shuttled to different genome sites. This process promotes continual diversification and thereby thwarts the host's immune system. Therefore, the genetic drift in viruses that may lead to antigenic diversity due to selective pressures is not the same as the true immune evasion processes observed in other parasites.

## 9 The Role of the Canyon

As discussed above, one way to test the canyon hypothesis is to ascertain the main function of the canyon. Recent results on the mechanism of action of an antiviral compound have shed some light on the role of the canyon.

A hydrophobic cavity lies directly beneath the canyon floor into which some hydrophobic compounds bind (Smith et al. 1986). WIN compounds induce relatively small changes in the canyon floor upon binding within the hydrophobic pocket. The largest changes observed occur when these hydrophobic compounds create an opening to the binding pocket by displacing M1221 (Badger et al. 1988, 1989; Smith et al. 1986). However, the conformational changes are limited to residues immediately surrounding the drug, and thus it is difficult to speculate as to a possible structural mechanism of stabilization other than to invoke the notion of a global thermodynamic effect.

Subsequently, WIN compounds were shown to affect the receptor-virus interactions of the major but not the minor serotypes of HRV (Pevear et al. 1989). On this basis, conformational changes in the canyon floor were thought to be responsible for abrogation of cell attachment and also used as evidence to ascribe a role for these canyon-floor residues in the binding of ICAM-1 to the major serotypes of rhinoviruses (Oliveira et al. 1993). While this hypothesis is attractive because of its conceptual simplicity, it is also possible that drug-induced abrogation of receptor binding could be due to allosteric effects on the canyon region. For example, binding of substrates to an enzyme active site can often be affected by ligands bound to distant sites (e.g., Peterson and Smith 1999). In this case, we would propose that the WIN compounds affect receptor binding by altering the ability of the canyon to undergo the conformational changes that occur subsequent to receptor binding rather than causing steric interference of ICAM-1 attachment.

This latter contention is supported by a number of studies demonstrating that mutations in and near the canyon region affect the uncoating process and that drug resistance can arise via mutations distal to the drug-binding region. While some of the mutations that confer resistance in HRV14 make direct contact with the bound drug, many of the other residues point away from the bound WIN compounds and most lie on strands that remain relatively unchanged by drug binding. Therefore, these latter mutations may facilitate the canyon-associated conformational changes in the presence of drug rather than simply affecting receptor or drug binding.

Drug-dependent mutants of HRV16 were isolated when the virus was grown in the presence of WIN compounds. Interestingly, these mutations (V1210A, M1103T, T1208A) occurred in residues somewhat distal to the bound drug and located midway up the south canyon wall. These mutations destabilize virions, and hence, virions can only assemble in the presence of drug. This is an important finding in that it demonstrates that crucial residues are not only confined to the very bottom of the canyon and that the canyon region is very sensitive to mutations. This shows that this region, which is in contact with Fab17 in the case of HRV14, is functionally crucial and unlikely to be the site of naturally occurring escape mutations.

Drug-resistant mutations are located in three regions of poliovirus: the drug-binding pocket itself, the north wall of the canyon (~17Å away from the bound drug), and the at the viral capsid inner surface. A number of these latter mutations are clustered around the VP4/capsid interface. Again, these results suggest that the neutralizing effects of the drug are not immediately localized to the bottom of the canyon.

The presence of electron density in the WIN drug-binding cavity of a number of picornaviruses has led to an hypothesis that this region and the bound "pocket factors" may play an important role in the viral life cycle (Oliveira et al. 1993). It was proposed that host cells produce various pocket factors that stabilize virions in the extracellular milieu, and these factors compete with receptor for binding to the virion. It was further proposed that WIN compounds and receptor bind better than the pocket factor. However, there are some data inconsistent with this model. Firstly, the $t_{1/2}$ of WIN compounds disassociating from HRV14 is on the order of 30–90min (Fox et al. 1991). If pocket factors have weaker binding constants (smaller $t_{1/2}$ of disassociation) (Oliveira et al. 1993), it is unclear how these compounds can remain bound to virions during lengthy purification procedures.

Secondly, the structure of the Fab17/HRV14 complex (Smith et al. 1996) showed clear evidence for a bound pocket factor (Fig. 9) even though the structure of HRV14 alone did not (Rossmann et al. 1985). The only difference in the preparation of virus samples in these two studies was the addition of Fab fragments and PEG 400 as a cryo-solvent to freeze the crystals of the complex. This suggests that the previously observed pocket-factors may not have been host-derived but rather could be attributed to low molecular weight solutes in crystallization liquors or from additives used in the purification protocol. While these results suggest that the observed factors may not be from the cell, it is still possible that pocket factors exist in vivo. However, results from naturally occurring, drug-resistant mutants suggest that, when the pocket is altered in a manner which blocks binding of WIN compounds, the infectivity of these mutant virions are not demonstrably different than that of native virions. Also, since purified HRV14 lacks a pocket factor yet is still infectious, it seems likely that these compounds are dispensable in the viral life cycle. In addition, receptor binding to some major group rhinoviruses is extremely insensitive to WIN compounds (e.g., HRV89; Pevear et al. 1992).

We would propose that the drug-binding cavity has evolved to maintain flexibility of this region. Empty cavities keep proteins at a higher energy state, where it will take less energy to cause structural changes (Matthews 1993). Indeed, a direct relationship between protein stability and function has been demonstrated (Shoichet et al. 1995) in T4 lysozyme. If this cavity itself plays a role in the viral life cycle, then it would explain why other picornaviruses (e.g., FMDV and mengovirus) have highly homologous structures in this region but do not have bound pocket factors.

Recent mass spectroscopic analysis of HRV14 trypsin digestion products have provided evidence that the drug binding cavity does indeed play a role in capsid dynamics (Lewis et al. 1998). As expected, trypsin cleaves HRV14 adjacent to the highly exposed NIm-IA site (Fig. 10). However, trypsin also cleaves the VP1 and VP4 N-termini that lie buried at the capsid/RNA interface. Use of immobilized trypsin gives similar results, hence cleavage of buried residues cannot be attributed to penetration of the capsid by trypsin. One might imagine cleavage occurring if viral proteins transiently "lift" and expose the rest of the capsid, but it is unclear how this could happen without exposing the ssRNA genome to RNAse digestion (RNAse is ubiquitous and it is present in high concentrations during purification protocols). Alternatively, the N-termini of VP1 and VP4 transiently and dynamically extrude from the capsid. Such a hypothesis was first proposed for poliovirus, in which it was found that antibodies raised against peptides representing VP1 and VP4 N-termini only bound to poliovirus when the antibodies and virions were added together at 37°C (Li et al. 1994). In addition, the N-terminus of poliovirus VP1 was shown to extrude from the capsid and interact with liposomes (Fricks and Hogle 1990). Evidence for peptide extrusion was subsequently obtained for trypsin digestion experiments with flockhouse virus, which gave a mass spectroscopic analysis similar to that for HRV14 (Bothner et al. 1998).

A particularly noteworthy result of the trypsin digestion experiments with HRV14 is that WIN compounds completely blocked cleavage at all of the sites

(Fig. 5). HRV14 alone was cleaved by trypsin within 5min, whereas the presence of WIN compounds delayed cleavage for more than 18h. Though this result was not completely unexpected, because WIN compounds are known to stabilize HRV capsids, it is difficult to reconcile how trypsin cannot access an exposed NIm-IA site that is more than 30Å away from the pocket. An interesting corollary is that only NIm-IA antibodies stabilize HRV14 against low pH inactivation. This growing body of results clearly indicates that the canyon is highly dynamic and can easily be stabilized by small hydrophobic compounds and antibodies. Regardless of whether N-termini cleavage accompanies extrusion or transient disassociation, WIN binding in the pocket at the base of the canyon clearly abrogates these conformational changes.

We would therefore argue that all of these results are elucidating the real role of the canyon region in rhinovirus. The virus is binding to receptor in a two-step process (SMITH and BAKER 1999). The first step is a weak association with the south wall of the canyon. Subsequent to this, in a temperature-dependent manner, some conformational process occurs that increases the affinity of the virus for the receptor (HOLLAND 1962; HOLLAND and HOYER 1962). Indeed, the "breathing" observed in poliovirus (LI et al. 1994) and rhinovirus (LEWIS et al. ) is probably the first stage of viral uncoating which is promoted and extended by interactions with the receptor. In a "snake-like" manner, these viruses may be constantly probing the environment by extending their myristylated N-termini. When the receptors are bound, the virion undergoes a major conformational change that accentuates this probing, allowing the termini to find their membrane target. It has been suggested that this expulsion occurs about the fivefold axes (KOLATKAR et al. 1999) or at the quasi threefold axes between VP1, VP2, and VP3 (BELNAP et al. 2000a). While only future structural studies can discern whether either model is correct, it is interesting to note that the location of the digested termini are closer to the icosahedral threefold and twofold axes than the fivefold axes (LEWIS et al. 1998).

## 10 Conclusions

It appears that the mechanism of antibody-mediated neutralization may be simpler than previously envisioned. While antibodies might induce conformational changes, it appears that one can produce strongly neutralizing antibodies that do not cause structural changes. Instead, we have found that increasing the valency of antibody binding can increase the neutralization efficacy, and this is most likely due to an increase in antibody avidity. At least for this system, it appears that the main effect of antibody in vitro is the abrogation of cellular attachment. In vivo, it is likely that components of innate immunity play a synergistic role in neutralization. If this is true, then the development of vaccines is greatly simplified in that one needs only to elicit antibodies that can recognize the native virions and not be concerned with antibody-mediated induction of secondary effects.

The previously proposed role of the canyon region is as follows. The dimensions of the canyon are supposed to prohibit penetration of the antibodies. Therefore, the canyon apparently evolved to protect essential residues. These essential residues are on the floor of the canyon under which a hydrophobic pocket lies that contains factors that, via steric effects, control the binding of receptor to the canyon.

We would argue instead that the canyon has evolved to interact with receptor and to facilitate subsequent conformational changes that are necessary for injection of viral genome into the host cell. Indeed, our results clearly demonstrate that the canyon does not offer protection from immune recognition. We would further suggest that the drug-binding region is an empty cavity that affords the region the necessary conformational flexibility to respond to receptor binding. We conclude that since the canyon has such a pivotal role in the viral life cycle, and it does not hide crucial residues from antibodies, that it is not the evolutionary consequence of host immunity evasion. Therefore, we propose that the major evolutionary force in all viruses is the recognition of a receptor and subsequent conformational processes that facilitate genome release.

## References


Acharya R, Fry E, Stuart E, Fox G, Rowlands E, Brown F (1989) The three-dimensional structure of foot-and-mouth disease virus at 2.9Å resolution. Nature 327:709–716

Badger J, Minor I, Kremer MJ, Oliveira MA, Smith TJ, Griffith JP, Guerin MA, Krishnaswamy S, Luo M, Rossmann MG, McKinlay MA, Dlana GD, Kutko FJ, Fancher M, Rueckert RR, Heinz BA (1988) Structural analysis of a series of antiviral agents complexed with human rhinovirus 14. Proc Natl Acad Sci USA 85:3304–3308

Badger J, Minor I, Oliveira MA, Smith TJ, Rossmann MG (1989) Structural analysis of antiviral agents that interact with the capsid of human rhinoviruses. Proteins 6:1–19

Belnap DM, Filman DJ, Trus BL, Cheng N, Booy FP, Conway JF, Curry S, Hiremath CN, Tsang SK, Steven AC, Hogle JM (2000a) Molecular tectonic model of virus structural transitions: the putative cell entry states of poliovirus. J Virol 74:1342–1354

Belnap DM, McDermott BM, Filman DJ, Cheng N, Trus BL, Zuccola HJ, Racaniello VR, Hogle JM, Steven AC (2000b) Three-dimensional structure of poliovirus receptor bound to poliovirus. Proc Natl Acad Sci USA 97:73–78

Bizebard T, Gigant B, Rigolet P, Rasmussen B, Diat O, Bosecke P, Wharton SA, Skehel JJ, Knossow M (1995) Structure of influenza virus haemagglutinin complexed with a neutralizing antibody. Nature 376:92–94

Bothner B, Dong XF, Bibbs L, Johnson JE, Siuzdak G (1998) Evidence of viral capsid dynamics using limited proteolysis and mass spectrometry. J Biol Chem 273:673–676

Brioen P, Dekegel D, Boeyé A (1983) Neutralization of poliovirus by antibody-mediated polymerization. Virology 127:463–468

Brioen P, Thomas AAM, Boeyé A (1985) Lack of quantitative correlation between the neutralization of poliovirus and the antibody-mediated pI shift of the virions. J Gen Virol 66:609–613

Che Z, Olson NH, Leippe D, Lee W-M, Mosser A, Rueckert RR, Baker TS, Smith TJ (1998) Antibody-mediated neutralization of human rhinovirus 14 explored by means of cryo-electron microscopy and X-ray crystallography of virus-Fab complexes. J Virol 72:4610–4622

Colman PM (1997) Virus versus antibody. Structure 5:591–593

Colman PM, Varghese JN, Laver WG (1983) Structure of the catalytic and antigenic sites in influenza virus neuraminidase. Nature 303:41–44

Colonno RJ, Callahan PL, Leippe DM, Rueckert RR (1989) Inhibition of rhinovirus attachment by neutralizing monoclonal antibodies and their Fab fragments. J Virol 63:36–42

Colonno RJ, Condra JH, et al. (1988) Evidence for the direct involvement of the rhinovirus canyon in receptor binding. Proc Natl Acad Sci USA 85:5449–5453

Delaet I, Boeye A (1993) Monoclonal antibodies that disrupt poliovirus only at fever temperatures. J Virol 67:5299–5302

Diez J, Davila M, Escarmis C, Mateu MG, Dominguez J, Perez JJ, Giralt E, Melero JA, Domingo E (1990) Unique amino acid substitutions in the capsid proteins of foot-and-mouth disease virus from a persistent infection in cell culture. J Virol 64:5519–28

Dulbecco R, Vogt M, Strickland AGR (1956) A study of the basic aspects of neutralization of two animal viruses, western equine encephalitis and poliomyelitis virus. Virology 2:162–205

Emini EA, Kao S, Lewis AJ, Crainic R, Wimmer E (1983a) Functional basis of poliovirus neutralization determined with monospecific neutralizing antibodies. J Virol 46:466–474

Emini EA, Ostapchuk P, Wimmer E (1983b) Bivalent attachment of antibody onto poliovirus leads to conformational alteration and neutralization. J Virol 48:547–550

Fox MP, McKinlay MA, Diana GD, Dutko FJ (1991) Binding affinities of structurally related human rhinovirus capsid-binding compounds are related to their activities against human rhinovirus type14. Antimicrob Agents Chemother 35:1040–1047

Fricks CE, Hogle JM (1990) Cell-induced conformational change in poliovirus; externalization of the amino terminus of VP1 is responsible for liposome binding. J Virol 64:1934–1945

Fry EE, Lea SM, Jackson T, Newman JWI, Ellard FM, Blakemore WE, Abu-Ghazaleh R, Samuel A, King AMQ, Stuart DI (1999) The structure and function of a foot-and-mouth disease virus–oligosaccharide receptor complex. EMBO J 18:543–554

Harber J, Bernhardt G, Lu HH, Sgro JY, Wimmer E (1995) Canyon rim residues, including antigenic determinants, modulate serotype-specific binding of polioviruses to mutants of the poliovirus receptor. Virology 214:559–570

Heinz BA, Rueckert RR, Shepard DA, Dutko FJ, McKinlay MA, Francher M, Rossmann MG, Badger J, Smith TJ (1989) Genetic and molecular analysis of spontaneous mutants of human rhinovirus 14 resistant to an antiviral compound. J Virol 63:2476–2485

Hewat EA, Blaas D (1996) Structure of a neutralizing antibody bound bivalently to human rhinovirus 2. EMBO J 15:1515–1523

Hewat EA, Verdaguer N, Fita I, Blakemore W, Brookes S, King A, Newman J, Domingo E, Mateau MG, Stuart DI (1997) Structure of the complex of an Fab fragment of a neutralizing antibody with foot-and-mouth disease virus: positioning of a highly mobile antigenic loop. EMBO J 16:1492–1500

Holland JJ (1962) Irreversible eclipse of poliovirus by Hela cells. Virology 16:163–176

Holland JJ, Hoyer BH (1962) Early stages of enterovirus infection. Cold Spring Harbor Symp Quant Biol 27:101–112

Hornick CL, Karush F (1972) Antibody affinity-III. The role of multivalence. Immunochemistry 9:325–340

Jackson T, Ellard FM, Abu Ghazaleh R, Brookes SM, Blakemore WE, Corteyn AH, Stuart DI, Newman JWI, King AMQ (1996) Efficient infection of cells in culture by type O foot-and-mouth disease virus requires binding to cell surface heparan sulfate. J Virol 70:5282–5287

Kilbourne ED, Laver WG, Shulman JL, Webster RG (1968) Antiviral activity of antiserum specific for an influenza virus neuraminidase. J Virol 2:281–288

Kolatkar PR, Bella J, Olson NH, Bator CM, Baker TS, Rossmann MG (1999) Structural studies of two rhinovirus serotypes complexed with fragments of their cellular receptor. EMBO J 18:6249–6259

Leippe DM (1991) Stoichiometry of picornavirus neutralization by murine monoclonal antibodies. University of Wisconsin, Ph.D. thesis

Lewis JK, Bothner B, Smith TJ, Siuzdak G (1998) Antiviral agent blocks breathing of the common cold virus. Proc Natl Acad Sci USA 95:6774–6778

Lewis JK, Bothner B, Smith TJ, Siuzdak G (1998) Antiviral agent blocks breathing of the common cold virus. Proc Natl Acad Sci USA 95:6774–6778

Li Q, Yafal AG, Lee YMH, Hogle J, Chow M (1994) Poliovirus neutralization by antibodies to internal epitopes of VP4 and VP1 results from reversible exposure of the sequences at physiological temperatures. J Virol 68:3965–3970

Liu H, Smith TJ, Lee WM, Mosser A, Rueckert RR, Olson NH, Cheng RH, Baker TS (1994) Structure determination of an fab fragment that neutralizes human rhinovirus 14 and analysis of the fab-virus complex. J Mol Biol 240:127–137

Mandel B (1967) The interaction of neutralized poliovirus with HeLa cells. II. Elution, penetration, uncoating. Virology 31:247–259

Mandel B (1976) Neutralization of poliovirus: A hypothesis to explain the mechanism and the one-hit character of the neutralization reaction. Virology 69:500–510

Mason PW, Baxt B, Brown F, Harber J, Murdin A, Wimmer E (1993) Antibody-complexed foot-and-mouth disease virus, but not poliovirus, can infect normally insusceptible cells via the Fc receptor. Virology 192:568–577

Matthews BW (1993) Structural and genetic analysis of protein folding and stability. Curr Opin Struct Biol 3:589–593

McCullough KC, De Simone F, Brocchi E, Capucci L, Crowther JR, Kihm U (1992) Protective immune response against foot-and-mouth disease. J Virol 66:1835–1840

Mosser AG, Leippe DM, Rueckert RR (1989) Neutralization of picornaviruses: Support for the pentamer bridging hypothesis. In: Semler BL, Ehrenfeld E (eds) Molecular aspects of picornavirus infection and detection. ASM, Washington, DC, pp 155–167

Mosser AG, Rueckert RR (1993) WIN 51711-dependent mutants of poliovirus type 3: evidence that virions decay after release from cells unless drug is present. J Virol 67:1246–1254

Mosser AG, Sgro JY, Rueckert RR (1994) Distribution of drug resistance mutations in type 3 poliovirus identifies three regions involved in uncoating functions. J. Virol 68:8193–8201

Oliveira AC, Ishimaru D, Goncalves RB, Smith TJ, Mason P, Sá-Carvalho D, Silva JL (1999) Low temperature and pressure stability of picornaviruses: implications for virus uncoating. Biophys J 76:1270–1279

Oliveira MA, Zhao R, Lee W-M, Kremer MJ, Minor I, Rueckert RR, Diana GD, Pevear DC, Dutko FJ, McKinlay MA, Rossmann MG (1993) The structure of human rhinovirus 16. Structure 1:51–68

Olson NH, Kolatkar PR, Oliveira MA, Cheng RH, Greve JM, McClelland A, Baker TS, Rossmann MG (1993) Structure of a human rhinovirus complexed with its receptor molecule. Proc Natl Acad Sci USA 90:507–511

Parren PW, Mondor I, Naniche D, Ditzel HJ, Klasse PJ, Burton DR, Sattentau QJ (1998) Neutralization of human immunodeficiency virus type 1 by antibody to gp120 is determined primarily by occupancy of sites on the virion irrespective of epitope specificity. J Virol 72:3512–3519

Peterson PE, Smith TJ (1999) The structure of bovine glutamate dehydrogenase provides insights into the mechanism of allostery. Structure 7:769–782

Pevear DC, Diana GD, McKinlay MA (1992) Molecular targets for antiviral therapy of the common cold. Sem Virol 2:41–55

Pevear DC, Fancher FJ, Feloc PJ, Rossmann MG, Miller MS, Diana G, Treasurywala AM, McKinlay MA, Dutko FJ (1989) Conformational change in the floor of the human rhinovirus canyon blocks adsorption to HeLa cell receptors. J Virol 63:2002–2007

Porta C, Cheng RH, Chen Z, Baker TS, Johnson JE (1994) Direct imaging of interactions between an icosahedral virus and conjugate Fab fragments by cryoelectron microscopy and X-ray crystallography. Virology 204:777–788

Rieder E, Baxt B, Mason PW (1994) Animal-derived antigenic variants of foot-and-mouth disease virus type A12 have low affinity for cells in culture. J Virol 68:5296–5299

Rossmann MG (1989) The canyon hypothesis. J Biol Chem 264:14587–14590

Rossmann MG, Arnold E, Erickson JW, Frankenberger EA, Griffith JP, Hecht HJ, Johnson JE, Kamer G, Luo M, Mosser AG, Rueckert RR, Sherry B, Vriend G (1985) Structure of a human common cold virus and functional relationship to other picornaviruses. Nature 317:145–153

Rueckert RR (1996) Picornaviridae and their replication. In: Fields BN, Knipe DM (eds) Fundamental virology. Raven, New York, pp 609–654

Schmaljohn AL, Johnson ED, Dalrymple JM, Cole GA (1982) Nonneutralizing monoclonal antibodies can prevent lethal alphavirus encephalitis. Nature 297:70–72

Schulman JL (1975) Immunology of influenza. In: Kilbourne ED (ed) The influenza viruses and influenza. Academic, New York, pp 373–393

Sherry B, Mosser AG, Colonno RJ, Rueckert RR (1986) Use of monoclonal antibodies to identify four neutralization immunogens on a common cold picornavirus, human rhinovirus 14. J Virol 57:246–257

Sherry B, Rueckert RR (1985) Evidence for at least two dominant neutralization antigens on human rhinovirus 14. J Virol 53:137–143

Shoichet BK, Baase WA, Kuroki R, Matthews BW (1995) A relationship between protein stability and protein function. Proc Natl Acad Sci USA 92:452–456

Smith TJ, Baker TS (1999) Picornaviruses: epitopes, canyons, pockets. Adv Virus Res 52:1–23

Smith TJ, Chase ES, Schmidt TJ, Olson NH, Baker TS (1996) Neutralizing antibody to human rhinovirus 14 penetrates the receptor-binding canyon. Nature 383:350–354

Smith TJ, Cheng RH, Olson NH, Peterson P, Chase E, Kuhn RJ, Baker TS (1995) Putative receptor binding sites on alphaviruses as visualized by cryoelectron microscopy. Proc Natl Acad Sci USA 92:10648–10652

Smith TJ, Kremer MJ, Luo M, Vriend G, Arnold E, Kamer G, Rossmann MG, McKinlay MA, Diana GD, Otto MJ (1986) The site of attachment in human rhinovirus 14 for antiviral agents that inhibit uncoating. Science 233:1286–1293

Smith TJ, Olson NH, Cheng RH, Liu H, Chase E, Lee WM, Leippe DM, Mosser AG, Ruekert RR, Baker TS (1993) Structure of human rhinovirus complexed with Fab fragments from a neutralizing antibody. J Virol 67:1148–1158

Thomas AA, Vrijsen R, Boeye A (1986) Relationship between poliovirus neutralization and aggregation. J Virol 59:479–485

Thomas AAM, Brioen P, Boeyé A (1985) A monoclonal antibody that neutralizes poliovirus by cross-linking virions. J Virol 54:7–13

Thouvenin E, Laurent S, Madelaine MF, Rasschaert D, Vautherot JF, Hewat EA (1997) Bivalent binding of a neutralising antibody to a calicivirus involves the torsional flexibility of the antibody hinge. J Mol Biol 270:238–246

Verdaguer N, Mateu MG, Reu D, Giralt E, Domingo E, Fita I (1995) Structure of the major antigenic loop of foot-and-mouth disease virus complexed with a neutralizing antibody: direct involvement of the Arg-Gly-Asp motif in the interaction. EMBO J 14:1690–1696

Vrijsen R, Mosser A, Boeye A (1993) Post-absorption neutralization of poliovirus. J Virol 67:3126–3133

Wade RH, Taveau JC, Lamy JN (1989) Concerning the axial rotational flexiblity of the Fab regions of immunoglobulin G. J Mol Biol 206:349–356

Wang K-S, Schmaljohn AL, Kuhn RJ, Strauss JH (1991) Antiidiotypic antibodies as probes for the Sindbis virus receptor. Virology 181:694–702

Wang P, Porta C, Chen Z, Baker TS, Johnson JE (1992) Identification of a Fab interaction site (footprint) on an icosahedral virus by cryo-electron microsocpy and X-ray crystallography. Nature 355:275–278

Watson JD, Tooze J, Kurtz DT (1983) Recombinant DNA: a short course. Scientific American Books, New York, pp 147–148

Wien MW, Curry S, Filman DJ, Hogle JM (1997) Structural studies of poliovirus mutants that overcome receptor defects. Nat Struct Biol 4:666–674

Wien MW, Filman DJ, Stura EA, Guillot S, Delpeyroux F, Crainic R, Hogle JM (1995) Structure of the complex between the Fab fragment of a neutralizing antibody for type 1 poliovirus and its viral epitope. Nat Struct Biol 2:232–243

Wikoff WR, Wang G, Parrish CR, Cheng RH, Strassheim ML, Baker TS, Rossmann MG (1994) The structure of a neutralized virus: canine parvovirus complexed with neutralizing antibody fragment. Structure 2:595–607

Yongning H, Bowman VD, Mueller S, Bator CM, Bella J, Peng X, Baker TS, Wimmer E, Kuhn RJ, Rossmann MG (2000) Interaction of the poliovirus receptor with poliovirus. Proc Natl Acad Sci USA 97:79–84

# Structural Studies on Antibody Interacting with Viruses

E. Hewat[1] and D. Blaas[2]



## 1 Introduction

One of the responses of the immune system to an invading virus is the production of a large number of antibody clones that bind to well-defined areas on the viral capsid. These areas, epitopes, are usually formed by spike proteins or loops protruding from the surface of the viral capsid. Immunogenicity correlates with the hydrophilic character of the amino acid sequence (Hopp and Woods 1981), its flexibility (Karplus and Schulz 1985), and other parameters influencing the establishment of a tight contact between the viral epitope and the antigen-binding site

[1] Institut de Biologie Structurale J-P Ebel, 41 Avenue rue Jules Horowitz, 38027, Grenoble, Cedex 1, France
[2] Institute of Medical Biochemistry, Division Biochemistry, University of Vienna (VBC), Dr. Bohr Gasse 9/3, 1030 Vienna, Austria

of the antibody, the paratope. The polyclonal nature of the antiserum prohibits a detailed analysis of the mechanisms of viral neutralization. Monoclonal antibodies (mAbs), however, are ideally suited for the analysis of the topology of attachment, and the identification and description of the interactions between virus and antibody. The mechanisms of viral neutralization can be summarized as: (1) aggregation (leading to a reduction of free viral particles), (2) inhibition of receptor binding by steric hindrance, (3) physical disruption of the virus particle, and (4) inhibition of release of the viral genome. Any one mAb may invoke a subtle combination of these mechanisms.

The site of antibody attachment can be determined by molecular biology methods, in particular, by analysis of mutations in the genome of virus variants that have escaped neutralization. This method has proved very useful in defining amino acid residues involved in the interaction between virus and antibody, and has yielded important information on antibody-binding sites in the absence of structural data.

Understanding how antibodies neutralize viruses at the atomic level is a major challenge in structural virology. Antibodies bind to viruses, forming large molecular complexes that are both flexible and fleeting, so the structures of such complexes are not easily amenable to solution by X-ray crystallography. However, the developing methodology of combining structural data from cryoelectron microscopy and X-ray crystallography to give a pseudo-atomic model is proving highly appropriate in this context (Porta et al. 1994; Wang et al. 1992; Baker and Johnson 1996). The structure of virus/antibody complexes provides information not only on how antibodies neutralize virus, but also on how they interact with antigen in general; for example, data about the flexibility of antibodies and the allowable distance between two binding sites of a bivalently attaching immunoglobulin molecule might be deduced.

## 1.1 Combining Cryoelectron Microscopy and X-Ray Data of Virus/Antibody Complexes

The cryoelectron microscopy map of the virus-Fab complex alone provides some information about the relative positions and orientation of the Fabs on the viral surface, their flexibility, the occupancy of the Fab sites, and the distance between binding sites. The pseudo-atomic structure derived from the combined data goes further in revealing if there is any major rearrangement in the viral capsid on binding of the antibody and whether the cell receptor site is obscured. It can also verify whether the antibody binding is mono- or bivalent. The value and the extent of the density attributed to the antibody gives some indication of its occupancy and movement (or flexibility). In every reconstruction in which a whole IgG is attached to an icosahedral virus, the Fc fragment is either missing or only very weakly represented. This is taken to indicate movement of the Fc portion associated with the highly flexible nature of the antibody hinge region.

Fitting X-ray structures into cryoelectron microscopy maps of virus-antibody complexes, in the absence of steric hindrance, is one of the more tractable fitting problems. The characteristic shape of an Fab fragment facilitates the task, so that even at low resolution a fit is possible since the binding site (paratope) of the Fab is known. It should be noted, however, that there may be a certain ambiguity in the fit due to the presence of a pseudo twofold symmetry axis coincident with the long axis of the Fab. This ambiguity can be resolved even for a resolution of 30Å if the elbow angle of the Fab is large (HEWAT et al. 1997). Additional information is often useful in confirming the fit or resolving the ambiguity. For example, when the structure of an Fab fragment co-crystallized with a peptide corresponding to the viral epitope is known, the position of the peptide must be compatible with the position of the corresponding sequence on the viral surface, although substantial rearrangements might occur due to induced fit (HEWAT and BLAAS 1996).

The current precision of the rigid body fit of roughly 4Å is not quite sufficient to determine detailed amino acid interactions, but it is certainly sufficient to suggest useful point mutation experiments. Nor can adaptive movements of interacting loops be determined except in special cases, as evoked above, in which a peptide that mimics the epitope is co-crystallized with the Fab.

## 1.2 Steric Hindrance of Neighboring Antibody

Many of the spherical viruses display icosahedral symmetry, such that the symmetry-related binding sites are often very close together. Thus, there is often steric hindrance between the Fabs on neighboring epitopes resulting in partial occupancy. This complicates the interpretation of the density map. The Fab occupancy will reflect the underlying differences in the conformation and accessibility of the viral proteins in different symmetry environments. This is a factor that can influence the neutralizing capacity of the antibody since not all possible epitopes can be covered.

An example of steric hindrance of neighboring antibodies is seen in the complex of rabbit hemorrhagic disease virus (RHDV), a calicivirus, and the strongly neutralizing mAb-E3 (THOUVENIN et al. 1997). The RHDV virion is 40nm in diameter and consists of 180 copies of a single 60-kDa protein arranged as dimeric arches on a $T = 3$ icosahedral lattice. The electron microscope map of the RHDV capsid complexed with the whole mAb-E3 reveals that the density in the Fab domains is only about half of that in the capsid and hence the occupation of the Fab sites is incomplete. Further, there is no density that can be attributed to the Fc fragment. The Fc fragment is apparently quite free to move about the hinge. From the lack of appreciable aggregation of the complexes and from the form of the density, it follows that the mAbs attach bivalently to the capsid. There are three epitopes on neighboring arches, related by a local threefold icosahedral symmetry axis, which compete for each mAb. Any given mAb may thus bind to any pair of the three epitopes (Fig. 1). In this model each dimeric arch has two epitopes such that the bound Fabs overlap slightly. It is notable that the mAbs are bound to

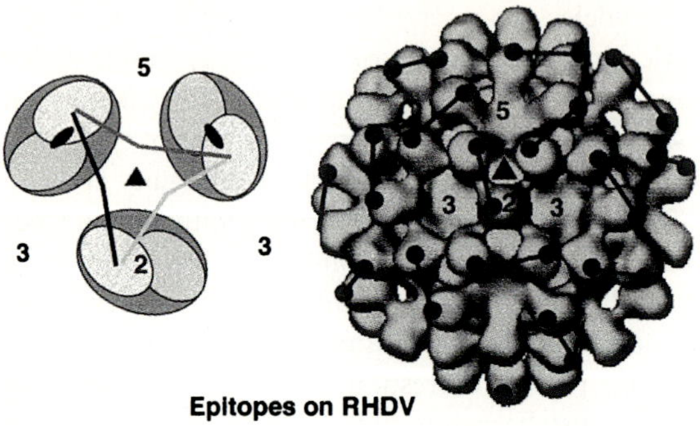


**Fig. 1.** The epitopes of the neutralizing monoclonal antibody (mAb)-E3 on the rabbit hemorrhagic disease virus (RHDV) capsid. The icosahedral axes are marked *2*, *3* and *5* and the local *2* and *3*-fold axes are marked with *symbols*. *Left*: Three dimeric arches of the capsid are represented in *dark gray*, and possible pairs of binding sites for one mAb are shown linked by lines in different shades of *gray*. The overlapping epitopes on each arch are indicated. *Right*: one of the many possible distributions of mAb-E3 on the capsid. (Adapted from Fig. 6 in Thouvenin et al. 1997)

epitopes that are related by an approximate threefold symmetry axis, rather than a twofold axis: This is a striking example of the torsional flexibility of the hinge region of an immunoglobulin molecule.

It is natural to suppose that the density in electron microscopy maps is correlated to the occupancy and mobility of the corresponding molecule. However, quantifying this relation, even when the molecules are fixed relative to one another, is not without error at the resolution of 20–30Å. At best the precision is of the order of ±15%.

## 1.3 Antibodies as Markers

While neutralizing mAbs are of particular interest, antibodies that simply mark a given viral protein or that modify the viral activity (e.g., inhibit RNA transcription) also merit structural study. An interesting example of antibodies used as markers is the herpes simplex virus, in which distinct mAbs separately labeled the hexons or the pentons of the capsid (Trus et al. 1992). This was interpreted as indicating that, although the same protein forms both hexons and pentons, it adopts different conformations that present different antigenic characters. There is also the case of a neutralizing mAb used as a marker for the immunogenic spike protein of rotavirus (Prasad et al. 1990). Two Fabs were seen bound to each spike, thus identifying the spike as a dimer. An example of antibodies that modify viral activity is the case of mAbs directed against the double-layered particle of rotavirus. Certain of these mAbs inhibit the RNA transcription activity of the viral particle (Lawton et al. 1999).

In the following, we will concentrate on two members of the picornavirus family and their interactions with neutralizing mAbs. After a short introduction to the picornaviridae, we will consider the minor group of human rhinoviruses, represented by serotype 2 (HRV2), and the aphtoviruses, represented by foot-and-mouth disease virus serotype C-S8c1 (FMDV C). Complementary results on the major group of HRVs are presented elsewhere in this volume. It will be recalled that HRVs are classified into two groups depending on their cellular receptor. HRVs that bind to ICAM-1 are assigned to the major group, and HRVs that bind to members of the low-density lipoprotein receptor (LDL-r) family are assigned to the minor group. Currently, there is more structural information available on the complexes formed by picornaviruses and their neutralizing mAbs than for any other virus family.

## 2 Picornaviruses

Picornaviruses are small (~30nm) icosahedral particles with an RNA genome of positive (messenger sense) polarity and a capsid composed of 60 copies each of the viral proteins VP1–VP4. This large virus family includes enterovirus, rhinovirus, cardiovirus, aphtovirus, hepativirus A, and some others which are less well-characterized (Knowles 1999; Rueckert 1996). X-ray structure analysis of a number of picornaviruses showed that loops connecting the anti-parallel β-barrels of VP1, 2, and 3 are exposed, whereas VP4 is internal and in close association with the genomic RNA (Acharya et al. 1989; Hogle et al. 1985; Muckelbauer et al. 1995; Rossmann et al. 1985). The N-terminus of VP1 and parts of VP4 appear to become temporally exposed by "viral breathing", as seen from antibodies reacting with internal epitopes of poliovirus and coxsackievirus (Li et al. 1994; Roivainen et al. 1993) and from VP4 peptides released upon digestion of HRV14 with trypsin (Lewis et al. 1998). These findings clearly indicate that the virus structure, as deduced from the crystal, does not exactly reflect the real situation, as the capsid appears to be much more plastic and flexible. Nevertheless, the disorder seen upon X-ray analysis at some positions, in particular at the N-termini of VP4 and in VP1 of HRV2, might be taken as an indication for the presence of more than one conformation (Verdaguer et al. 2000).

The receptor-binding site of poliovirus and major-group rhinoviruses lies in the canyon, a cleft encircling the fivefold axes of icosahedral symmetry (Belnap et al. 2000; He et al. 2000; Olson et al. 1993). For minor-group HRVs, the receptor-binding site is not in the canyon but rather consists of two surface exposed loops (BC and HI) of VP1 close to the 5-fold axes of symmetry (Hewat et al. 2000). FMDVs have a remarkably smooth surface, except for a long flexible loop connecting β-strands G and H of VP1. This GH-loop is highly immunogenic and contains an Arg-Gly-Asp (RGD) sequence that is recognized by the integrin αvβ3 (Acharya et al. 1989; Berinstein et al. 1995).

Certain neutralizing immunogenic sites and receptor-binding sites partially overlap. It is remarkable that, in general, a one-amino-acid change in the epitope leads to escape from neutralization by a given mAb (Smith et al. 1996; Verdaguer et al. 1995), whereas receptor-binding can support quite some variation in the amino acids contacted by the receptor.

## 2.1 Human Rhinoviruses

### 2.1.1 Neutralization Epitopes

Analysis of escape mutants of the major-group virus HRV14 (Sherry et al. 1985) and of the minor-group virus HRV2 (Appleyard et al. 1990) indicated the presence of four neutralizing immunogenic sites (NIm-1a, NIm-1b, NIm-2, and NIm-3) for HRV14 and three NIm sites (1, 2, and 3, also called A, B, and C) for HRV2 (Fig. 2). For HRV14, these sites were mapped onto the viral capsid structure obtained by X-ray crystallography (Rossmann et al. 1985). A similar mapping was carried out for HRV2 using the X-ray structure of HRV1-A (Kim et al. 1989), a close relative of HRV2, with 95% sequence identity of the capsid proteins. The assumption that the location of the sites was equivalent in these two serotypes (Appleyard et al. 1990) has now been confirmed with the resolution of the atomic structure of HRV2 by X-ray crystallography (Verdaguer et al. 2000).

### 2.1.2 The Structure of Human Rhinovirus/Antibody Complexes

The pseudo-atomic structures of five different HRV/antibody complexes have been determined by combining cryoelectron microscopy and X-ray data. Three of the complexes involve mAbs directed against the NIm-1a site of the major-group HRV14 and substantially overlap the receptor site of ICAM-1 (Che et al. 1998). Two of the mAbs bind bivalently across the twofold axes and are strongly neutralizing. The third binds monovalently and is weakly neutralizing. The other two complexes, which will be described here, involve mAbs directed against site B of the minor-group virus HRV2. Both mAbs are weakly neutralizing, although one, mAb-8F5, is bound bivalently (Hewat and Blaas 1996) and the other, mAb 3B10, is bound monovalently (Hewat et al. 1998). Site B of HRV2 is equivalent to the NIm-2 site of the major-group viruses; as viral binding to the cell is not grossly impaired, neither apparently overlaps the receptor-binding site.

mAb-8F5 binds to a sequential epitope on VP2 of HRV2 that lies between amino acids 153 and 164 (Barnett et al. 1995; Skern et al. 1987). Since it recognizes a linear sequence, it was possible to determine the structure of the Fab fragment in complex with a synthetic 15-mer peptide (that mimics the viral epitope) and to compare it with the structure of the Fab fragment alone (Tormo et al. 1994). Thus, it appeared that binding of the peptide to the antibody causes some structural rearrangement of the paratope. The elbow angle of the Fab also differs in the presence and absence of the peptide, but this may be dictated by the crystallization conditions rather than by the influence of the bound peptide. Several lines of evidence suggested that the IgG 8F5 binds bivalently to HRV2. Biochemical data

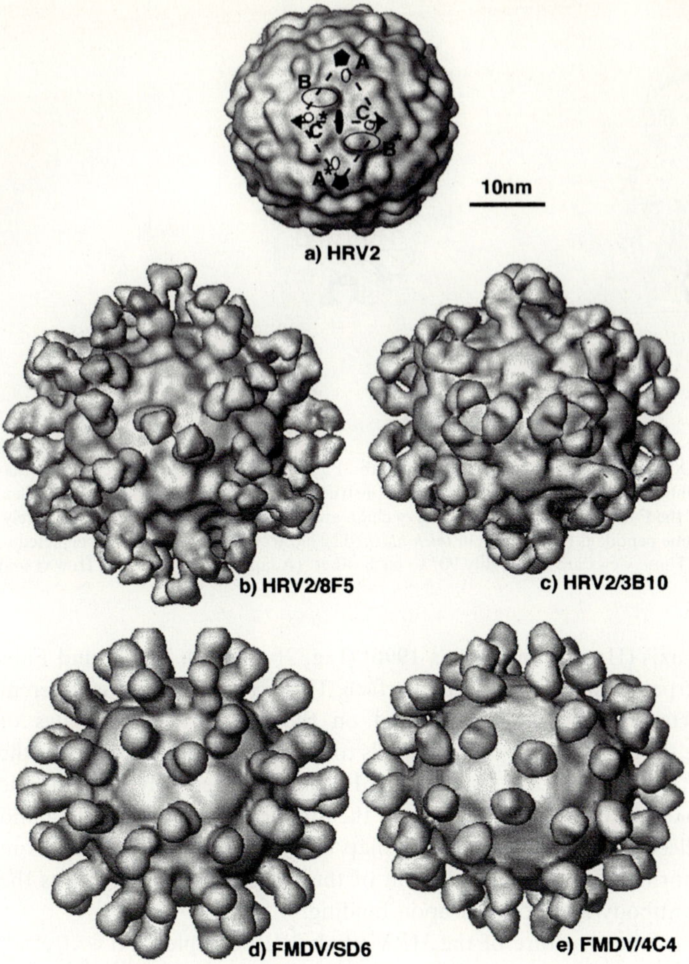


**Fig. 2a–e.** Isosurface representations of the X-ray map of HRV2 limited to 25Å resolution (**a**), the cryoelectron microscope maps of the complex HRV2/MAb-8F5 (**b**), HRV2/Fab-3B10 (**c**), FMDV C-S8c1/Fab-SD6 (**d**), and FMDV C-S8c1/Fab-4C4 (**e**). In **a**, two asymmetric unit triangles and the three immunogenic sites of HRV2 are marked *A*, *B*, and *C*. mAb 8F5 binds to sites *B* and *B**. All representations are viewed down the twofold axis

indicated a maximum of 30 molecules of IgG bound per virion and the absence of any appreciable aggregation. Also, a model of Fab-8F5 bound to HRV2 (Tormo et al. 1995), produced by superposing the peptide backbone onto the corresponding region in HRV1-A, suggested bivalent binding of the antibody over the twofold icosahedral axes. However, the center-to-center distance between the epitopes spanned by the two arms of the IgG molecule is only 6nm, which is slightly less than was previously thought possible (Mosser et al. 1989). The pseudo-atomic structure of the complex of whole 8F5 IgG and HRV2 confirmed bivalent binding

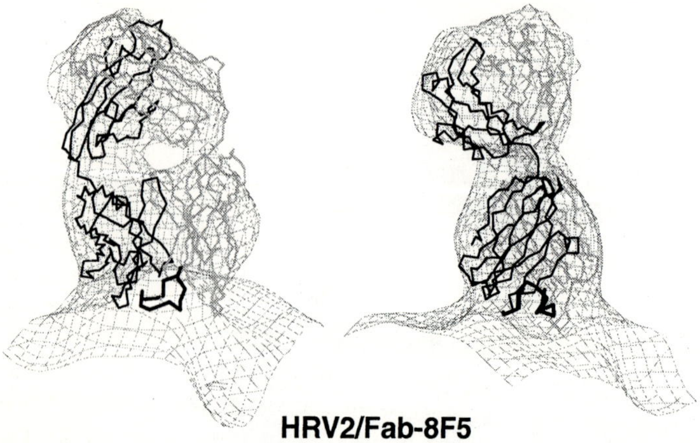


**Fig. 3.** The fit of the X-ray crystallographic structure of Fab-8F5 (plus the 15-mer synthetic peptide that mimics the epitope) into the cryoelectron microscope reconstruction of the HRV2/MAb-8F5 complex. The Cα backbones of the Fab-8F5 light chain and heavy chain are shown in *gray* and *black*, respectively, and the 15-mer synthetic peptide is represented in *thick black*. The electron microscope map is depicted in *light-gray wire mesh*. These views are at roughly 90° to each other. (Adapted from Fig. 5 of HEWAT and BLAAS 1996)

over the twofold axis (HEWAT and BLAAS 1996) (Fig. 2b, Fig. 3). The bound Fabs project almost perpendicularly to the viral surface. This Fab orientation is different from that predicted by the modeling based on the structural coordinates of HRV1A, because the epitope loop of VP2 bends towards the twofold axis when the antibody is bound. Recently, the structure of HRV2 has been solved to atomic resolution (VERDAGUER et al. 2000). Fitting of the atomic structures of HRV2 and 8F5 to the cryoelectron microscopy density map of the virus/mAb complex is in progress and confirms that the conformation of the antigenic loop as well as the paratope of the antibody are modified upon binding.

The pseudo-atomic structure of the HRV2/Fab-3B10 complex (HEWAT et al. 1998) corroborates that the mAb has an area epitope (Fig. 2c). It shows clearly why mAb-3B10 can only be bound monovalently to HRV2. Although mAb-3B10 attaches to the same site B as 8F5, it is inclined at an angle of 45° to the virus surface pointing away from the twofold axis. Even taking into account the flexibility of the elbow angle, it is not possible to model a bivalently bound mAb into the density map.

## 2.2 Foot-and-Mouth Disease Viruses

### 2.2.1 Neutralization Epitopes

Four or five immunogenic sites involving VP1, VP2, and VP3 have been detected on the seven serotypes of FMDV (CROWTHER et al. 1993; MATEU et al. 1995). However, FMDV differs from the other picornaviruses in having a relatively smooth surface with one exceptionally long exposed loop, the GH-loop of VP1 (residues 134–160), which not only constitutes most of the major antigenic site (Bittle et al. 1982; Pfaff et al. 1982; Strohmaier et al. 1982), but also contains a conserved RGD (Arg-Gly-Asp) sequence that binds to the αvβ3 integrin receptor (Berinstein et al. 1995). This loop is highly disordered in the native virus structure of all serotypes yet examined including FMDV-C-S8c1 (Lea et al. 1994). When the disulfide bond linking the base of the loop (Cys-134 to Cys-130) of VP2 is reduced in FMDV-O1, however, the loop adopts a more highly ordered position lying along the viral surface (Logan et al. 1993). Nevertheless, the reduced virus retains most of its infectivity. While most escape mutations involve residues that directly take part in binding antibody, several escape mutations on FMDV-O1 have been shown to occur at positions remote from the antibody-binding site. Observed conformational changes at a distance, in the GH-loop, are believed to destroy the epitope (Parry et al. 1990). The mutations outside the GH-loop appear to switch the population of conformations sampled by this loop from a predominantly "up" position close to the capsid fivefold axes (where those mutations were found) to a predominately "down" position closer to the capsid twofold axes (Logan et al. 1993; Parry et al. 1990).

### 2.2.2 The Structure of FMDV/Antibody Complexes

Direct information on how mAbs interact with FMDV is available from the structural studies of two FMDV-C/Fab complexes. The strongly neutralizing mAbs SD6 and 4C4, elicited against FMDV-C, bind to continuous epitopes on the GH-loop. The crystal structure of a synthetic peptide representing the GH-loop of FMDV-C complexed with Fab fragments of mAb-SD6 (Verdaguer et al. 1995) and mAb-4C4 (Verdaguer et al. 1998) have been determined. Both the complexed peptides show considerable structural similarity to the VP1 GH-loop of the reduced form of FMDV-O, and the RGD motif interacts directly with the Fabs. These structures, with the RGD triplet forming an open-turn flanked by a β-strand and a helical region, reinforces the picture of the GH-loop of FMDV VP1 acting as a mobile structural module.

In the pseudo-atomic structures of both FMDV-C/Fab complexes, the Fabs are seen to project almost radially from the viral surface in an orientation that precludes bivalent binding (Hewat et al. 1997; Verdaguer et al. 1999) (Fig. 2d,e). While Fab-SD6 occupies a well-defined position in the complex, Fab-4C4 remains mobile and is thus represented by a lower density smeared out over a larger volume. This difference in mobility of the bound Fab correlates with the observed difference in the number of amino acids of the GH-loop that interact with the Fab: ten for Fab-SD6 and eight for Fab-4C4. Despite this difference in mobility of the bound Fabs, the neutralization curves for the Fab and the IgG in both cases are essentially identical (Verdaguer et al. 1997, 1999).

The location of the GH-loop in the complexes is defined by the best-fit position of the Fabs in the cryoelectron microscopy maps (Fig. 4). In the FMDV/Fab-4C4 complex, the GH-loop is fully extended so that the RGD motif lies at

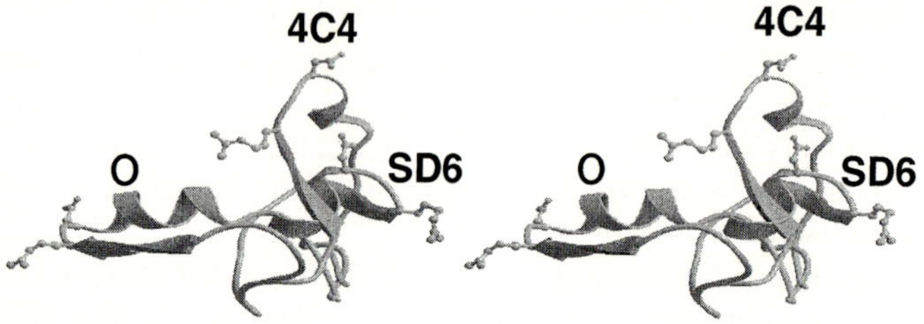


**Fig. 4.** Stereo view of ribbon diagrams of the VP1 GH-loop of foot-and-mouth disease virus (*FMDV*), as seen in the crystallographic structure of the reduced FMDV-O1 BFS (*O*), and in the pseudo-atomic models obtained from fitting the Fab atomic structures into the FMDV-C/ Fab complexes for Fabs SD6 and 4C4. The capsid surface is horizontal and the loop (O) in the reduced virus lies flat onto the surface in the "down position". The GH-loops bound to the Fab-SD6 (SD6) and Fab-4C4 (4C4) are both in an "up" position. The side chains of arginine and aspartic acid from the essential RGD are shown explicitly. (Adapted from Fig. 5b in Verdagurer et al. 1999)

its maximal distance from the capsid surface. A simple hinge rotation at the surface of the capsid would allow the transition between this extended position and the loop position in reduced FMDV type O. When the Fab SD6 is bound, the loop is also in an extended position, but it is twisted around its base by roughly 180° and so remains closer to the capsid surface. The extended position of the loop seen in these FMDV/Fab complexes was proposed earlier as the "up" conformation of the loop, in contrast to the "down" conformation with the loop lying flat against the capsid surface, as seen in the structure of reduced FMDV-O1 (Parry et al. 1990).

## 3 Mechanisms of Neutralization

### 3.1 Structural Changes

Early work on poliovirus by Mandel (1976), followed by Emini et al. (1983a), suggested that bivalent binding of antibodies induces a reduction in isoelectric point from neutral to around 4. This led to the hypothesis that neutralization and a change of isoelectric point, associated with a structural modification, were strongly correlated. This was also supported by the finding that infectivity and the normal isoelectric point were reestablished upon cleavage of attached IgG molecules with papain. Furthermore, upon cross linking of the Fab fragments remaining on the viral capsid with anti-mouse IgG antibodies, the pI and infectivity were once again reduced (Emini et al. 1983b). This was taken to indicate that bivalent antibody binding led to structural changes of the virions. In another investigation, however, no quantitative correlation between a shift in isoelectric point and neutralization of poliovirus 1 was observed (Brioen et al. 1985a). Using a panel of mAbs directed against each NIm site of HRV14, Colonno et al. (1989) demonstrated that neutralization was mainly due to inhibition of cell attachment. This work also indicated that the neutralizing capacity was correlated with a substantial decrease in the isoelectric point (to between 1.8 and 3.2) of the virus upon attachment of the immunoglobulins. However, the concept of structural change as a function of bivalent attachment had to be dismissed, as even the monovalent Fab fragments led to a similar decrease in the isoelectric point, although a 13- to 61-fold higher concentration was needed for a neutralization capacity similar to that of the IgG molecules. Analysis of complexes between HRV14 and Fab fragments of mAb 17A-I by X-ray crystallography failed to reveal any gross structural rearrangement of the virus structure that might have resulted from antibody binding (Smith et al. 1996). Similarly, for all picornavirus/antibody complexes for which pseudo-atomic models have been obtained by combining X-ray crystallography and electron cryomicroscopy data, no modifications of the viral capsid at around 2.5-nm resolution have been detected upon antibody attachment. However, capsid modifications may occur in a low ionic strength medium in which a "hit and run" mechanism of viral neutralization was demonstrated for poliovirus. Attachment of one particular mAb, which under physiologic conditions led to aggregation, literally disrupted and thus inactivated the virus when incubated in low salt buffer (Brioen et al. 1985b). Thus, at least under physiologic conditions, it is highly unlikely that major structural changes are responsible for neutralization of these picornaviruses.

However, from data on 8F5 bound bivalently to HRV2, it became clear that the epitope loop of VP2, encompassing amino acids G153 to Q167, becomes rearranged to better fit the paratope of 8F5 (Verdaguer et al. 2000), and it is also bent over towards the twofold axis. Whether these minor modifications of the structure or the shielding of the exposed R161 from solvent (Tormo et al. 1994) is sufficient to explain the decrease of the isoelectric point to around 3.2 (unpublished observations) is not clear.

## 3.2 Aggregation

Structural work on seven HRV/mAb and FMDV/mAb complexes presented in this volume shows a roughly even distribution between monovalent and bivalent antibody binding. It is clear that monovalent antibody binding opens the possibility for aggregation as a neutralization mechanism. Neutralization curves for poliovirus and HRV14 revealed that some weakly neutralizing antibodies display a U-shaped dependency of surviving virus vs the antibody to virus ratio. Surviving virus was minimal at about 100 antibodies per virion, whereas at higher antibody/virus ratios the number of surviving virus increased (Mosser et al. 1989). The U-shaped curve was interpreted as indicating that at a low antibody to virus ratio (such that not all epitopes are occupied by antibody) aggregation is the predominant means of neutralization. The increase in infectivity at higher antibody to virus ratio (at which the virion is likely to be saturated with antibody and little or no aggregation occurs) indicates that the antibodies do not prevent attachment to the receptor, nor do they inhibit uncoating. It is notable that the neutralization curves for the two strongly neutralizing antibodies (SD6 and 4C4) against FMDV-C display the standard rapid drop of more than 5 logs in surviving virus (Verdaguer et al. 1999), both for the mAb and the Fab fragment. Thus, although these mAbs bind monovalently to FMDV-C, there is no indication that aggregation plays a role in neutralization.

## 3.3 Inhibition of Receptor Binding

Any antibody bound to the viral surface is a potential cause of interference with receptor binding. However, several cases of viral infectivity being unaffected by antibody binding have been reported. This may be due either to a low affinity, resulting in displacement of the antibodies by the receptors, or to binding at sites remote from the receptor attachment site.

The pseudo-atomic structures of major-group HRV14 or HRV16 and recombinant ICAM-1 confirmed that the receptor binds in the canyon. Thus, the mAbs that bind to the NIm-1a site clearly obstruct the -binding site (Che et al. 1998; Smith et al. 1996). Similarly, for FMDV-C the mAbs SD6 and 4C4 compete with the integrin receptor for the binding site on the GH-loop (Verdaguer et al. 1995). All structural and biochemical evidence indicates that inhibition of receptor binding is the principal mechanism of neutralization for these two mAbs. In the case of minor group HRVs, recombinant very-low density lipoprotein receptor fragments N-terminally fused to maltose-binding-protein (MBP) bound virus as efficiently as the recombinant protein without its fusion partner. This suggested that receptor binding does not occur in the canyon, as MBP would probably sterically hinder entry of the receptor moiety of the fusion protein (Ronacher et al. 2000). Indeed, recent structure determination revealed binding of the receptors to the BC and the HI loops of VP1 (Hewat et al. 2000).

## 3.4 Inhibition of Uncoating

There is some indirect experimental evidence that antibodies attached bivalently over the twofold axes of icosahedral symmetry stabilize the virus against structural modifications necessary for RNA release. For example, poliovirus is easily inactivated in hypotonic buffer at 37°C, but this inactivation is prevented by incubation with neutralizing antibodies and infectious virus is recovered upon dissociation of the IgG at low pH (Wetz 1993). Also, "postabsorption neutralization" has been observed for antibodies binding either bivalently or rendered bivalent by cross-linking the antibodies with anti-mouse Fc-specific sera. Virus already attached to the cells could still be neutralized, but infectious virions could be recovered upon dissociation of the antibodies (Vrijsen et al. 1993). This indicates that the structural modification from 160S native particles to subviral particles, the remainder of the uncoating process, was prevented by the bound antibodies.

# 4 Discussion

What conclusions can be drawn from the seven picornavirus-antibody complexes for which the pseudo-atomic structure is known? In this rather limited collection, there are examples of both strongly and weakly neutralizing mAbs that invoke each of three mechanisms of neutralization, namely, (1) obstruction and inhibition of receptor binding, (2) aggregation by cross-linking virus particles and (3) inhibition of virus uncoating by bivalent binding and cross-linking of pentamers. The four strongly neutralizing mAbs, two against HRV14, and two against FMDV, each inhibit receptor binding and bind strongly to the virus (they each form a virus-Fab complex with 100% occupancy). Thus, it appears that a mAb that binds strongly to the receptor-binding site will be a strong neutralizer. However, bivalent binding is neither necessary (e.g., the two monovalently bound, strongly neutralizing mAbs against FMDV) nor sufficient (e.g., the bivalently bound, weakly neutralizing mAb against HRV2) to ensure strong neutralization. It is the affinity (or avidity) of the antibody that matters. The cooperativity of bivalent mAb binding can result in an affinity constant 100–1000 times higher than for the Fab alone. Modification of the virus capsid on antibody binding leading to inhibition of one or more steps in the infection cycle has been evoked, particularly in the case of polioviruses (Emini et al. 1983b; Wien et al. 1995), but there is no evidence for this in any of the mAbs considered here. In fact, the X-ray structure of an HRV14/Fab complex (Smith et al. 1996) shows no modification of the virus structure. The mAbs with a high affinity, irrespective of their mode of binding, will probably also be the most effective in vivo in marking the virus for destruction by other pathways of the immune system (McCullough et al. 1992).

*Acknowledgement.* D.B. was supported by the Austrian Science Foundation (FWF) grant 12269-MOB.

# References

Acharya R, Fry E, Stuart D, Fox G, Rowland D, Brown F (1989) The three-dimensional structure of foot-and-mouth disease virus at 2.9Å resolution. Nature 337:709–716

Appleyard G, Russell SM, Clarke BE, Speller SA, Trowbridge M, Vadolas J (1990) Neutralization epitopes of human rhinovirus type 2. J Gen Virol 71:1275–1282

Baker TS, Johnson JE (1996) Low resolution meets high: towards a resolution continuum from cells to atoms. Curr Opin Struct Biol 6:585–594

Barnett PV, Rowlands DJ, Campbell RO, Parry NR (1995) Monoclonal antibodies to a peptide of human rhinovirus type 2 with different specificities recognize the same minimum sequence. J Gen Virol 76:1255–1261

Belnap DM, McDermott BM Jr, Filman DJ, Cheng N, Trus BL, Zuccola HJ, Racaniello VR, Hogle JM, Steven AC (2000) Three-dimensional structure of poliovirus receptor bound to poliovirus. Proc Natl Acad Sci USA 97:73–78

Berinstein A, Roivainen M, Hovi T, Mason PW, Baxt B (1995) Antibodies to the vitronectin receptor (integrin alpha(v)beta(3)) inhibit binding and infection of foot-and-mouth disease virus to cultured cells. J Virol 69:2664–2666

Bittle JL, Houghten RA, Alexander H, Shinnick TM, Sutcliffe JG, Lerner RA, Rowlands DJ, Brown F (1982) Protection against foot-and-mouth disease by immunization with a chemically synthesized peptide predicted from the viral nucleotide sequence. Nature 298:30–33

Brioen P, Thomas AA, Boeye A (1985a) Lack of quantitative correlation between the neutralization of poliovirus and the antibody-mediated pI shift of the virion. J Gen Virol 66:609–613

Brioen P, Rombau B, Boeye A (1985b) Hit-and-run neutralization of poliovirus. J Gen Virol 66:2495–2499

Che ZW, Olson NH, Leippe D, Lee WM, Mosser AG, Rueckert RR, Baker TS, Smith TJ (1998) Antibody-mediated neutralization of human rhinovirus 14 explored by means of cryoelectron microscopy and X-ray crystallography of virus-fab complexes. J Virol 72:4610–4622

Colonno RJ, Callahan PL, Leippe DM, Rueckert RR, Tomassini JE (1989) Inhibition of rhinovirus attachment by neutralizing monoclonal antibodies and their Fab fragments. J Virol 63:36–42

Crowther JR, Farias S, Carpenter WC, Samuel AR (1993) Identification of a fifth neutralizable site on type O foot-and-mouth disease virus following characterization of single and quintuple monoclonal antibody escape mutant. J Gen Virol 74:1547–1553

Emini EA, Kao SY, Lewis AJ, Crainic R, Wimmer E (1983a) Functional basis of poliovirus neutralization determined with monospecific neutralizing antibodies. J Virol 46:466–474

Emini EA, Ostapchuk P, Wimmer E (1983b) Bivalent attachment of antibody onto poliovirus leads to conformational alteration and neutralization. J Virol 48:547–550

He Y, Bowman VD, Mueller S, Bator CM, Bella J, Peng X, Baker TS, Wimmer E, Kuhn RJ, Rossmann MG (2000) Interaction of the poliovirus receptor with poliovirus. Proc Natl Acad Sci USA 97:79–84

Hewat EA, Blaas D (1996) Structure of a neutralizing antibody bound bivalently to human rhinovirus. 2 EMBO J 15:1515–1523

Hewat EA, Marlovits TC, Blaas D (1998) Structure of a neutralizing antibody bound monovalently to human rhinovirus 2. J Virol 72:4396–4402

Hewat EA, Neumann E, Conway JF, Moser R, Ronacher B, Marlovits TC, Blaas D (2000) The cellular receptor to human rhinovirus 2 binds around the 5-fold axis and not in the canyon: a structural view. EMBO J 19:6317–6325

Hewat EA, Verdaguer N, Fita I, Blakemore W, Brookes S, King A, Newman J, Domingo E, Mateu MG, Stuart DI (1997) Structure of the complex of an Fab fragment of a neutralizing antibody with foot-and-mouth disease virus: Positioning of a highly mobile antigenic loop. EMBO J 16:1492–1500

Hogle JM, Chow M, Filman DJ (1985) Three-Dimensional Structure of Poliovirus at 2.9Å Resolution Science. 229:1358–1365

Hopp TP, Woods KR (1981) Prediction of protein antigenic determinants from amino acid sequences. Proc Natl Acad Sci USA 78:3824–3828

Karplus PA, Schulz GE (1985) Prediction of chain flexibility in proteins: a tool for the selection of peptide antigens. Naturwissenschaften 72:212–213

Kim S, Smith TJ, Chapman MS, Rossmann MG, Pevear DC, Dutko FJ, Felock PJ, Diana GD, McKinlay MA (1989) Crystal structure of human rhinovirus serotype-1A (Hrv1A). J Mol Biol 210:91–111

Knowles N (1999) The Picornavirus Home Page [http://www.iah.bbsrc.ac.uk/virus/Picornaviridae/SequenceDatabase/Index.html]

Lawton JA, Estes MK, Prasad BV (1999) Comparative structural analysis of transcriptionally competent and incompetent rotavirus-antibody complexes. Proc Natl Acad Sci USA 96:5428–5433

Lea S, Hernandez J, Blakemore W, Brocchi E, Curry S, Domingo E, Fry E, Abughazaleh R, King A, Newman J, Stuart D, Mateu MG (1994) The structure and antigenicity of a type-C foot-and-mouth disease virus. Structure 2:123–139

Lewis JK, Bothner B, Smith TJ, Siuzdak G (1998) Antiviral agent blocks breathing of the common cold virus. Proc Natl Acad Sci USA 95:6774–6778

Li Q, Yafal AG, Lee YM, Hogle J, Chow M (1994) Poliovirus neutralization by antibodies to internal epitopes of VP4 and VP1 results from reversible exposure of these sequences at physiological temperature. J Virol 68:3965–3970

Logan D, Abu Ghazaleh R, Blakemore W, Curry S, Jackson T, King A, Lea S, Lewis R, Newman J, Parry N, Rowlands D, Stuart D, Fry E (1993) Structure of a major immunogenic site on foot-and-mouth disease virus. Nature 362:566–568

Mandel B (1976) Neutralization of poliovirus: a hypothesis to explain the mechanism and the one-hit character of the neutralization reaction. Virology 69:500–510

Mateu MG, Camarero JA, Giralt E, Andreu D, Domingo E (1995) Direct evaluation of the immunodominance of a major antigenic site of foot-and-mouth disease virus in a natural host. Virology 206:298–306

McCullough KC, De Simone F, Brocchi E, Capucci L, Crowther JR, Kihm U (1992) Protective immune response against foot-and-mouth disease. J Virol 66:1835–1840

Mosser AG, Leippe DM, Rueckert RR (1989) Neutralization of picornaviruses: support for the pentamer bridging hypothesis. In: Semler B, Ehrenfeld E (eds) Molecular aspects of picornavirus infection and detection. American Society for Microbiology, Washington DC, pp 155–167

Muckelbauer JK, Kremer M, Minor I, Diana G, Dutko FJ, Groarke J, Pevear DC, Rossmann MG (1995) The structure of coxsackievirus B3 at 3.5Å resolution. Structure 3:653–667

Olson NH, Kolatkar PR, Oliveira MA, Cheng RH, Greve JM, McClelland A, Baker TS, Rossmann MG (1993) Structure of a human rhinovirus complexed with its receptor molecule. Proc Natl Acad Sci USA 90:507–511

Parry N, Fox G, Rowlands D, Brown F, Fry E, Acharya R, Logan D, Stuart D (1990) Structural and serological evidence for a novel mechanism of antigenic variation in foot-and-mouth disease virus. Nature 347:569–572

Pfaff E, Mussgay M, Bohm HO, Schulz GE, Schaller H (1982) Antibodies against a preselected peptide recognize and neutralize foot and mouth disease virus. EMBO J 1:869–874

Porta C, Wang G, Cheng H, Chen Z, Baker TS, Johnson JE (1994) Direct imaging of interactions between an icosahedral virus and conjugate F(ab) fragments by cryoelectron microscopy and X-ray crystallography. Virology 204:777–788

Prasad BV, Burns JW, Marietta E, Estes MK, Chiu W (1990) Localization of VP4 neutralization sites in rotavirus by three-dimensional cryo-electron microscopy. Nature 343:476–479

Roivainen M, Piirainen L, Rysa T, Narvanen A, Hovi T (1993) An immunodominant N-terminal region of VP1 protein of poliovirion that is buried in crystal structure can be exposed in solution. Virology 195:762–765

Ronacher B, Marlovits TC, Moser R, Blaas D (2000) Eukaryotic and prokaryotic expression and folding of human very-low density lipoprotein receptor fragments: neutralization capacity towards human rhinovirus HRV2. Virology 278:541–550

Rossmann MG, Arnold E, Erickson JW, Frankenberger EA, Griffith JP, Hecht HJ, Johnson JE, Kamer G, Luo M, Mosser AG, Rueckert RR, Sherry B, Vriend G (1985) Structure of a human common cold virus and functional relationship to other picornaviruses. Nature 317:145–153

Rueckert RR (1996) Picornaviridae: the viruses and their replication. In: Fields BN, Kipe DM, Howley PM (eds) Fields' Virology. Lippincott-Raven, Philadelphia, pp 609–654

Sherry B, Mosser AG, Colonno J, Rueckert RR (1985) Use of monoclonal antibodies to identify four neutralization immunogens on a common cold picornavirus human rhinovirus 14. J Virol 57:246–257

Skern T, Neubauer C, Frasel L, Gruendler P, Sommergruber W, Zorn W, Kuechler E, Blaas D (1987) A neutralizing epitope on human rhinovirus type 2 includes amino acid residues between 153 and 164 of virus capsid protein VP2. J Gen Virol 68:315–323

Smith TJ, Chase ES, Schmidt TJ, Olson NH, Baker TS (1996) Neutralizing antibody to human rhinovirus 14 penetrates the receptor-binding canyon. Nature 383:350–354

Strohmaier K, Franze R, Adam KH (1982) Location and characterization of the antigenic portion of the FMDV immunizing protein. J Gen Virol 59:295–306

Thouvenin E, Laurent S, Madelaine MF, Rasschaert D, Vautherot JF, Hewat EA (1997) Bivalent binding of a neutralising antibody to a calicivirus involves the torsional flexibility of the antibody hinge. J Mol Biol 270:238–246

Tormo J, Blaas D, Parry NR, Rowlands D, Stuart D, Fita I (1994) Crystal structure of a human rhinovirus neutralizing antibody complexed with a peptide derived from viral capsid protein VP2. EMBO J 13:2247–2256

Tormo J, Centeno NB, Fontana E, Bubendorfer T, Fita I, Blaas D (1995) Docking of a human rhinovirus neutralizing antibody onto the viral capsid. Proteins: Struct Funct Genet 23:491–501

Trus BL, Newcomb WW, Booy FP, Brown JC, Steven AC (1992) Distinct monoclonal antibodies separately label the hexons or the pentons of herpes simplex virus capsid. Proc Natl Acad Sci USA 89:11508–11512

Verdaguer N, Blaas D, Fita I (2000) Structure of human rhinovirus serotype 2. (HRV2). J Mol Biol 300:1179–1194

Verdaguer N, Fita I, Domingo E, Mateu MG (1997) Efficient neutralization of foot and mouth disease virus by monovalent antibody binding. J Virol 71:9813–9816

Verdaguer N, Mateu MG, Andre D, Giralt E, Domingo E, Fita I (1995) Structure of the major antigenic loop of foot-and-mouth disease virus complexed with a neutralizing antibody: Direct involvement of the Arg-Gly-Asp motif in the interaction. EMBO J 14:1690–1696

Verdaguer N, Schoehn G, Ochoa WF, Fita I, Brookes S, King A, Domingo E, Mateu MG, Stuart D, Hewat EA (1999) Flexibility of the major antigenic loop of foot-and-mouth disease virus bound to a fab fragment of a neutralising antibody: structure and neutralisation. Virology 255:260–268

Verdaguer N, Sevilla N, Valero ML, Stuart D, Brocchi E, Andreu D, Giralt E, Domingo E, Mateu MG, Fita I (1998) A similar pattern of interaction for different antibodies with a major antigenic site of foot-and-mouth disease virus: implications for intratypic antigenic variation. J Virol 72:739–748

Vrijsen R, Mosser A, Boeye A (1993) Postabsorption neutralization of poliovirus. J Virol 67:3126–3133

Wang GJ, Porta C, Chen ZG, Baker TS, Johnson JE (1992) Identification of a fab interaction footprint site on an icosahedral virus by cryoelectron microscopy and X-ray crystallography. Nature 355: 275–278

Wetz K (1993) Attachment of neutralizing antibodies stabilizes the capsid of poliovirus against uncoating. Virology 192:465–472

Wien MW, Filman DJ, Stura EA, Guillot S, Delpeyroux F, Crainic R, Hogle JM (1995) Structure of the complex between the Fab fragment of a neutralizing antibody for type 1 poliovirus and its viral epitope. Nat Struct Biol 2:232–243

# Shape and Electrostatic Complementarity at Viral Antigen-Antibody Complexes

V.C. Epa and P.M. Colman



## 1 Introduction

This chapter discusses the results of applying to the interfaces of viral antigen-antibody complexes metrics for shape (Lawrence and Colman 1993) and electrostatic (McCoy et al. 1997) complementarity that were developed in this laboratory. Characterization and quantification of interfaces formed by protein association are useful steps towards understanding processes such as neutralization of viral antigen by antibodies, and this work discusses some trends and implications discernible from the application of these particular metrics.

## 2 Computational Methods

Briefly, the shape complementarity metric Sc is determined by computing a modulated scalar product between the two unit vectors which are normal to the molecular surfaces and are located on the points closest to each other on the two surfaces buried from the solvent by the two complexing proteins. Two distributions

---

Biomolecular Research Institute, 343 Royal Parade, Parkville, Victoria 3052, Australia of S values result, one each from the discrete set of points on the two surfaces, and taking the average of the median value of each distribution gives Sc for that particular complex. The surface is sampled at 15 dots/Å$^2$, and those portions of the surface that lie within 1.5Å of the non-buried surface are excluded from the calculation.

The electrostatic complementarity measure EC is determined by computing the negative of the Pearson and the Spearman (rank) correlation coefficients between the electrostatic potentials due to each of the proteins on the interface surfaces. Only the protein atoms in the crystal structure are included in the calculation. Hydrogen atoms are added and limited energy minimization is carried out prior to the coordinates of the points mapping the interface surfaces being determined by the program MS (Connolly 1983). Aspartate, glutamate, arginine, and lysine residue side-chains as well as the N- and C-termini side-chains are considered ionized, while chain breaks resulting from disorder in the protein structure are kept neutral. Electrostatic potentials are computed by numerically solving the linear Poisson–Boltzmann equation, as implemented in the program DelPhi (Gilson and Honig 1988). The electrostatic potentials are calculated for what is called the "partially solvated" case, i.e. point atomic charges and radii from the PARSE parameter set (Sitkoff et al. 1994) are assigned to the atoms of one protein, while the other protein atoms are assigned zero charges.

Readers are directed to the original references, Lawrence and Colman (1993) and McCoy et al. (1997), for the complete computational details.

The results of applying these calculations on 20 viral antigen-antibody complexes from the Protein Data Bank (http://www.rcsb.org/pdb/) are shown in Table 1. In this table, $EC^P$ and $EC^S$ are the Pearson and Spearman correlation coefficients, respectively, for the partially solvated cases multiplied by −1.0, Sc and D are the median shape complementarity metric and distance (in Å), respectively, between the two surfaces. Area is the total buried surface area (in Å$^2$), $K_d$ is the experimentally determined dissociation constant (in $M^{-1}$) for the complex, and the "Charge" column gives the total formal charge for the complex, with the formal charges for the antigen and antibody components indicated in parentheses.

## 3 General Trends

All the viral antigen-antibody complexes studied, with one exception, exhibit electrostatic (potential) complementarity (regardless of whether the Pearson or the Spearman correlation coefficient is considered) and significant shape complementarity.

There is a much smaller range in the magnitude of Sc than there is among the EC values, with a significantly high shape complementarity exhibited by all the complexes, regardless of their components. Optimization of shape complementarity is presumably easier to achieve by an "induced fit" mechanism during complexation than is optimization of electrostatic complementarity.

Maps of the local shape correlation scalar function $S(x)$ (LAWRENCE and COLMAN 1993) on the interface surfaces show that the interface consists of numerous small patches of high and low shape complementarity, all of which have an additive effect on the final value of the metric Sc.

Likewise, analysis of the maps of electrostatic potential on the interface surfaces show the cumulative effect of local patches of complementary as well as anti-complementary electrostatic potential on the final value of the metric EC. Local "hot spots" of very high electrostatic complementarity on the interface can play a dominant role in determining the value of EC for an interface, and these hot spots need not be located centrally in the interface.

The one exception to the above pattern of electrostatic complementarity in viral antigen-antibody complex interfaces is 2HMI, the HIV-1 reverse transcriptase-Fab 28 complex, which shows moderate electrostatic *anti*-complementarity (i.e. a negative value of EC) at the interface. (2HMI does have high shape complementarity.) It should be noted, however, that in this complex the reverse transcriptase chain had many incomplete side-chains, and a total of 42 side-chains were built and subjected to limited energy minimization before the complementarity analysis.

There is another HIV-1 reverse transcriptase-Fab complex, 1C9R, in the Protein Data Bank, solved at the lower resolution of 3.5Å. Here the missing side-chains are represented by alanines. Computed EC values for this complex also gives moderate electrostatic anti-complementarity (results not shown).

## 4 Peptides vs Proteins

Keeping in mind that only a limited number of complexes are available for this structural study and that the generalizing conclusions made here may not hold as more viral antigen-antibody complexes have their structures solved, it appears that antibody complexes with small viral peptides achieve higher electrostatic and shape complementarity than the complexes with larger protein antigens. Particularly in the case of shape complementarity (and to some extent in the case of electrostatic complementarity), this is easily understood – for the smaller, flexible peptides can more easily achieve closer contact with the paratope surface. It can be seen from Table 1 that D, the median distance between the antigen and antibody surfaces, is significantly smaller for the peptide complexes than for the larger protein complexes.

1GC1 (antibody complex of the HIV-1 GP120 protein) and 1ACY (an antibody complex of a 12-residue fragment peptide of GP120) are examples of the viral-peptide-antibody complex having significantly higher electrostatic complementarity at the interface than the integral viral-protein-antibody complex. (This peptide corresponds to a region that is missing in the GP120 structure in 1GC1.)

**Table 1.** Complementarity at viral-antigen-antibody interfaces[1,2]

| PDB ID | Description | $EC^P$ | $EC^S$ | SC | D | Area | Charge | $K_d$ | Resolution, R-factor |
|---|---|---|---|---|---|---|---|---|---|
| 1NCA[3] | N9 NA-NC41 Fab | +0.72 | +0.61 | 0.65 | 0.60 | 1207 | −3.0 (−3.0, 0.0) | $8.3 \times 10^{-8}$ | 2.50, 0.191 |
| 1NCB[4] | N9 NA*-NC41 Fab | +0.72 | +0.71 | 0.62 | 0.59 | 1147 | −5.0 (−4.0, −1.0) | | 2.50, 0.165 |
| 1NCC[4] | N9 NA*-NC41 Fab | +0.71 | +0.68 | 0.60 | 0.61 | 1297 | −2.0 (−2.0, 0.0) | | 2.50, 0.212 |
| 1NCD[3] | w N9 NA-NC41 Fab | +0.68 | +0.62 | 0.65 | 0.61 | 1173 | −5.0 (−4.0, −1.0) | $1.3 \times 10^{-6}$ | 2.90, 0.157 |
| 1A14[5] | N9 NA-NC10 scFv | +0.64 | +0.64 | 0.67 | 0.50 | 756 | −9.0 (−3.0, −6.0) | | 2.50, 0.200 |
| 1NMB[6] | w N9 NA-NC10 Fab | +0.60 | +0.52 | 0.66 | 0.51 | 742 | −7.0 (−4.0, −3.0) | $1.4 \times 10^{-8}$ | 2.50, 0.210 |
| 1NMC[5] | N9 NA-NC10 scFv | +0.66 | +0.68 | 0.65 | 0.54 | 741 | −9.0 (−3.0, −6.0) | $3.3 \times 10^{-8}$ | 2.50, 0.220 |
| 1QFU[7] | HA1-HC45 Fab | +0.45 | +0.47 | 0.65 | 0.57 | 953 | +8.0 (+4.0, +4.0) | $3.0 \times 10^{-9}$ | 2.80, 0.198 |
| 2VIR[8] | HA1-HC19 Fab | +0.45 | +0.37 | 0.67 | 0.50 | 641 | −1.0 (+2.0, −3.0) | $1.0 \times 10^{-9}$ | 3.25, 0.198 |
| 2VIS[8] | HA1*-HC19 Fab | +0.40 | +0.31 | 0.62 | 0.58 | 721 | +2.0 (+5.0, −3.0) | $4.0 \times 10^{-6}$ | 3.25, 0.188 |
| 2VIT[8] | HA1*-HC19 Fab | +0.48 | +0.47 | 0.72 | 0.46 | 644 | −1.0 (+2.0, −3.0) | | 3.25, 0.188 |
| 1AFV[9] | HIV-1 P24-25.3 Fab | +0.56 | +0.48 | 0.65 | 0.51 | 801 | −3.0 (−3.0, +0.0) | | 3.70, 0.217 |
| 1GC1[10] | HIV-1 GP120-17B Fab | +0.55 | +0.55 | 0.70 | 0.53 | 586 | +6.0 (+4.0, +2.0) | | 2.50, 0.210 |
| 1IFH[11] | HA1 peptide-17/9 Fab | +0.62 | +0.60 | 0.80 | 0.40 | 572 | −2.0 (−2.0, +0.0) | | 2.80, 0.170 |
| 1A3R[12] | Rhinovirus VP2 peptide-8F5 Fab | +0.75 | +0.80 | 0.76 | 0.47 | 1284 | −2.0 (−1.0, −1.0) | | 2.10, 0.171 |
| 1ACY[13] | HIV-1 GP120 peptide-59.1 Fab | +0.69 | +0.72 | 0.80 | 0.38 | 612 | +2.0 (+1.0, +1.0) | $9.3 \times 10^{-5}$ | 3.00, 0.210 |
| 1GGI[14] | HIV-1 GP120 peptide-50.1 Fab | +0.58 | +0.41 | 0.80 | 0.38 | 681 | +0.0 (+2.0, −2.0) | $2.2 \times 10^{-7}$ | 2.80, 0.188 |
| 1FPT[15] | Poliovirus type I (Mahoney) peptide-C3 Fab | +0.75 | +0.78 | 0.74 | 0.46 | 670 | +0.0 (+0.0, +0.0) | | 3.00, 0.230 |
| 2HRP[16] | HIV-1 Protease peptide-F11.2.32 Fab | +0.66 | +0.75 | 0.80 | 0.33 | 703 | +5.0 (+3.0, +2.0) | $5.0 \times 10^{-9}$ | 2.20, 0.198 |

[1] See text in Sect. 2, "Computational Methods", for details.
[2] For the HIV-1 reverse-transcriptase-Fab 28 complex [PDB ID: 2HMI, Ding et al. (1998)], the $EC^P$, $EC^S$, Sc, D, Area, Charge, and Resolution, R-factor column values are −0.23, −0.35, 0.71, 0.48, 626, −6.0 (+5.0, −11.0), and 2.80, 0.271, respectively.
[3] Tulip et al. (1992a); N9 NA is tern N9 neuraminidase (1NCA), while "w N9 NA" is whale N9 neuraminidase (1NCD); $K_d$ values from Gruen et al. (1993).
[4] Tulip et al. (1992b); N9 NA* is the mutant N329D neuraminidase in 1NCB and I368R mutant neuraminidase in 1NCC.
[5] Malby et al. (1998); scFv is single-chain Fv with a 5-mer polypeptide linker in 1A14 and a 15-mer polypeptide linker in 1NMC; $K_d$ value from Malby et al. (1993).
[6] Malby et al. (1994); $K_d$ value from Gruen et al. (1993).
[7] Fleury et al. (1999); HA1 is a fragment of the HA1 chain of influenza virus hemagglutinin; $K_d$ value (measured for binding of HA1 to HC45 IgG) from Fleury et al. (1999).
[8] Fleury et al. (1998); HA1* are the mutants T131I (2VIS) and T155I (2VIT) of the HA1 chain fragment of influenza virus hemagglutinin; $K_d$ values from Fleury et al. (1998).
[9] Momany et al. (1996).
[10] Kwong et al. (1998).
[11] Schulze-Gahmen et al. (1993).
[12] Tormo et al. (1994).
[13] Ghiara et al. (1994); $K_d$ values (measured between synthetic peptides and the Mabs) from White-Scharf et al. (1993).
[14] Rini et al. (1993); $K_d$ values (measured between synthetic peptides and the Mabs) from White-Scharf et al. (1993).
[15] Wien et al. (1995).
[16] Lescar et al. (1997); $K_d$ value from same reference.

The EC values for the rhinovirus- and poliovirus-peptide-Fab complexes, 1A3R and 1FPT, are higher than calculated previously for any complex (McCoy et al. 1997). The Sc values computed for the viral-protein-antibody complexes (with the possible exception of 2HMI) fall within the range found for antigen-antibody interfaces by Lawrence and Colman (1993), although the viral-peptide-antibody complexes (the structures of which were not available at the time of the study by Lawrence and Colman) show consistently much higher shape complementarity.

## 5 Effect of Different Viruses, Viral Proteins, and Antibodies on Complementarity

What kind of variation in complementarity is there between antibody complexes formed by different viruses or between different envelope glycoproteins from the same virus? It is obvious from the results that two different protein antigens from the same virus can form antibody complexes with quite different shape and electrostatic complementarity. For example, most of the influenza viral neuraminidase-Fab complexes show relatively high electrostatic complementarity, while all the influenza viral hemagglutinin-Fab complexes are only moderately electrostatic complementary at their interfaces. (We note that three of the hemagglutinin complex structures were solved at lower resolution.) The latter complexes (for example, 1QFU and 2VIR) all show EC values of such moderate and similar magnitudes, despite the fact that the antibodies complex with quite different antigenic sites of the HA1 chain of hemagglutinin. [In the case of 1QFU, this is at some distance from the virus receptor binding site (Fleury et al. 1999).]

On the other hand, the shape complementarity metric, Sc, exhibits considerably less variation between (and among) the neuraminidase and hemagglutinin antibody complexes.

In contrast, the complex between the hemagglutinin peptide and the Igg2 A 17/9 Fab, 1IFH, displays (in addition to very high shape complementarity) an electrostatic complementarity metric that is as high as some of the neuraminidase complexes. Clearly, the binding position of the peptide, the "cleft" between the H and the L chains of the Fab, affords not only a more intimate steric fit, but also much more complementary electrostatic potentials across the interface. It should be noted that this peptide is in a region distant from the epitopes of 1QFU and 2VIR, and also that the conformation of the peptide in 1IFH is quite different from the conformation it assumes as a portion of the HA1 chain in 1QFU and 2VIR.

Unlike the influenza virus, in the case of HIV-1, the antibody complexes formed by two different proteins, for example GP120 in 1GC1 and the viral capsid protein P24 in 1AFV, do not exhibit markedly different electrostatic complementarity (ignoring the unusual result in the case of 2HMI). Likewise, 1ACY and 2HRP, which are Fab complexes of a HIV-1 GP120 peptide fragment and a HIV-1 protease peptide fragment, respectively, show electrostatic and shape complementarity of comparable magnitude.

On the other hand, 1GGI, the complex between the nine-residue HIV-1 GP120 peptide and Fab 50.1, shows significantly lower electrostatic complementarity than 1ACY, the complex between the ten-residue HIV-1 GP120 peptide and Fab 59.1 (although their Sc values are comparable). The two peptides in this case have only four residues at their N-terminus (1ACY) and C-terminus (1GGI) in common, and the paratope regions in the heavy chains of the respective Fabs are quite different [the complementary determining regions of the two Fabs have only 46% sequence identity (GHIARA et al. 1994)], no doubt contributing to their different EC values.

The neuraminidase complexes with NC41 Fab appear to show greater electrostatic complementarity at the interface than do the complexes with NC10. (Although the Spearman and Pearson correlation coefficients do not always follow the same trends, complicating generalization.) The interfaces of the NC10 complexes are significantly smaller in area than the NC41 complexes, and the marked differences in the complementarity-determining regions (CDRs) of NC41 and NC10 [35% sequence identity (MALBY et al. 1994)] contribute to the differences in EC. [While the sequences of the H1 CDR are identical in NC41 and NC10, H1 makes no contact with the antigen in the case of NC10 (MALBY et al. 1994)]. Even in the common overlapping epitope region of the NC41 and NC10 complexes, the complementarity arises from different sections of the interface (MCCOY et al. 1997).The slightly lower electrostatic complementarity in 1NMB (whale N9 neuraminidase-NC10 complex) than in 1NMC and 1A14 (tern N9 neuraminidase-NC10 complexes) may be attributed to the differences between whale and tern N9 neuraminidase sequences as well as side-chain conformational differences at the interfaces (MCCOY et al. 1997). The same factors probably cause the slight differences in complementarity between the tern and whale N9 neuraminidase-NC41 Fab complexes, 1NCA and 1NCD.

We may further observe that complementarity in 1A14 and 1NMC complexes has not been compromised, despite having the $V_H$ and $V_L$ domains joined by a polypeptide linker (5-mer and 15-mer, respectively). This is not surprising, as it has been previously noted (MALBY et al. 1998) that the 5-mer and 15-mer linkers do not affect $V_L$-$V_H$ pairing in the complexes.

## 6 Escape Mutants

Small differences in EC and Sc values are observed among the neuraminidase point-mutant complexes 1NCA (wild-type tern N9 neuraminidase), 1NCB (N329 → D), and 1NCC (I368 → R) as well as among the hemagglutinin point-mutant complexes 2VIR (wild-type hemagglutinin), 2VIS (T131 → I), and 2VIT (T155 → I). (Although it should be noted that residues 46, 87, and 109 in the Fab heavy chain are Lys, Thr, and Val in 1NCA and 1NCC, while they are Glu, Lys, and Leu, respectively, in 1NCB; the residues 81, 165 and 285 in the HA1 chain are asparagines in 2VIS while they are aspartates in 2VIR and 2VIT.) These could be due to the chemical differences in the mutated residues as well as a cumulative effect of small conformational changes of the side-chains.

The neuraminidase mutant N329D (1NCB) is an escape mutant selected with the antibody NC10, while mutants I368R, A369D, S370L, and K432 N were selected by other antibodies. Examination of the wild-type neuraminidase-NC10 (ScFv) (i.e. in the complex 1A14) interface surface properties shows that residue N329 is located on a region of locally high shape and electrostatic potential complementarity, while I368, A369, S370 and K432 are located on surface regions of lesser complementarity.

On the other hand, in the hemagglutinin-HC19 antibody Fab complex 2VIR, residues T131, S157, G158 (T131I, S157L, G158 E being the escape mutants of hemagglutinin selected by the antibody HC19) are not located on interface surface regions of obviously higher shape or electrostatic potential complementarity than are T128, G135, S145, S193, A198, all residue locations of escape mutants selected by other antibodies.

## 7 Complementarity and Binding

There is no correlation between EC and the product of total formal charges of the components of the antigen-antibody complex (as is obvious from the fact that the only system that does not show complementarity, 2HMI, has a very large negative product, while the system with the highest EC, 1A3R, has a positive product).

Nor is there any obvious correlation between the complementarity metrics and the experimentally observed dissociation constant. It should be understood that the free energy of association (or dissociation) depends not only on the strength of interaction between the two components at their interface (represented by the metrics EC and Sc), but also on other factors such as the desolvation energies of the individual components.

## References

Connolly ML (1983) Analytical molecular surface recognition. J Appl Crystallog 16:548–558

Ding J, Das K, Yu H, Sarafianos SG, Clark Jr. AD, Jacobo-Molina A, Tantillo C, Hughes SH, Arnold E (1998) Structure and functional implications of the polymerase active site region in a complex of HIV-1 RT with a double-stranded DNA and an antibody Fab fragment at 2.8Å resolution. J Mol Biol 284:1095–1111

Fleury D,Wharton SA, Skehel JJ, Knossow W, Bizebard T (1998) Antigen distortion allows influenza virus to escape neutralization. Nat Struct Biol 5:119–123

Fleury D, Barrere B, Bizebard T, Daniels RS, Skehel JJ, Knossow M (1999) A complex of influenza hemagglutinin with a neutralizing antibody that binds outside the virus receptor binding site. Nat Struct Biol 6:530–534

Ghiara JB, Stura EA, Stanfield RL, Profy AT, Wilson IA (1994) Crystal structure of the principal neutralizing site of HIV-1. Science 264:82–85

Gilson MK, Honig B (1988) Calculation of the total electrostatic energy of a macromolecular system: solvation energies, binding energies, and conformational analysis. Proteins: Struct Funct Genet 4:7–18

Gruen L, McInerney TL, Webster RG, Jackson DC (1993) Binding affinity of influenza virus N9 neuraminidase with Fab fragments of monoclonal antibodies NC10 and NC41. J Protein Chem 12:255–259

Kwong PD, Wyatt R, Robinson J, Sweet RW, Sodroski J, Hendrickson WA (1998) Structure of an HIV GP120 envelope glycoprotein in complex with the CD4 receptor and a neutralizing human antibody. Nature 393:648–659

Lawrence MC, Colman PM (1993) Shape complementarity at protein/protein interfaces. J Mol Biol 234:946–950

Lescar J, Stouracova R, Riotott MM, Chitarra V, Brynda J, Fabry M, Horejsi M, Sedlacek J, Bentley GA (1997) Three-dimensional structure of an Fab-peptide complex: structural basis of HIV-1 protease inhibition by a monoclonal antibody. J Mol Biol 267:1207–1222

Malby RL, McCoy AJ, Kortt AA, Hudson PJ, Colman PM (1998) Three dimensional structure of single-chain Fv-neuraminidase complexes. J Mol Biol 279:901–910

Malby RL, Tulip WR, Harley VR, McKimm-Breschkin JL, Laver WG, Webster RG, Colman PM (1994) The structure of a complex between the NC10 antibody and influenza virus neuraminidase and comparison with the overlapping binding site of the NC41 antibody. Structure 2:733–746

Malby RL, Caldwell JB, Gruen LC, Harley VR, Ivancic N, Kortt AA, Lilley GG, Power BE, Webster RG, Colman PM, Hudson PJ (1993) Recombinant antineuraminidase single chain antibody: expression, characterization, and crystallization in complex with antigen. Proteins: Struct Funct Genet 16:57–63

McCoy AJ, Epa VC, Colman PM (1997) Electrostatic complementarity at protein/protein interfaces. J Mol Biol 268:570–584

Momany C, Kovari LC, Prongay AJ, Keller W, Gitti RK, Lee BM, Gorbalenya AE, Tong L, McClure J, Ehrlich LS, Summers MF, Carter C, Rossmann MG (1996) Crystal structure of dimeric HIV-1 capsid protein. Nat Struct Biol 3:763–770

Rini JM, Stanfield RL, Stura EA, Salinas PA, Profy AT, Wilson IA (1993) Crystal structure of an HIV type 1 neutralizing antibody, 50.1, in complex with its V3 loop peptide antigen. Proc Natl Acad Sci USA 90:6325–6329

Schulze-Gahmen U, Rini JM, Wilson IA (1993) A detailed analysis of the free and bound conformation of an antibody: X-ray structures of anti-peptide Fab 17/9 and three different Fab–peptide complexes. J Mol Biol 234:1098–1118

Sitkoff D, Sharp KA, Honig B (1994) Accurate calculation of hydration free energies using macroscopic solvent models. J Phys Chem 98:1978–1988

Tormo J, Blaas D, Parry NR, Rowlands D, Stuart D, Fita I (1994) Crystal structure of a human rhinovirus neutralizing antibody complexed with a peptide derived from viral capsid protein VP2. EMBO J 13:2247–2256

Tulip WR, Varghese JN, Laver WG, Webster RG, Colman PM (1992a) Refined crystal structure of the influenza virus N9 neuraminidase–NC41 Fab complex. J Mol Biol 227:122–148

Tulip WR, Varghese JN, Webster RG, Laver WG, Colman PM (1992b) Crystal structures of two mutant neuraminidase-antibody complexes with amino acid substitutions in the interface. J Mol Biol 227:149–159

White-Scharf ME, Potts BJ, Smith LM, Sokolowski KA, Rusche JR, Silver S (1993) Broadly neutralizing monoclonal antibodies to the V3 region of HIV-1 can be elicited by peptide immunization. Virology 192:197–206

Wien MW, Filman DJ, Stura EA, Guillot S, Delpeyroux F, Crainic R, Hogle JM (1995) Three-dimensional structure of the complex between the Fab fragment of an neutralizing antibody for type 1 poliovirus and its viral epitope. Nat Struct Biol 2:232–243

# Structural Studies on Viral Escape from Antibody Neutralization

T. Bizebard[1], C. Barbey-Martin[1], D. Fleury[1], B. Gigant[1], B. Barrère[2], J.J. Skehel[2], and M. Knossow[1]



## 1 Introduction

Antigenic variations allow viruses to reinfect their hosts. The recent determination of the three-dimensional structures of picornaviruses, of the surface antigens of enveloped viruses, and of their complexes with neutralizing antibodies has provided a structural interpretation for a large amount of data on antigenic variation and has prompted further investigation of the interactions between neutralizing antibodies and their viral targets. This chapter describes the nature and location in virus structures of antigenic variations that allow escape from neutralization of infectivity by antibodies and the effect of antigenic variations on antibody-virus interactions.

## 2 Escape Mutations Significantly Reduce Neutralizing Antibody Affinity for Viruses

Quantitative analyses of the changes that an antigen molecule needs to undergo in order for the virus to escape neutralization by an antibody mainly involve

[1] Laboratoire d'Enzymologie et Biochimie Structurales, UPR 9063, CNRS, Bât. 34, CNRS, 91198 Gif-sur-Yvette Cedex, France

[2] National Institute for Medical Research, Mill Hill, London NW7 1AA, UK

monoclonal-antibody-selected virus variants which most frequently differ from the original strain by a single amino acid substitution.

Escape is associated with significantly lower affinity of the antibody for the escape mutant compared with its affinity for the wild-type strain. The few quantitative measurements demonstrate that monoclonal antibodies (mAbs) still bind to the escape mutants that have been selected by growth of wild-type virus in their presence, albeit with a much reduced affinity. In the case of antibodies against the hemagglutinin (HA) of influenza virus, escape mutants were recognized with affinities between $2.5 \times 10^5 M^{-1}$ and $2.0 \times 10^6 M^{-1}$, significantly lower than the affinity of the antibody for wild-type HA ($10^9 M^{-1}$) (FLEURY et al. 1998).

## 3 The Sites of Escape Mutations Are on the Viral Surface

The location of the sites where escape mutations occur has been studied for influenza HA and neuraminidase, tick-borne encephalitis virus surface antigen, HIV Env glycoprotein, and the picornaviruses: poliovirus, rhinovirus, foot-and-mouth disease virus (FMDV), and mengovirus.

In the case of the surface antigens of enveloped viruses, location of the escape mutations with respect to the viral membrane is possible only if the orientation of the antigen with respect to the viral surface is known. This is the case for the two major surface glycoproteins of influenza virus (HA and neuraminidase). The atomic structures of these proteins are known (VARGHESE et al. 1982; WILSON et al. 1981) and their orientations on the virus have been established. In the case of HA, the sites of escape mutations coincide with surface residues located on the membrane-distal tip of the molecule (Fig. 1). Interestingly, variations observed between epidemic strains involve substitutions of buried residues as well as of surface residues. However, the former are not retained in the HAs of viruses isolated from subsequent epidemics, suggesting that they are not antigenically significant. By contrast, most substitutions retained in the HA of successive epidemic strains (43 out of 45 between 1968 and 1995) involve surface residues, and their locations coincide with those of amino acid substitutions found in mAb-selected mutants, strongly suggesting that these mutations have antigenic significance. Amino acids of the neuraminidase molecule that change during antigenic drift are also located on the membrane-distal surface of the molecule, close to the active site of this enzyme (VARGHESE et al. 1982).

Antigenic mutations of viruses are located at solvent-exposed positions on their surface, presumably because the corresponding residues are accessible to antibodies. This presumption has been used to orient the tick-borne encephalitis virus E antigen with respect to the viral surface. Mutations that allow escape of the virus from neutralization are located on one face of this flat molecule, suggesting that this is the outer face (REY et al. 1995). Similar arguments have led to a proposed orientation of the crystallized HIV gp120 fragment with respect to the gp120-gp41 trimer and to the viral membrane (KWONG et al. 1998).

The surface of picornaviruses is irregular, with many protrusions and invaginations. The major antigenic sites have also been mapped by locating on the three-dimensional structures the sites of escape mutations selected by growing these viruses in the presence of mAbs. In the case of rhinovirus serotype 14, four sites have been identified corresponding to protruding regions on the viral surface, facing towards the exterior of the virus (ROSSMANN et al. 1985). Similarly, sites where escape mutations of poliovirus or FMDV serotypes O and C occur are in exposed loops on the surface of the virus (ACHARYA et al. 1989; HOGLE et al. 1986; LEA et al. 1995). In a systematic study of FMDV antigenic mutants, Lea and co-workers (LEA et al. 1994) found that escape mutations more often correspond to residues with side-chains accessible to solvent, which are not involved in interactions with other amino acids of the viral capsid.

The extensive study of influenza HA antigenic variations allows a detailed description of the locations of positions where antigenic mutations occur. HA is a trimer of identical subunits; each subunit consists of a membrane-proximal helix-rich stem structure and a membrane-distal receptor-binding globular domain (WILEY and SKEHEL 1987). All antigenic mutations are located on the surface of the membrane-distal globular domain, from the membrane-distal tip of this domain to its base, closer to the membrane, and removed from the outer surface of the virus (Fig. 2). Apparently, the overall density of protein molecules on the viral surface allows penetration of antibody-combining sites to this depth, at the expense of some loss of their avidity for HA on the virus. When the neutralization titers of two antibodies having similar affinity for solubilized HA were compared, the neutralization titer of the antibody binding to the base of the membrane-distal domain was found to be lower, by an order of magnitude, than that of the antibody binding to the tip of the molecule (FLEURY et al. 1999). The avidities of both antibodies are nevertheless sufficient to neutralize virus infectivity. An analysis of the sites of substitutions in HA between the 1968 Hong Kong pandemic and 1995 shows that the steric hindrance brought about by the crowding of antigens on the viral surface is not sufficient to cause a major disproportion between the numbers of antigenic mutations occurring at the tip and at the base of the membrane-distal domain of HA, i.e., 16 positions where residues have been mutated are at the base of the HA globular domain, as compared to 24 at its tip.

## 4 A Small Number of Amino Acid Substitutions in an Epitope Allow a Virus to Escape from Neutralization

The decrease in affinity by a factor of $10^2$–$10^4$ which accompanies escape from neutralization by an antibody corresponds to a 2.5–5kcal/mol difference in the interaction energy between the antibody and the antigen. This is in the upper range of the stabilization energy provided by a hydrogen bond between charged residues (CREIGHTON 1993). The loss of one or a few interactions in the interface between a

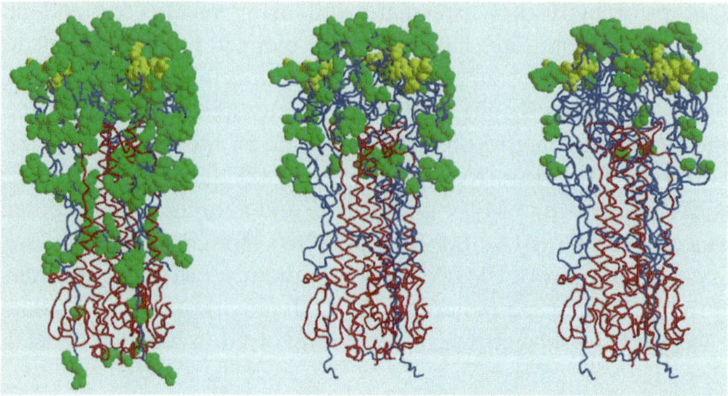

**Fig. 1.** (Caption opposite)

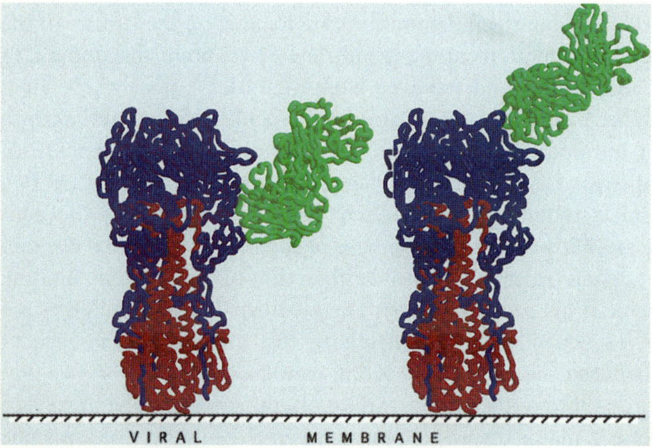


**Fig. 2.** (Caption opposite)

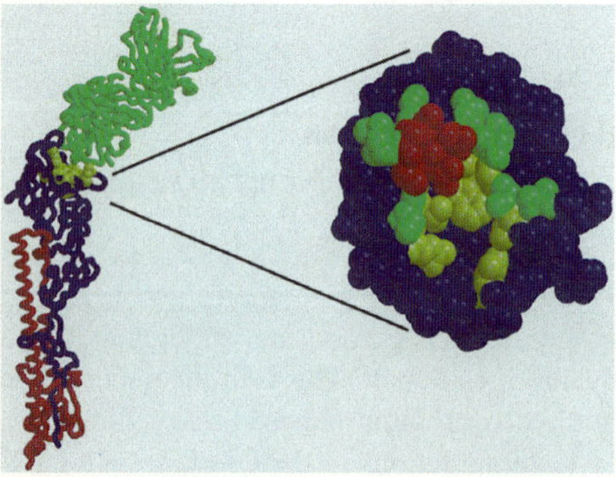

**Fig. 3.** (Caption opposite)

viral antigen and an antibody could therefore allow the virus to escape from neutralization.

Systematic investigation of the sites of mutations that cause a decrease in affinity and lead to escape from neutralization by an antibody indicate that only a few positions in the antigen-antibody interface are involved, much less than the number of residues in the interface. Two antibody-viral antigen complexes have been studied. In the case of influenza neuraminidase, when 15 out of the 19 residues in the interface with an antibody were mutated, only three were found to cause a complete loss of interaction (NUSS et al. 1993). Similarly, mutations that allow influenza virus to escape from neutralization were identified at only four of the 13 positions in the epitope of a neutralizing antibody on influenza HA (BIZEBARD et al. 1995).

These results are consistent with observations on the contribution of individual residues to the interaction energy between two proteins, i.e., that only a few residues in the interface contribute to most of the binding affinity between the two partner proteins (PRASAD et al. 1993; CLACKSON and WELLS 1995; DALL'ACQUA et al. 1998). For example, in a classical study of the interaction between human growth hormone and its receptor, CLACKSON and WELLS (1995) showed that mutations at only six positions of human growth hormone out of thirty three in the interface lead to a decrease of the interaction energy greater than 1.5kcal/mol.

## 5 Structural Changes in an Antibody/Antigen Interface that Allow Escape from Neutralization

Steric hindrance as the structural basis for escape has been proposed from comparison of the structures of mutant and wild-type antigens when the wild-type

**Fig. 1.** Distribution of sequence changes in hemagglutinins (HAs) of the Hong Kong pandemic era during the 1968–1995 period. The space-filling models represent, in *yellow*, the virus-receptor binding site and, in *green*, substituted amino acids. *Left*: All substitutions in HAs of viruses isolated between 1968 and 1995; *middle*: amino acid substitutions between 1968 and 1995 that were retained in subsequent years; *right*: amino acid substitutions detected in monoclonal-antibody-selected variants of A/Hong Kong/68 HA. The α-carbon tracings of the HA trimers are colored *blue* and *red* to denote the HA1 and HA2 polypeptide chains

**Fig. 2.** Ribbon models of Fab fragments (*green*) of neutralizing antibodies bound to the HA trimer. In both models, only one of the three Fabs bound is displayed. The viral membrane is at the *bottom*, and the spacing of HAs indicates the average distance of these molecules on the viral surface (CUSACK et al. 1985). The Fab fragment on the *left* binds at the base of the membrane-distal domain of HA, that on the *right* binds close to its tip

**Fig. 3.** *Left*: Ribbon model of the Fab fragment of a neutralizing antibody bound to an HA monomer. The virus-receptor binding site is highlighted in *yellow*. *Right*: Close-up of a space-filling model of HA, as seen from the Fab. HA is in *blue*, except residues contacted by the antibody, which are in *green*; *yellow*, Residues that also belong to the virus-receptor binding site; *red*, locations of mutations that allow the virus to escape from neutralization by the antibody amino acid changed to a bulkier residue in the escape mutant (Bizebard et al. 1995; Knossow et al. 1984; Colman et al. 1994; Nuss et al. 1993; Lea et al. 1995). The consequences of structural changes that cause escape from neutralization have been suggested based on antigen-antibody structures (Bizebard et al. 1995; Fleury et al. 1999; Tulip et al. 1992) and structures of viral peptide-antibody (Tormo et al. 1994; Verdaguer et al. 1995; Wien et al. 1995) complexes. The deduction is that the change of an antigen side-chain buried in the antigen-antibody interface to a bulkier one causes steric hindrance in the interface, which rearranges to form less favorable interactions than in the wild-type complex.

A second kind of perturbation of antigen structure in escape mutants entails the breaking of at least one interaction (hydrogen bond, salt bridge) crucial for complex stability. An investigation of the effects of influenza virus neuraminidase point mutations on antibody binding provides a possible example of such a situation: the replacement of a serine residue by a glycine is detrimental to complex stability, most likely because of the loss of a hydrogen bond between the serine hydroxyl and the side-chain of an antibody aspartate (Nuss et al. 1993).

The two types of structural changes in an antibody-antigen interface that allow escape from neutralization do not need to be exclusive of each other. Pathogens will be selected with any type of structural change, albeit very subtle, if it allows them to escape from antibody-mediated neutralization. Indeed, the combined contribution of steric hindrance and of a broken hydrogen bond has been illustrated by the structure of a complex between the HA of an escape mutant and the neutralizing antibody that was used to select it. The mutation involved is a threonine to isoleucine change and the larger, mutated side chain is accommodated in the antibody-mutant HA interface without any apparent unfavorable interaction. However, because of the breaking of a hydrogen bond made by the threonine hydroxyl in wild-type HA, the structure of the uncomplexed variant HA changes at locations close to but distinct from the mutated side chain. This new HA conformation would lead to steric hindrance in the complex to a point where it could not be accommodated by the antibody: the complex is formed with an energetically unfavorable, conformationally wild-type-like mutated antigen (Fleury et al. 1998). In this case, two types of changes in HA have led to escape: the breaking of a hydrogen bond and steric hindrance between the antibody and mutated HA.

## 6 Antibody Response and the Functional Elements of Viruses

Escape mutations change the antigenicity of viruses without loss of the function of the antigens. Escape mutations occur at positions exposed on the virus outer surface, and residues involved in virus binding to their cellular receptors or residues of the active sites of extracellular viral enzymes are also exposed but functionally constrained. Do antibodies directed against virus functional sites exist, and how do viruses escape neutralization by these antibodies? Plausible answers have been given (Rossmann et al. 1985; Varghese et al. 1982; Wilson et al. 1981).

One proposition is that critical elements present on viral surface proteins are not accessible to antibodies, because they are located in concave sites of the virus structure and therefore presumably accessible to receptors but not to antibodies. This possibility was emphasized when the rhinovirus receptor-binding site was found to lie in a deep "canyon" of the capsid and has been commonly quoted as the "canyon hypothesis" (Rossmann et al. 1985). However, it has also been observed for influenza virus HA (Bizebard et al. 1995) and for rhinovirus (Smith et al. 1996) that antibody epitopes can include such concave sites.

Extrapolating from the size of the influenza virus neuraminidase active site and from the known structures of antibody-hapten complexes (Colman et al. 1983), and based on the sizes of influenza HA virus-receptor binding site (Weis et al. 1988) and of antibody-antigen interfaces (Amit et al. 1986), a second possibility was suggested (Colman et al. 1983; Weis et al. 1988). This was that when epitopes of neutralizing antibodies are larger than viral functional sites, the "antibody footprint" will include residues that are not involved in the function. Mutations at these positions will allow the virus to escape from neutralization without compromising its viability. Support for this possibility is provided by the finding that all mutations that allow influenza virus to escape from neutralization by an antibody whose epitope partially overlaps the virus receptor-binding site are at positions in the epitope outside this site (Bizebard et al. 1995) (Fig. 3).

A variation on this theme, that residues dispensable for virus viability are embedded in the virus-receptor interface, is illustrated by analysis of HIV gp160 surface-antigen variations. HIV gp120 binds to the primary cellular receptor CD4 with an association constant of $\sim 10^8 M^{-1}$ (Wu et al. 1996), compared to $10^3 M^{-1}$ in the case of HA-influenza virus-receptor interaction (Sauter et al. 1989). In this case, as in those of viruses that bind their cellular receptor with high affinity, a difficulty with the antibody footprint hypothesis is encountered: the size of the virus-receptor interface may be similar to that of an antibody-viral antigen interface and the two interfaces might coincide. Indeed, the structure of the complex formed between a fragment of HIV gp120 surface antigen and CD4 (Kwong et al. 1998) demonstrates that the surface area of the virus-receptor interface ($800 Å^2$) is in the range of those of antigen-antibody interfaces (Janin and Chothia 1990). However, the complementarity of HIV gp120 and CD4 is imperfect. As a consequence, there is a "variational island" on the gp120 molecule where sequence changes do not affect CD4 binding. This variational island is a location where mutations that allow escape from neutralization by antibodies directed to the CD4 binding site may take place (Kwong et al. 1998). Further along this line, it has been proposed that even when the footprints of an antibody and of the virus receptor coincide, owing to differences between virus-receptor and virus-antibody interactions, mutations that compromise binding to the antibody do not necessarily compromise binding to the receptor and may be efficient escape mutations (Colman 1997). The only limitation stems from an observation made regarding influenza virus: in spite of the numerous antibodies that bind to this area of HA, some of the influenza virus receptor-binding-site residues are totally invariant among viruses that infect humans (Wilson and Cox 1990). For a virus to escape from neutralization by an antibody, the residues that most influence virus-antibody interaction and those required for any viral function, such as virus-receptor binding, will have to differ.

## 7 Conclusions

Molecular models of antigen-antibody complexes have provided an understanding of the mechanisms by which a virus escapes neutralization by an antibody; they also demonstrate that a small receptor-binding site is of some advantage to the virus, since escape from neutralization is possible while maintaining functionality. In addition to antigenic novelty, necessary for the existence of a threshold of susceptible hosts, other factors are important for the generation of new epidemics. One of them is transmissibility, as was illustrated by the influenza outbreak that occurred in Hong Kong in 1998 (Subbarao et al. 1998). Further insight into the molecular mechanisms that govern transmissibility will be required to understand the selection of emerging strains.

Two additional issues concerning virus escape from neutralization by antibodies remain poorly understood. Firstly, several hypotheses have been made as to how viruses with a large receptor-binding site escape neutralization by antibodies binding to this site, but additional structures of antibody complexes are required to support them. Secondly, the large antigenic surface of most viruses raises the question of how viruses vary sufficiently to escape from neutralization when several antibodies bind to them at non-overlapping sites. Knowledge of the spectrum of antibodies actually produced by infected individuals and of the range of mutants in circulation is required to address this question.

*Acknowledgement.* This work was supported by C.N.R.S., M.R.C., and by a grant from the EU (Contract BIOMED BMH4-97-2393).

## References

Acharya R, Fry E, Stuart D, Fox G, Rowlands D, Brown F (1989) The three-dimensional structure of foot-and-mouth disease virus at 2.9Å resolution. Nature 337:709–716

Amit AG, Mariuzza RA, Phillips SEV, Poljak RJ (1986) Three-dimensional structure of an antigen-antibody complex at 2.8Å resolution. Science 233:747–753

Bizebard T, Gigant B, Rigolet P, Rasmussen B, Diat O, Bösecke S, Wharton S, Skehel JJ, Knossow M (1995) Structure of influenza hemagglutinin complexed with a neutralizing antibody. Nature 376: 92–94

Clackson T, Wells JA (1995) A hot spot of binding energy in a hormone-receptor interface. Science 267:383–386

Colman PM (1994) Influenza virus neuraminidase: structure, antibodies and inhibitors. Protein Sci 3:1687–1696

Colman PM (1997) Virus versus antibody. Structure 5:591–593

Colman PM, Varghese JN, Laver WG (1983) Structure of the catalytic and antigenic sites in influenza virus neuraminidase. Nature 303:41–44

Colman PM, Air GM, Webster RG, Varghese JN, Baker AT, Lentz MR, Tulloch PA, Laver WG (1987) How antibodies recognize virus proteins. Immunol Today 8:323–326

Creighton TE (1993) Proteins: structures and molecular properties, 2nd edn. Freeman, New York

Cusack S, Ruigrok RWH, Krygsman PCJ, Mellema JE (1985) Structure and composition of influenza virus. J Mol Biol 186:565–582

Dall'Acqua W, Goldman ER, Lin W, Teng C, Tsuchiya D, Li H, Ysern X, Braden BC, Li Y, Smith-Gill SJ, Mariuzza RA (1998) A mutational analysis of binding interactions in antigen-antibody protein-protein complexes. Biochemistry 37:7981–7991

Fleury D, Wharton SA, Skehel JJ, Knossow M, Bizebard T (1998) Antigen distortion allows influenza virus to escape neutralization. Nat Struct Biol 5:119–123

Fleury D, Barrere B, Bizebard T, Daniels RS, Skehel JJ, Knossow M (1999) A complex of influenza hemagglutinin with a neutralizing antibody that binds outside the virus receptor binding site. Nat Struct Biol 6:530–534

Hogle JM, Chow M, Filman DJ (1986) Three-dimensional structure of poliovirus at 2.9Å resolution. Science 229:1358–1365

Janin J, Chothia C (1990) The structure of protein-protein recognition sites. J Biol Chem 265:16027–16030

Knossow M, Daniels RS, Douglas AR, Skehel JJ, Wiley DC (1984) Three-dimensional structure of an antigenic mutant of the influenza virus haemagglutinin. Nature 311:678–680

Kwong PD, Wyatt R, Robinson J, Sweet RW, Sodroski J, Hendrickson WA (1998) Structure of an HIV gp120 envelope glycoprotein in complex with the CD4 receptor and a neutralising human antibody. Nature 393:648–659

Lea S, Hernandez J, Blakemore W, Brocchi E, Curry S, Domingo E, Fry E, Abu-Ghazaleh R, King A, Newman J, Stuart D, Mateu MG (1994) The structure and antigenicity of a type C foot-and-mouth disease virus. Structure 2:123–139

Lea S, Abu-Ghazaleh R, Blakemore W, Curry S, Fry E, Jackson T, King A, Logan D, Newman J, Stuart D (1995) Structural comparison of two strains of foot-and-mouth disease virus subtype O1 and a laboratory variant, G67. Structure 3:571–580

Logan D, Abu-Ghazaleh R, Blakemore W, Curry S, Jackson T, King A, Lea S, Lewis R, Newman J, Parry N, Rowlands D, Stuart D, Fry E (1993) Structure of a major immunogenic site on foot-and-mouth disease virus. Nature 362:566–568

Nuss JM, Whitaker PB, Air GM (1993) Identification of critical contact residues in the NC41 epitope of a subtype N9 influenza virus neuraminidase. Proteins 15:121–132

Parry N, Fox G, Rowlands D, Brown F, Fry E, Acharya R, Logan D, Stuart D (1990) Structural and serological evidence for a novel mechanism of antigenic variation in foot-and-mouth disease virus. Nature 347:569–572

Prasad L, Sharma S, Vandonselaar M, Quail JW, Lee JS, Waygood EB, Wilson KS, Dauter Z, Delbaere LTJ (1993) Evaluation of mutagenesis for epitope mapping. J Biol Chem 268:10705–10708

Rey FA, Heinz FX, Mandl C, Kunz C, Harrison SC (1995) The envelope glycoprotein from tick-borne encephalitis virus at 2Å resolution. Nature 375:291–298

Rossmann MG, Arnold E, Erickson JW, Frankenberger EA, Griffith JP, Hecht H-J, Johnson JE, Kamer G, Luo M, Mosser AG, Rueckert RR, Sherry B, Vriend G (1985) Structure of a human common cold virus and functional relationship to other picornaviruses. Nature 317:145–153

Sauter NK, Bednarski MD, Wurzburg BA, Hanson JE, Whitesides GM, Skehel JJ, Wiley DC (1989). Haemagglutinins from two influenza virus variants bind to sialic acid derivatives with millimolar dissociation constants: A 500MHz proton nuclear magnetic resonance study. Biochemistry 28:8388–8396

Smith TJ, Chase ES, Schmidt TJ, Olson NH, Baker TS (1996) Neutralizing antibody to human rhinovirus 14 penetrates the receptor-binding canyon. Nature 383:350–354

Subbarao K, Klimor A, Karz J, Regnery H, Lim W, Hall H, Purdue M, Swayne D, Bender C, Huang J, Hemphill M, Rose T, Shaw M, Xu X, Fukuda K, Cox N (1998). Characterization of an avian influenza A (H5N1) virus isolated from a child with a respiratory illness. Science 279: 393–396

Takemoto DK, Skehel JJ, Wiley DC (1996) A surface plasmon resonance assay for the binding of influenza virus hemagglutinin to its sialic acid receptor. Virology 217:452–458

Tormo J, Blaas D, Parry NG, Rowlands D, Stuart D, Fita I (1994) Crystal structure of a human rhinovirus neutralizing antibody complexed with a peptide derived from viral capsid protein VP2. EMBO J 13:2247–2256

Tulip WR, Varghese JN, Baker AT, van Donkelaar A, Laver WG, Webster RG, Colman PM (1991) Refined atomic structures of N9 subtype influenza virus neuraminidase and escape mutants. J Mol Biol 221:487–497

Tulip WR, Varghese JN, Laver WG, Webster RG, Colman PM (1992) Refined structure of the influenza virus neuraminidase-NC41 Fab complex. J Mol Biol 227:122–148

Varghese JN, Laver WG, Colman PM (1982) Structure of the influenza virus glycoprotein antigen neuraminidase at 2.9Å resolution. Nature 303:35–40

Verdaguer N, Mateu MG, Andreu D, Giralt E, Domingo E, Fita I (1995) Structure of the major antigenic loop of foot-and-mouth disease virus complexed with a neutralizing antibody: direct involvement of the Arg-Gly-Asp motif in the interaction. EMBO J. 14:1690–1696

Weis W, Brown JH, Cusack S, Paulson JC, Skehel JJ, Wiley DC (1988) Structure of the influenza virus hemagglutinin complexed with its receptor, sialic acid. Nature 333:426–431

Wien MW, Filman DJ, Stura EA, Guillot S, Delpeyroux F, Crainic R, Hogle JM (1995) Structure of the complex between the Fab fragment of a neutralizing antibody for type 1 poliovirus and its viral epitope. Nat Struct Biol 2:232–242

Wiley DC, Skehel JJ (1987) The structure and function of the hemagglutinin membrane glycoprotein of influenza virus. Ann Rev Biochem 56:365–394

Wilson IA, Cox NJ (1990) Structural basis of immune recognition of influenza virus haemagglutinin. Annu Rev Immunol 8:737–771

Wilson IA, Skehel JJ, Wiley DC (1981) Structure of the hemagglutinin membrane glycoprotein of influenza virus at 3Å resolution. Nature 289:366–373

Wu H, Myszka DG, Tendian SW, Brouillette CG, Sweet RW, Chaiken IM, Hendrickson WA (1996) Kinetic and structural analysis of mutant CD4 receptors that are defective in HIV gp120 binding. Proc Natl Acad Sci USA 93:15030–15035

# Affinity, Complementarity, Cooperativity, and Specificity in Antibody Recognition

N.S. GREENSPAN



## 1 Introduction

The recognition of antigens by antibodies provides one of the best-studied paradigms of molecular specificity in all of biology. Consequently, there is an extensive repository of experimental information available for use in exploring the numerous connections among the four concepts highlighted in the title of this chapter. First, each of the four concepts is examined individually, in sequence, in the context of monovalent recognition. The concepts are then revisited in the more challenging context of multivalent recognition. In the latter setting, the concepts are treated two-at-a-time. I emphasize some points that are insufficiently appreciated by immunologists, even if they are familiar to most individuals from other disciplines, such as biophysics.

Institute of Pathology, Biomedical Research Building, Room 927, Case Western Reserve University, 10900 Euclid Avenue, Cleveland, OH 44106-4943, USA

## 2 Monovalent Interactions

### 2.1 Affinity

Affinity is a measure of the strength of the noncovalent interaction between two molecules. In immunology, it is customary to distinguish between *intrinsic affinity*, which applies to monovalent interactions, and *functional affinity*, which applies to multivalent interactions. Some prefer the term *avidity* to refer to the strength of a multivalent interaction. A disadvantage of the term "avidity" is that it was used historically in the context of assays for the strength of multivalent interactions that were not pure measures of binding between antigen and antibody. Examples of such assays are those based on precipitation or agglutination (Karush 1976). In this section, the focus will be on intrinsic affinity.

Intrinsic affinity is defined as the equilibrium association constant characterizing an interaction between one antibody paratope (combining site) and one epitope (antigenic determinant). Therefore, if monovalent paratopes are represented by $P$, monovalent epitopes by $E$, paratope–epitope complexes by $PE$, and the equilibrium association constant by $K_A$, then, by the Law of Mass Action:

$$K_A = [PE]/[P][E]$$

The above equation indicates that the intrinsic affinity is the reciprocal of the concentration of monovalent antigen at which half of the paratopes will be occupied. One can regard this measure of the strength of the attraction between paratope and epitope as relating to the relative likelihood that epitope and paratope will be found in complex, as opposed to being found in the unbound state in solution. The equilibrium association constant can also be related to the free energy change that accompanies complex formation.

The term "intrinsic", in "intrinsic affinity", can be taken to refer to the fact that the magnitude of a monovalent affinity is a measure of how well the paratope fits the epitope (complementarity), while the functional affinity or avidity can reach relatively high values merely as a consequence of multivalence, even if each individual interaction is relatively weak or involves a relatively poor fit. However, the precise value of an intrinsic affinity is not an attribute that is truly intrinsic to the paratope, the epitope, or even the paratope–epitope pair. In fact, the exact value of the intrinsic affinity will depend on many factors, in addition to the degree of complementarity (as defined by any measure other than the intrinsic affinity itself) between paratope and epitope. If significant conformational changes in antibody or antigen accompany the formation of the antibody-antigen complex (Rini et al. 1992; reviewed in Wilson and Stanfield 1994; Davies and Cohen 1996), the intrinsic affinity will reflect this fact. The other factors that influence the value of the intrinsic affinity include variables that are extrinsic to the antibody and the antigen: temperature, pressure, pH, ionic strength, the concentration of other macromolecules in the solution (even if these macromolecules exhibit no significant tendency to react preferentially with either reactant or the complex), and the presence of any structures (e.g., membranes, fibers) that confine the interacting species (ZIMMERMAN and MINTON 1993). These latter two factors will now be considered in more detail.

In biological systems, intrinsic affinities between molecules are likely to involve what have been referred to as volume-exclusion (MINTON 1993; reviewed in ZIMMERMAN and MINTON 1993) and confinement effects (MINTON 1992; reviewed in ZIMMERMAN and MINTON 1993). Volume-exclusion effects apply when total solute (e.g., protein) concentration is high. The free energy changes associated with the formation of molecular complexes under these conditions can be substantially altered relative to the situation in which the same molecular complexes form in an environment characterized by a low overall protein or solute concentration.

In evaluating the potential significance of these effects, it is worth noting that total protein concentration is typically in the range of 65–80mg/ml in blood and can be over 300mg/ml intracellularly. The contrast with the usual protein concentrations in typical in vitro media of approximately 10mg/ml (or less) is marked. These effects on affinities are enhanced when the predominant background solute species is smaller than the receptors or ligands of interest. In a provocative hypothetical scenario discussed by MINTON (1993), a site (e.g., a paratope) that, in the absence of crowding effects binds more strongly to an epitope that is part of a smaller ligand A than to an epitope associated with a larger ligand B, may reverse this preference as the concentration of background species increases (Fig. 1). Thus, despite the fact that the structures involved in these latter interactions have not been altered in any way by the perturbations in the system, the specificity, as defined in thermodynamic terms, has been substantially modified by altering the concentrations of molecular species that a purely structural perspective would suggest are irrelevant to the interaction.

Similarly, the presence of large numbers of fibers or of membranes, creating compartments which confine macromolecules to regions only slightly larger than the molecules themselves, can substantially alter the chemical potentials of these molecules (MINTON 1992). Consequently, the free energy changes associated with complex formation can be substantially altered in comparison to the same interactions occurring in bulk solution. In general, confinement, like molecular crowding, tends to enhance molecular association. Calculations suggest that, particularly inside of cells where fibrous elements have been estimated to occupy approximately 20% of the total volume of the cytoplasm, such effects can be quantitatively significant. The magnitude of the confinement-mediated enhancement of molecular association will be influenced by the shape and size of the confining compartment and by the shape and size of the resulting aggregates (MINTON 1992).

The key conclusion to be drawn from the existence of volume-exclusion and confinement effects is that a free energy change of complex formation reflects many factors that cannot be assessed from knowledge, however detailed, of the structural features of the interface between receptor and ligand (Fig. 2). Another way of stating this point is that the magnitude of the free energy change associated with the union of receptor and ligand takes into account the composition and attributes of a

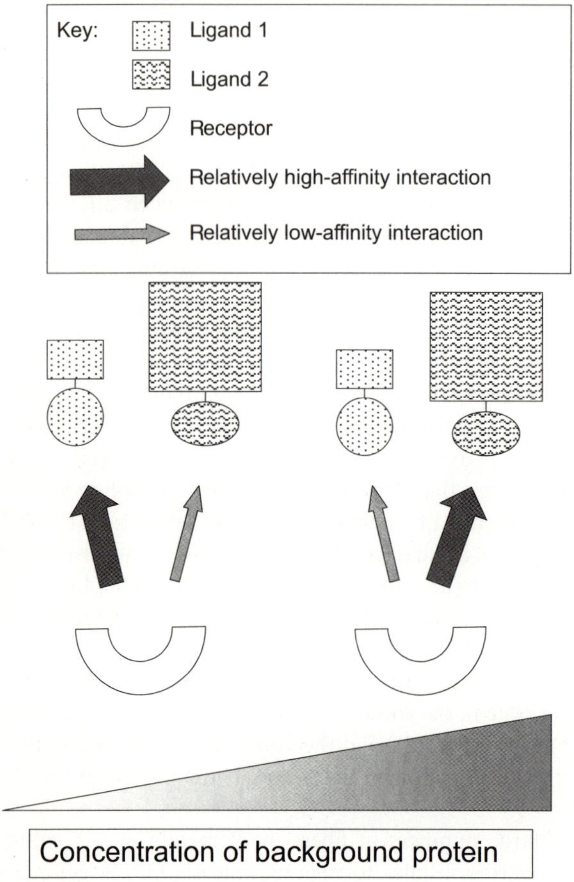


**Fig. 1.** The hypothetical influence of molecular crowding on the thermodynamics of molecular discrimination exhibited by a receptor (MINTON 1993). At low background protein concentration, the monovalent receptor exhibits a higher intrinsic affinity for ligand 1 than for ligand 2. When the background protein concentration passes a threshold (i.e., under conditions of molecular crowding), the absolute value of the magnitude of the free energy of complex formation becomes greater for ligand 2 than for ligand 1. Ligand 2 has less shape complementarity than ligand 1 with respect to the receptor but is larger. (Reproduced with permission from GREENSPAN 2001)

molecular system, which includes receptor and ligand but which also includes many other molecular species. Standard thinking about molecular specificity frequently engenders indifference to these other molecular species. Yet, the concentrations of these other molecules can have effects on the specific interaction under consideration of a magnitude approximating the magnitude of the free energy change attributable to the intermolecular forces that are typically considered (ZIMMERMAN and MINTON 1993).

Further evidence that molecular context can affect the binding properties of a binding site is provided by FOOTE and RAMAN (2000). These authors reported a strong correlation between the locations of ligand-binding sites and the locations of the principal axes of inertia for proteins selected for ligand binding function. Specifically, they found that at least one principal axis of inertia passes through the protein surface at a site used for molecular recognition. They also suggested that location of a binding site, such as an antibody paratope, around a principal axis of inertia would contribute to the stability of a complex formed between that binding

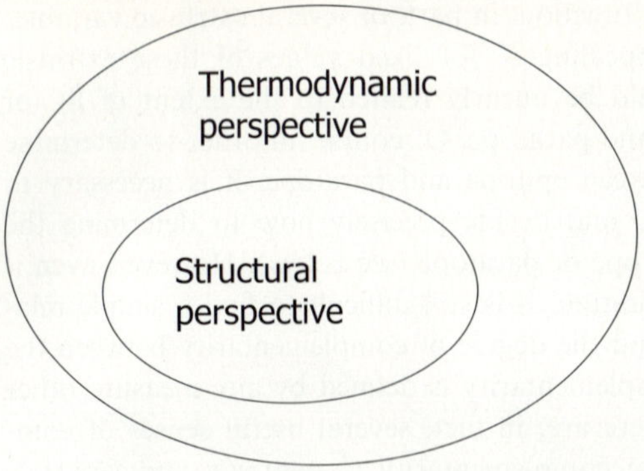


**Fig. 2.** Illustration of the relationship between thermodynamic and structural perspectives with respect to the factors determining the affinity between an antibody paratope and an epitope on an antigen molecule. Of necessity, a thermodynamic perspective must include all of the forces and effects operative at the intermolecular interface, but the free energy of complex formation can also be influenced by factors that cannot be assessed solely by reference to the structural details of the interacting surfaces

site and a ligand. While it remains to be determined how quantitatively significant these effects are, their existence serves to emphasize the lack of a one-to-one correspondence between binding site structure and noncovalent binding properties.

The range of intrinsic affinities characterizing biologically meaningful interactions between antibodies and antigens is large, extending over at least seven orders of magnitude, from a minimum of about $10^4M^{-1}$ up to at least $10^{11}M^{-1}$. While affinities higher than $10^{11}M^{-1}$ have been reported for antibodies from sheep, murine monoclonal antibodies derived from hybridomas do not generally bind cognate antigens with intrinsic affinities above this figure (Webster et al. 1994). The nature of the antigen can influence where in this broad range the intrinsic affinities of the corresponding antibodies tend to fall. For instance, on average, antibodies specific for protein antigens tend to bind their cognate ligands with higher intrinsic affinities than do antibodies specific for carbohydrate antigens (Karush 1978). On average, it would be expected that more rigid antigens would bind to antibodies with higher intrinsic affinities than would more flexible antigens due to the entropic costs, associated with complex formation, of restricting the motions of previously mobile molecular moieties.

## 2.2 Complementarity

The essence of the concept of complementarity, considered from a general perspective, is that two entities fit together in some sense to create a whole. In science and mathematics, the terms "complementary" and "complement" have many distinct technical definitions and uses. For example, two mutant bacteriophage genomes can complement one another in an infected bacterial cell if the mutations in the respective genomes inactivate genes at distinct loci; together the wild-type genes in *trans*, each residing at a locus corresponding to the mutant gene in the other genome, can restore a "whole" phenotype. In mathematics, a set and its complement constitute a whole universe of discourse. Similarly, a whole bimolecular complex results from the noncovalent association of complementary molecules that correspond to antibody and antigen or receptor and ligand.

Given that intrinsic affinity is a function, in part, of several extrinsic variables (e.g., temperature), it would be appealing if, for fixed values of these extrinsic variables, the intrinsic affinity would be linearly related to the extent of fit, or complementarity, between epitope and paratope. Of course, in order to determine the extent of complementarity between epitope and paratope, it is necessary to define the term "complementarity" and decide precisely how to determine the composition and boundaries of epitope or paratope (see below). However, even if one puts aside this issue of site definition, it is still difficult to find a simple relationship between intrinsic affinity and the degree of complementarity between the interacting surfaces, as long as complementarity is defined by any measure other than the intrinsic affinity itself. There are, in fact, several useful senses of complementarity or fit, including shape complementarity (LAWRENCE and COLMAN 1993; JONES and THORNTON 1996), electrostatic complementarity (MCCOY et al. 1997) or other forms of chemical complementarity, and various amalgamations of these. As discussed below, each of these senses of complementarity admits of more than one precise formulation.

The most obvious sense of complementarity pertains to the shapes of the interacting sites on antibody and antigen. While it is easy to imagine roughly what is meant by "shape complementarity", it is less straightforward to decide how to assign a single numerical value for the fit, solely with respect to shape, between two molecules. In fact, a number of distinct criteria have been considered for the assessment of shape complementarity, including the extent of buried surface, the absence of buried water molecules, the packing density of interface atoms, and the average distance between contact atoms on receptor and ligand. There are also several schemes for the precise calculation of shape complementarity, and a couple of examples of such scheme will illustrate the diversity of approaches. One proposal for quantitating shape complementarity relies on the correspondence between the directions of, respectively, outwardly directed and inwardly directed normal unit vectors originating at corresponding sites of closest approach randomly chosen from the interface between the two molecules (LAWRENCE and COLMAN 1993). Another scheme, from JONES and THORNTON (1996), referred to as the gap index, assesses shape complementarity by dividing the volume of the intermolecular interface by the solvent accessible area of the interface. It is not yet clear how closely these different measures of shape complementarity correlate with one another, or if any of the available algorithms for calculating shape complementarity is truly superior to the others in correlating with intrinsic affinity.

In fact, there are strong reasons to expect that no measure of shape complementarity would reliably correlate with intrinsic affinity even under conditions in which the extrinsic variables, such as temperature and ionic strength, are fixed. Molecular shape can remain relatively constant despite variations in atomic identity that dictate affinity-altering differences in chemical complementarity. An additional consideration also suggests that a simple relationship between shape complementarity and affinity is unlikely. It is sometimes the case, as noted above, that the conformation of the antibody or the antigen in the unbound state is altered in the process of forming the complex (WILSON and STANFIELD 1994; DAVIES and COHEN

1996). Such conformational change will generally entail an energetic penalty that would be difficult to discern, let alone estimate, by simply inspecting the structures at the intermolecular interface. The inevitable variation, from complex to complex, in the magnitudes of such energetic costs means that, for any given extent of shape complementarity characterizing a complex (final complementarity), there will be a range of free energies of complex formation. A corollary of this conclusion can be stated as follows. Given two analogues of a ligand, the analogue with greater structural similarity to the cognate ligand in the unbound state (initial shape complementarity to the receptor) may not bind better to the receptor of interest or elicit the greater functional effect (Hua et al. 1991; Greenspan 1997).

In considering the relationship between intrinsic affinity and epitope-paratope complementarity, it was intimated that it would be necessary to precisely define the sites referred to as epitope or paratope. This task turns out to be more challenging in practice than the conventional definitions would lead one to expect (Greenspan 1992; Greenspan and Di Cera 1999). The difficulties arise because the conventional definition is based on contact residues (or atoms). There is one great advantage in using physical contact as the only criterion for including an amino acid, atom, or other structural element in a paratope or epitope. If the three-dimensional structure of the complex is known at high resolution, the constituents of either epitope or paratope can be clearly delineated.

However, a major disadvantage of the contact-based definition for epitopes (paratopes) is that many, perhaps most, investigators are primarily interested in knowing which structural elements account for the affinity or differential affinities that characterize a given interaction between an antibody and, respectively, a given antigen or set of antigens. Of course, kinetic as well as equilibrium parameters may also be of interest. The relevant atoms or amino acids are not necessarily limited to contact residues, and the contact residues are not all necessarily of importance in contributing to affinity or differential affinity (or kinetics) (Tulip et al. 1992; Lavoie et al. 1992; Hawkins et al. 1993; Greenspan 1992; Greenspan and Di Cera 1999; Fig. 3).

In addition to structural techniques, investigators interested in characterizing epitopes or paratopes may use techniques such as systematic substitution of wild-type amino acid residues with alanine (alanine scanning mutagenesis; Cunningham and Wells 1989). This approach determines those positions in the primary structure of a protein or peptide at which alanine substitutions substantially alter affinity or differential affinity. Some investigators may then feel obliged to suggest that the wild-type residues for which alanine substitutions have a major effect are part of the epitope or paratope as conventionally defined (i.e., make contact with the molecular partner). In this case, the conventional definition of epitope or paratope serves to confuse more than clarify because the set of contact elements is not likely to be completely identical with the sets of affinity-determining or specificity-determining elements in spite of the high likelihood of overlap among these categories (Greenspan 1997; Greenspan and Di Cera 1999). Furthermore, the quantitative interpretation of results from alanine scanning mutagenesis studies, for the purpose of attributing energetic contributions (to binding) to specific amino acids, is

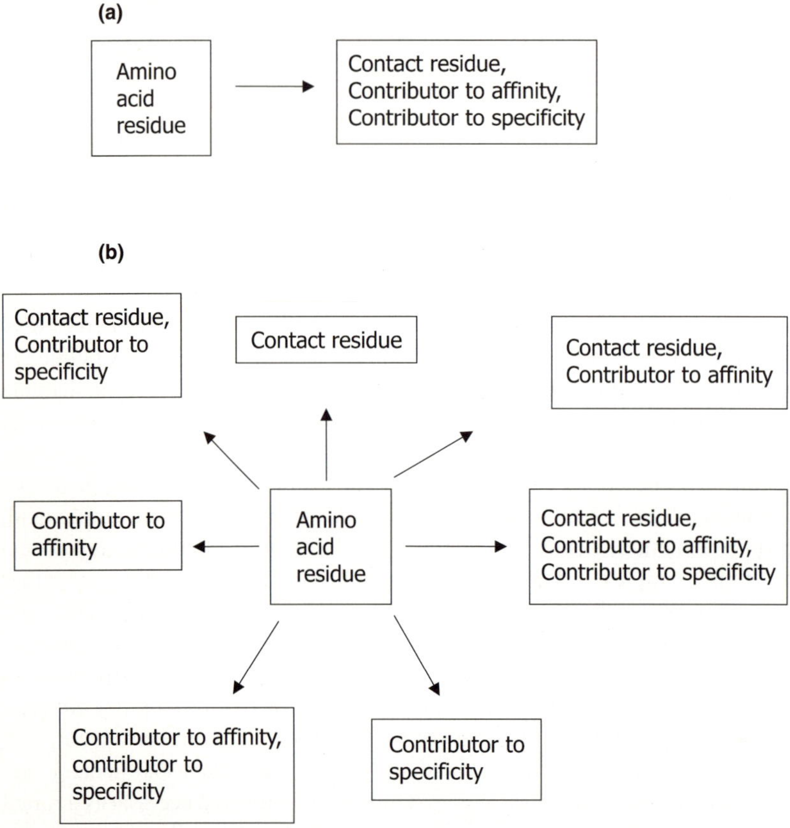


**Fig. 3A,B.** Two possible extremes in the relationships among the structural correlates of intermolecular contact, affinity, and specificity. **A** The simplest situation is that every structural element (e.g., amino acid residue) of an antibody that directly contacts the cognate epitope is also an important contributor to the free energy of complex formation with that cognate ligand and an important contributor to the differential free energy of complex formation that permits the antibody to distinguish between cognate and non-cognate antigens. **B** A more realistic scenario is that in which a given amino acid or atom (of the antibody) can participate in a noncovalent interaction in any one of the seven logically possible patterns. Experimental results suggests that the situation depicted in **A** is not general, and that most, probably all, of the patterns pictured in **B** are populated in actual biomolecular interactions. For example, contact residues might not be energetically-important residues, and conversely, important affinity-determining or specificity-determining residues might not be contact residues

frequently based on invalid assumptions of additivity and is consequently unreliable (Di Cera 1998a,b).

Investigators typically attribute the difference in the free energy of complex formation (with a given molecular partner) between wild-type and alanine-substituted molecules to the wild-type amino acid. This assumption is valid only if the removal of the wild-type side chain and its replacement with the (typically) smaller alanine side chain have no other effects on the energetics of the protein. However, it is often the case that such additional perturbations exist. This assertion is supported by the fact that, for the studies published so far, the free energy of complex formation for the wild-type receptor and the ligand of interest is not equal in magnitude to the sum of the individual contributions attributed to the amino acids for which alanine substitutions were made (Di Cera 1998a,b). Another infrequently noted subtlety of the interpretation of alanine mutations is that such mutations do not necessarily reveal the contributions to a bimolecular interaction of main chain, as opposed to side chain, atoms of the wild-type amino acid.

Another factor that complicates discussion of the complementarity characterizing the interaction between receptor and ligand, and the relationship between complementarity and intrinsic affinity or specificity, is the role of the solvent. In immunological (biological) contexts, water is obviously the dominant solvent and associations of water molecules with proteins are characterized by half-times of from $10^{-10}$ to $10^{-2}$s (Levitt and Park 1993). Consequently, the more tightly bound protein-associated water molecules may serve, for some purposes, as functional components of the protein. Furthermore, the binding of ligand to a site generally requires the displacement of some water molecules, but other water molecules may play crucial roles in the formation of the intermolecular interface (Quiocho et al. 1989). In a crystallographic study of the interaction between hen egg-white lysozyme and an Fv fragment derived from the monoclonal antibody D1.3, it was found that the number of bound water molecules associated with the antibody-antigen complex was greater than the sum of the bound waters found associated with the free reactants (Bhat et al. 1994). Furthermore, it was concluded that the bound water molecules played a significant role in the energetics of the interaction between D1.3 Fv and hen egg-white lysozyme.

The implication for measures of complementarity is that for some purposes the extent of shape complementarity between receptor and ligand should be calculated with bound water molecules included in the analysis. It is also conceivable that for some purposes it may be reasonable to exclude the bound water molecules from the determination of the degree of complementarity. These considerations suggest that there may be no single best approach for assessing the magnitude of shape complementarity for a complex between two biological molecules.

## 2.3 Cooperativity

The essence of cooperativity is non-additivity. If the effect of two interactions is more (positive cooperativity) or less (negative cooperativity) than the sum of the respective individual interactions, then cooperativity is present. Most discussions of molecular recognition in immunological contexts refer rarely, if at all, to the notion of cooperativity, at least in an explicit manner, despite the evidence that suggests that cooperativity is central to molecular recognition (Greenspan 1992; Di Cera 1998b). It is also noteworthy that, while the presence of cooperativity is frequently regarded as a special case, measures of cooperativity can range over a continuum from negative to positive values, and in this context, a value of zero would appear to be the special case.

The existence of cooperativity in the context of multivalent interactions, both intramolecular and intermolecular, is certainly acknowledged in the immunological literature, but it is less often noted that even in the case of a monovalent interaction between biomolecules there is almost certain to be a degree of cooperativity. Such cooperativity arises in part out of the fact that monovalent receptor-ligand interactions inevitably involve multiple individual atomic interactions. The energetic contribution of any atom (from epitope or paratope) is influenced by the presence or absence of contributions from other atoms. At an even more fundamental level, biological macromolecules such as proteins and nucleic acids are composed of many subunits that interact in attaining native conformation. Consequently, the process by which macromolecules achieve native conformation is intrinsically highly cooperative (LIM and SAUER 1989). GREGORIO WEBER (1975) noted that the strengths of the individual interactions between receptor and ligand are of the same order as those within the interacting molecular species. Consequently, "internal protein interactions cannot be separated from protein-ligand interactions, and when more than one ligand is bound the modification of the protein structure makes it impossible for the binding of the ligands to be independent in any rigorous sense" (WEBER 1975).

## 2.4 Specificity

In reference to monovalent interactions, specificity can have at least two distinguishable, if related, meanings. First, stating that a given antibody has specificity for a particular antigen can mean simply that the antibody exhibits an intrinsic affinity above some threshold for the antigen. Left implicit is the expectation that the antibody will be unlikely to exhibit such an affinity for a molecule chosen at random. For this sense of specificity, it can be argued that intrinsic affinity between antibody and antigen represents the ultimate measure. Alternatively, specificity can explicitly refer to the extent of discrimination between a cognate ligand and a non-cognate ligand (or ligands). In this case, the difference between two intrinsic affinities provides the most appropriate measure. So, whereas specificity in the second sense explicitly refers to discrimination, specificity in the first sense relates only implicitly to the notion of discrimination between molecules. Thus, these two senses of specificity overlap to some degree and both are useful, but it is frequently the case that the two senses are used interchangeably without explicit acknowledgment of their differences.

Complementarity plays a role in both types of specificity. As emphatically pointed out by Linus Pauling (PAULING and DELBRUCK 1940; PAULING 1962), significant free energies favoring complex formation can arise from the short-range forces operative at the intermolecular interface only if the components of the interacting species are in close proximity. These inter-atomic attractive forces decrease dramatically as a function of distance on the scale of Å. However, as noted above, since variable amounts of energy may be expended in attaining such proximity, one should not expect any simple relationship between final shape complementarity, final chemical complementarity, or some combination of the two, and intrinsic affinity. Since specificity in the second sense focuses on the specificities in the first sense (i.e., intrinsic affinities) for two or more ligands, the relative extents of final shape or chemical complementarity between one antibody and each of two antigens are critically involved, although not necessarily completely decisive, in determining the degree of discrimination.

Cooperativity is also crucial to understanding specificity in either sense. As mentioned earlier, the magnitude of the contribution attributable to any given noncovalent contact is dependent on the context created by all of the other contacts that constitute a given intermolecular interaction. This point is dramatically illustrated by the fact that modifying a single amino acid side chain by as little as a single methyl group or a single atom can alter the intrinsic affinity by one, two, or even three orders of magnitude. If ten to twenty amino acid residues of an antibody were directly involved in a noncovalent interaction with an antigen, as is often the case for protein antigens (Davies and Cohen 1996), then the absence of cooperativity would imply that changing any given residue could decrease the affinity by no more than 10–20%. The fact that changing one residue can alter the intrinsic affinity by orders of magnitude strongly implicates the cooperative nature of recognition. Thus, both intrinsic affinity for a cognate ligand and the difference in intrinsic affinities for cognate and non-cognate ligands are crucially dependent on cooperativity.

The existence and impact of molecular flexibility (atomic mobility) provides a basis for distinguishing between affinity and specificity (discrimination among ligands). If a binding region, such as an antibody paratope, is flexible, then it stands to reason that it would be better able to accommodate ligands with relatively dissimilar structures than would a paratope that is relatively rigid. On the other hand, the existence of greater atomic mobility among the constituents of the paratope prior to ligand binding would tend to diminish the intrinsic affinity in comparison with a paratope that has less atomic mobility in the unbound state but that can achieve the same final complementarity with respect to the epitope. Thus, molecular flexibility may give rise to a trade-off between maximal intrinsic affinity for one ligand vs reactivity, above some intrinsic affinity threshold, with the greatest number of ligands.

# 3 Multivalent Interactions

## 3.1 Functional Affinity and Cooperativity in Multivalent Binding

Multivalent interactions, in which two or more sites on a receptor simultaneously contact two or more sites on a ligand, are of importance for a vast range of biological phenomena. These include many antibody-antigen interactions, activation of the classical pathway of complement, interactions between immune complexes and Fc receptors on leukocytes, cell-cell interactions, cell-extracellular matrix interactions, signal transduction, and the regulation of gene transcription in lymphocytes and other cells. Analysis of multivalent interactions is inherently more complex than analysis of monovalent interactions (CROTHERS and METZGER 1972; DELISI 1976; KARUSH 1976; JENCKS 1981). A major source of this complexity is the existence of cooperative effects involving distinct binding sites.

As noted by DI CERA (1998b), there are two ways to describe binding equilibria involving soluble receptor molecules with two or more binding sites for a ligand. A global perspective emphasizes the overall behavior of ligand binding to the receptor molecule without explicitly distinguishing between the two (or more) sites. A site-specific perspective follows the binding occurring at each unique site. The conditions for zero cooperativity in the two perspectives do not necessarily coincide (DI CERA 1998b) if the two sites of the receptor molecule exhibit different intrinsic affinities for ligand. While, as noted above in the discussion of monovalent interactions, zero cooperativity is a formal possibility, a finite degree of cooperativity (positive or negative) is common. In cases of multivalent recognition, in which positive and negative cooperativity can be present together, the net effect depends on factors such as the relative spatial distributions of paratopes and epitopes.

Consider the case in which a soluble receptor molecule, such as an IgG antibody, with two identical binding domains interacts with a ligand (antigen) with two identical epitopes. It may appear obvious that each paratope contributes equally to the binding. This statement is true in one sense, since the intrinsic affinities of the two paratopes for the individual epitopes on the ligand are identical. Averaged over many encounters between the same antibody and the same bivalent antigen, the two paratopes of any individual antibody molecule would be expected to contribute equally. However, in another sense (see below) it is not true that the two paratopes contribute equivalently (KARUSH 1976).

In a single antibody-antigen interaction, the equilibrium association constants characterizing, respectively, the initial paratope–epitope interaction and the (temporally) second, structurally identical, paratope–epitope interaction will frequently be unequal. In the (temporally) second interaction, the linkage between the two molecules provided by the first paratope–epitope bond increases the effective concentrations of both paratope and epitope and diminishes the entropic penalty to be attributed to the second paratope on binding. On this basis, the equilibrium constant of the second interaction can correspond to a higher affinity, which contributes to intramolecular, positive cooperativity between the sites (CROTHERS and METZGER 1972; KARUSH 1976). If the orientation and the distance between the epitopes are non-optimal for the paratopes, then the equilibrium association constant for the temporally second reaction may be decreased, relative to the first reaction (KARUSH 1976). On that basis, a form of negative cooperativity between the paratopes can be exhibited.

JENCKS (1981) suggested that one can define intrinsic free energies of complex formation characterizing the interactions between distinct receptor sites and distinct ligand sites and a coupling free energy, which is derived primarily from alterations in rotational and translational entropy. WILLIAMS and WESTWELL (1998) have pointed out that, in addition to the entropic basis for positively cooperative binding of a bivalent or multivalent molecule, there is also an enthalpic contribution. Their argument is that the restriction of motion consequent to the binding of the first paratope will assist the binding (greater enthalpic contribution) by the temporally second paratope. They further suggest that this enthalpic effect is particularly pronounced for a binding site of relatively low intrinsic (monovalent) affinity when the other paratope, to which it is physically linked, binds relatively more tightly to its corresponding epitope (Williams and Westwell 1998).

Another key point is that, in analyzing a multivalent interaction, particularly an interaction in which two or more qualitatively different sites are involved, one must keep track of how the coupling free energy term is accounted for. It can be attributed to one or the other of the sites, or it can be kept separate, depending on the purpose of the analysis, but correct conclusions require that it be clear how the coupling free energy is being handled.

It is clear that structural determinants of the coupling free energy can be located at sites physically separated from the actual contact-mediating sites on the receptor or the ligand. Consequently, variation in structural elements distant from the interface-forming elements can exert substantial influence on both the functional affinity characterizing multivalent interactions and on the selectivity of such interactions (Greenspan and Cooper 1995). Furthermore, alterations in the binding of ligands at one site when a ligand binds to a distant site are not necessarily mediated through readily visualized conformation-altering signals transmitted to the contact residues (of the site with modified reactivity). Instead, such changes in binding activity may represent structurally subtle (verging on undetectable) pathways of energetic linkage (Williams and Westwell 1998; Freire 1999) or energetic costs for aligning sites for simultaneous engagement (Del Rio et al. 1993). As noted by Williams and Westwell (1998), the mere "tightening" (or "loosening") of the internal structure of a macromolecule upon binding by one ligand can be sufficient to account for altered affinities for additional ligands. Gross changes in molecular shape, as are frequently invoked by immunologists, are not essential, although dramatic shape changes can contribute to such functional linkage in a proportion of cases.

Several mechanisms of cooperative interactions between monoclonal antibodies (mAbs) binding to the same antigen or antigen complex have been described (reviewed in Thompson and Jackson 1984; Greenspan and Cooper 1993). Reports describing the ability of unlabeled mAb to enhance the binding of radiolabeled mAb to a protein or glycoprotein antigen in solution or attached to a solid surface began appearing as early as 1979 (Oi and Herzenberg 1979). Typically, the unlabeled and the labeled mAb were different mAbs, known or presumed to bind to different epitopes. Two basic mechanisms were thought to account for these instances of cooperativity between mAbs of differing specificity (Fig. 4A,B). The first mechanism involved the cross-linking of two soluble antigen molecules by one bivalent mAb, increasing the probability that a second mAb, specific for an independent epitope, could bind to two sites, instead of to only one site (Ehrlich et al. 1982; Holmes and Parham 1983; Moyle et al. 1983). The second form of co-

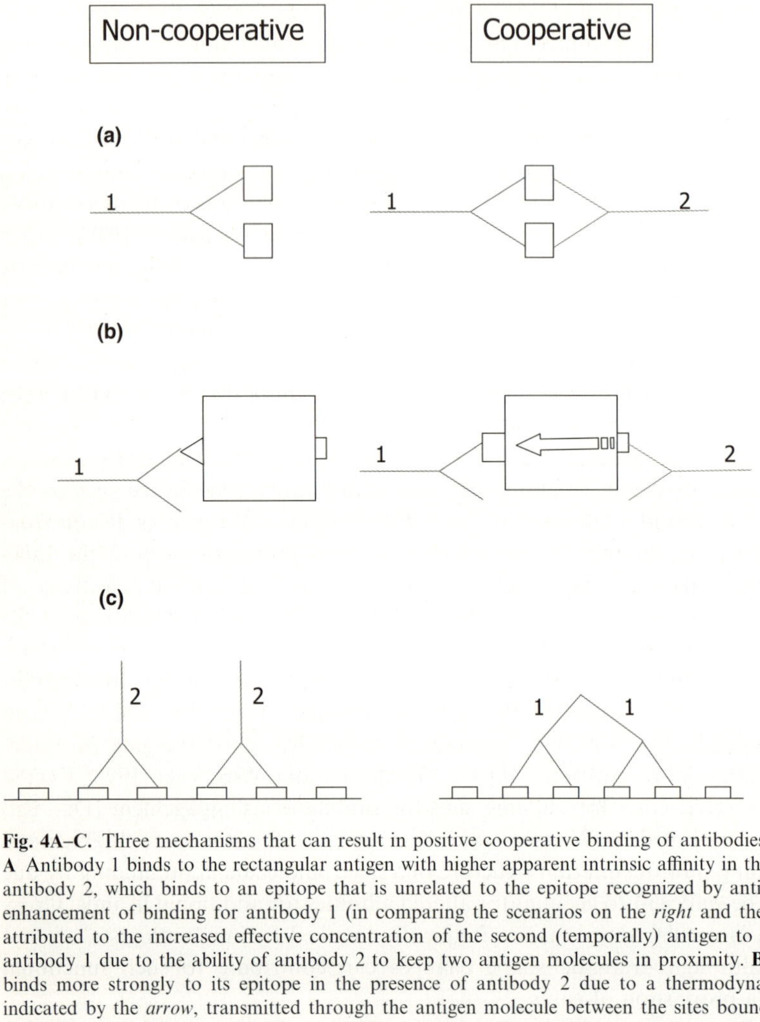


**Fig. 4A–C.** Three mechanisms that can result in positive cooperative binding of antibodies to antigens. **A** Antibody 1 binds to the rectangular antigen with higher apparent intrinsic affinity in the presence of antibody 2, which binds to an epitope that is unrelated to the epitope recognized by antibody 1. This enhancement of binding for antibody 1 (in comparing the scenarios on the *right* and the *left*) can be attributed to the increased effective concentration of the second (temporally) antigen to be bound by antibody 1 due to the ability of antibody 2 to keep two antigen molecules in proximity. **B** Antibody 1 binds more strongly to its epitope in the presence of antibody 2 due to a thermodynamic linkage, indicated by the *arrow*, transmitted through the antigen molecule between the sites bound by the two antibodies. Such linkage could reflect a significant conformational change or merely a subtle tightening of the antigen structure, as discussed in the text. **C** The antibodies labeled *1* (cooperative, *right*) bind more strongly than the antibodies labeled *2* (non-cooperative, *left*) to a multivalent antigen, displaying many copies of the same epitope, due to the presence of noncovalent interactions between the Fc regions of the antibodies labeled *1*. This cooperative binding mechanism is employed by murine IgG3 antibodies in binding to the group A streptococcal cell wall polysaccharide. Although the antibodies labeled *2* do not exhibit intermolecular cooperativity, if they use two paratopes to bind simultaneously to two epitopes displayed on the same surface, they are exploiting intramolecular cooperativity, as are the antibodies labeled *1*. Each of the antibodies in **C** could be depicted as binding to the antigen through only a single Fab arm. Then, only the antibodies labeled *1* would be binding cooperatively

operativity was believed to result from the elicitation of a conformational change in the antigen, associated with the binding of the first mAb, such that the intrinsic affinity of the interaction between the second mAb and the antigen was increased (Heinz et al. 1984; Parham et al. 1986). As noted above, such intramolecular (within the antigen) effects do not necessarily require a gross alteration of molecular shape.

A third mechanism of cooperative binding by antibodies, further described in the next section, involves noncovalent Fc–Fc interactions between antibodies binding to an antigenic surface expressing many copies of the relevant epitope(s) (Fig. 4C). This mechanism of cooperative binding, first observed with mAbs specific for the *N*-acetyl-glucosamine residues of the *Streptococcus pyogenes* cell wall polysaccharide, can be distinguished in several respects from the mechanisms described in the reports cited above. First, the antigen with which this phenomenon was initially discovered was a bacterial cell wall polysaccharide possessing a highly repetitive structure (Coligan et al. 1978), and not a protein or glycoprotein. Second, the unlabeled mAb and the labeled mAb could express identical or non-identical epitope specificities, as long as the relevant epitopes were present on the same surface (Greenspan et al. 1987, 1988). Third, cooperating mAbs could bind to different antigen molecules if both were attached to the same surface (Greenspan et al. 1988). Fourth, the occurrence of cooperative binding depended critically on the isotypes of the mAbs involved (Greenspan et al. 1987, 1988; Cooper et al. 1991). Fifth, the cooperativity between mAbs required intact Fc regions (Greenspan et al. 1987, 1988; Cooper et al. 1991).

Functional affinities characterizing antibody–antigen interactions will generally be greater than the corresponding intrinsic affinities. The classic studies of Karush and colleagues, in the era before monoclonal and recombinant antibodies, suggested that the ratio of functional affinity to intrinsic affinity could be greater than $10^4$ for IgG antibodies and greater than $10^5$ for IgM antibodies (Hornick and Karush 1972; Gopalakrishnan and Karush 1974). Enhancements factors for functional affinity vs intrinsic affinity for monovalent vs multivalent antibody fragments, such as single-chain Fv fragments, have also been measured (Pack et al. 1995; Rheinnecker et al. 1996) using ELISA and surface plasmon resonance. Although Rheinnecker et al. (1996) do not provide direct quantitative comparisons of monovalent and multivalent equilibrium constants, they do note that the bivalent and tetravalent forms they studied exhibited increases in their dissociation half-times, relative to the monovalent Fv fragment, of at least 20-fold and 160-fold, respectively. However, these larger species had slower association rate constants than the monomeric form. Pack et al. (1995) also find slower on-rates and slower off-rates for a tetrameric single-chain Fv construct than for the dimeric single-chain Fv construct or the monomeric single-chain Fv. They also note that the effect of multivalency on functional affinity is dependent on epitope density.

It is to be expected that both intrinsic and functional affinity of antibody for antigen will influence biological activities of antibody such as the neutralization of toxins or viruses, complement activation, and opsonization (Steward 1981). The biological significance of antibody affinity was apparent in a study of antibody-mediated protection to *Streptococcus pneumoniae* using human post-vaccination sera (Usinger and Lucas 1999). In this study, an inverse correlation was observed between apparent anti-pneumococcal functional affinity and the quantity of antibody required to protect mice from bacteremia. Higher-affinity antibody was also more effective in vitro in mediating opsonophagocytosis of pneumococci. In a model of vesicular stomatitis virus (VSV) infection, BACHMANN et al. (1997) found that mAb avidity correlated with neutralization in vitro. However, protection in vivo was not strongly influenced by antibody avidity above a threshold of about $\sim 2–5 \times 10^7 M^{-1}$.

Bivalence or multivalence not only contributes to stronger binding, it can also enhance the functional consequences of antibody binding (METZGER 1992). Such functional consequences can contribute to protection against pathogens or to immunopathology. One example relates to an unusual model system in which antibodies can clear Sindbis virus infection from cells of neural origin both in vivo and in vitro (LEVINE et al. 1991). It was found that intact virus glycoprotein-specific IgG and $F(ab')_2$ fragments effectively eliminate virus from an infected cell line in vitro, while Fab fragments are ineffective (UBOL et al. 1995). Adding anti-immunoglobulin specific, respectively, for Fab, immunoglobulin heavy chain, or κ-light chain to cross-link the Fab fragments restored the antiviral effect. In conjunction with data indicating comparable binding of Fab and $F(ab')_2$ fragments to viral antigen, these results suggest that bivalency is crucial to the biological effect.

## 3.2 Complementarity and Specificity in Multivalent Binding

In discussing monovalent recognition, it was pointed out that there are several useful definitions of complementarity, and that there is no simple relationship between the extent of epitope-paratope complementarity (defined by any measure other than the intrinsic affinity itself) and intrinsic affinity. The situation is both more interesting and more complicated for multivalent recognition. All of the factors that influence the relationship between complementarity and intrinsic affinity still apply to functional affinity, but in considering the relationship between epitope-paratope complementarity and functional affinity, additional factors become relevant. Among these additional factors are: (1) the number of paratopes that can interact simultaneously with epitopes, (2) statistical considerations relating to the number of ways that a fixed number of paratopes can interact with a fixed number of epitopes, (3) the spatial distribution of epitopes and the physical properties of the inter-epitope regions, and (4) the geometry and rigidity of the connections between binding sites (paratopes). The influence of multiple factors, unrelated to the details of paratope and epitope structure, on the functional affinity of a multivalent antibody for a multivalent antigen makes precise predictions of such functional affinities uncertain, even if the intrinsic affinities of the relevant monovalent interactions and the functional valences are known.

We have compared the binding patterns of a murine mAb of the IgG3 subclass to the binding patterns of IgG1 and IgG2b mAbs expressing the identical heavy and light chain variable domains. All three of these mAbs have specificity for the terminal, non-reducing *N*-acetyl-glucosamine (GlcNAc) residues of the cell wall polysaccharide of group A streptococci. They also bind to both naturally occurring

(Turner et al. 1990) and intentionally modified (Greenspan 1988) proteins displaying terminal GlcNAc moieties.

Using three strains of streptococci expressing high, intermediate, and low densities of cell wall GlcNAc, we found that the IgG3 mAb bound more effectively to all three bacterial strains than either the IgG1 or the IgG2b mAbs (Cooper et al. 1993). Furthermore, whereas the IgG3 bound best to the intermediate GlcNAc-density strain, the IgG1 and IgG2b mAbs both bound best to the high GlcNAc-density strain (Table 1). $F(ab')_2$ fragments from both IgG3 and IgG1, which retain bivalency but lack the Fc region, also behaved like the IgG1 and IgG2b mAbs. In general, comparable results were obtained using either flow cytometry or solid-phase enzyme-linked immunosorbent assay (ELISA) to make these comparisons. Somewhat more complicated differences in relative binding among the GlcNAc-specific IgG mAbs were also observed using GlcNAc-protein conjugates in ELISA and in assays based on a real-time biosensor device (BIAcore). Results from the biosensor experiments suggested that the differences between the IgG3 mAb and the IgG1 and IgG2b mAbs also extended to kinetic as well as equilibrium parameters. The IgG3 appeared to have a slower dissociation rate and a slightly faster association rate than the other two mAbs in interactions with a solid-phase GlcNAc-protein conjugate (Cooper et al. 1994).

A study of the relative binding of three bivalent proteins, expressing identical DNA-binding domains linked to one another through distinct molecular structures, to cognate and non-cognate DNA sequences reached a conclusion similar to ours regarding the mAbs specific for the GlcNAc-containing epitopes (Cuenoud and Schepartz 1993). It was found that differences in the structures linking the two identical DNA-recognizing domains were correlated with striking differences in discrimination between the cognate and non-cognate nucleotide sequences. The context-dependence of multivalent specificity has also been demonstrated by Horan et al. (1999), who investigated binding of a multivalent

**Table 1.** Binding[a] of HGAC 39 mAb and myeloma proteins to two streptococcal strains by ELISA. The 95% confidence intervals of the OD 405nm (absorbance units) are shown (from Cooper et al. 1993)

| Antibody | High epitope-density bacteria | | Intermediate epitope-density bacteria | |
|---|---|---|---|---|
| | Mean | 95% | Mean | 95% |
| HGAC 39.G3 | 1.429 | ±0.248 | 2.488 | ±0.159 |
| HGAC 39.G1 | 0.829 | ±0.100 | 0.733 | ±0.078 |
| HGAC 39.G2b | 1.100 | ±0.095 | 0.847 | ±0.060 |
| FLOPC 21 | 0.015 | ±0.023 | 0.017 | ±0.017 |
| MOPC 21 | 0.013 | ±0.009 | 0.016 | ±0.015 |
| MOPC 141 | 0.024 | ±0.012 | 0.056 | ±0.029 |

Isotype controls are: FLOPC 21, IgG3; MOPC 21, IgG1; MOPC 141, IgG2b.

[a] Mouse antibody (10μg/ml) was incubated with bacteria adsorbed to the solid phase. Antibody binding was measured following addition of alkaline phosphatase-conjugated goat antibody specific for mouse κ-constant domains. Background binding of the label to the solid-phase bacterial strains, in the absence of mouse antibody, has been subtracted.

lectin to two different carbohydrate ligands displayed at varying densities. They found that the lectin bound more strongly to one carbohydrate ligand at lower ligand densities and bound better to the second carbohydrate ligand at higher epitope densities.

These results suggest that the concept of complementarity as it is applied to a single binding site could be generalized to a more global measure of the magnitude of complementarity between multivalent receptors and ligands. This extended form of complementarity would take into account not only the "molecular fit" of each of the individual sites that engage in direct physical contact, but also the positive and negative cooperative effects between sites, such as the costs of bringing these sites into proper alignment for simultaneous interaction.

In our example from above, the IgG3 mAb would be regarded as more globally complementary, in the extended sense, than the IgG1 or IgG2b mAbs with respect to the multivalent GlcNAc arrays on the streptococcal surfaces. By contrast, all three mAbs would be regarded as having identical complementarity, of the standard monovalent sort, for the GlcNAc epitopes. Ultimately, the free energy change associated with complex formation will be the most definitive measure for quantitating complementarity in the extended, multivalent sense, although other measures of multivalent complementarity might be devised.

Thus, results from our comparisons among the GlcNAc-specific murine mAbs (IgG3, IgG1, and IgG2b) and results from other systems indicate that identical complementarity, defined at the level of single sites (epitope or paratope), does not guarantee identical extent or specificity (discrimination among multivalent antigens) of multivalent binding. The possibility that immunological receptors can exhibit identical specificities (in the sense of molecular discrimination) at the level of monovalent recognition but different specificities in terms of multivalent recognition can be further generalized to recognition by the antigen-specific cells (B and T lymphocytes) of the immune system (Greenspan 2001). In other words, the binding specificity of the antigen-binding receptor may not be identical to the functional specificity of the lymphoid cell that synthesizes and displays that receptor. Such divergence in specificity between receptor and cell can be attributed to the influences of other cell surface molecules expressed by lymphocytes on cell triggering by antigen. These conclusions suggest that immunological specificity is not a single well-defined, concept, but is a family of concepts (Greenspan 2001).

## 4 Conclusions

The four terms highlighted in the title are integral to analysis and discussion of antibody recognition. Each term encompasses a family of precisely defined concepts that are related but non-identical. Greater affinity and specificity will often be associated with greater shape and chemical complementarity between antibody (receptor) and antigen (ligand). However, the relationship between receptor-ligand complementarity and affinity or differential affinity (between cognate and non-cognate ligands) is complex and not completely predictable. This lack of absolute correlation relates in part to the fundamentally cooperative nature of macromolecular structure and biomolecular recognition and in part to the influence of factors that originate independently of the intermolecular interface.

*Acknowledgement.* This research was supported in part by United States National Institutes of Health Grant AI41657.

*Note added in proof.* Recent studies, employing in vitro selection schemes for affinity maturation, suggest that in a context free of the constraints present in vivo (Foote and Eisen 1995), antibodies may attain intrinsic affinities that approach picomolar (Hanes et al. 2000) or even femtomolar (Boder et al. 2000) values. These affinities are several orders of magnitude greater than those typically observed for antibodies produced following in vivo immunization (Foote and Eisen 2000).

Boder ET, Midelfort KS, Wittrup KD (2000) Directed evolution of antibody fragments with monovalent femtomolar antigen-binding affinity. Proc Natl Acad Sci USA 97:10701–10705

Foote J, Eisen HN (1995) Kinetic and affinity limits on antibodies produced during immune responses. Proc Natl Acad Sci USA 92:1254–1256

Foote J, Eisen HN (2000) Breaking the affinity ceiling for antibodies and T cell receptors. Proc Natl Acad Sci USA 97:10679–10681

Hanes J, Schaffitzel C, Knappik A, Pluckthun A (2000) Picomolar affinity antibodies from a fully synthetic naive library selected and evolved by ribosome display. Nat Biotechnol 18:1287–1292

# References

Bachmann MF, Kalinke U, Althage A, Freer G, Burkhart C, Roost H, Aguet M, Hengartner H, Zinkernagel RM (1997) The role of antibody concentration and avidity in antiviral protection. Science 276:2024–2027

Bhat TN, Bentley GA, Boulot G, Greene MI, Tello D, Dall'Acqua W, Souchon H, Schwarz FP, Mariuzza RA, Poljak RJ (1994) Bound water molecules and conformational stabilization help mediate an antigen-antibody association. Proc Natl Acad Sci USA 91:1089–1093

Coligan JE, Kindt TJ, Krause RM (1978) Structure of the streptococcal groups A, A-variant, and C carbohydrates. Immunochemistry 15:755–760

Cooper LJN, Schimenti JC, Glass DD, Greenspan NS (1991) H chain C domains influence the strength of binding of IgG for streptococcal group A carbohydrate. J Immunol 146:2659–2663

Cooper LJN, Robertson D, Granzow R, Greenspan NS (1994) Variable domain-identical antibodies exhibit IgG subclass-related differences in affinity and kinetic constants as determined by surface plasmon resonance. Molec Immunol 31:577–584

Cooper LJN, Shikhman AR, Glass DD, Kangisser D, Cunningham MW, Greenspan NS (1993) Role of heavy chain constant domains in antibody-antigen interaction: apparent specificity differences among streptococcal IgG antibodies expressing identical variable domains. J Immunol 150:2231–2242

Crothers DM, Metzger H (1972) The influence of polyvalency on the binding properties of antibodies. Immunochemistry 9:341–357

Cuenoud B, Schepartz A (1993) Altered specificity of DNA-binding proteins with transition metal dimerization domains. Science 259:510–513

Cunningham BC, Wells JA (1989) High-resolution epitope mapping of hGH-receptor interactions by alanine-scanning mutagenesis. Science 244:1081–1085

Davies DR, Cohen GH (1996) Interactions of protein antigens with antibodies. Proc Natl Acad Sci USA 93:7–12

DeLisi C (1976) Antigen antibody interactions. Springer, Berlin Heidelberg New York

Del Rio S, Setzer D (1993) The function of individual zinc fingers in sequence-specific DNA recognition by transcription factor IIIA. J Mol Biol 233:567–579
Di Cera E (1998a) Site-specific analysis of mutational effects in proteins. Adv Protein Chem 51:59–119
Di Cera E (1998b) Site-specific thermodynamics: understanding cooperativity in molecular recognition. Chem Rev 98:1563–1591
Ehrlich PH, Moyle WR, Moustafa ZA, Canfield RE (1982) Mixing two monoclonal antibodies yields enhanced affinity for antigen. J Immunol 128:2709–2713
Foote J, Raman A (2000) A relation between the principal axes of inertia and ligand binding. Proc Natl Acad Sci USA 97:978–983
Forsén S, Linse S (1995) Cooperativity: over the Hill. TIBS 20:495–497
Freire E (1999) The propagation of binding interactions to remote sites in proteins: analysis of the binding of the monoclonal antibody D1.3 to lysozyme. Proc Natl Acad Sci USA 96:10118–10122
Gopalakrishnan PV, Karush F (1974) Antibody affinity. VII. Multivalent interaction of anti-lactoside antibody. J Immunol 113:769–778
Greenspan NS (1992) Epitopes, paratopes, and other topes: do immunologists know what they are talking about? Bull Inst Pasteur 90:267–279
Greenspan NS (1997) Conceptions of epitopes and paratopes and the ontological dynamics of immunological recognition. In: Rouvray DH (ed) Concepts in chemistry: a contemporary challenge. Research Studies, Taunton, Somerset, England
Greenspan NS (2001) Dimensions of antigen recognition and levels of immunological specificity. Adv Canc Res 80:147–187
Greenspan NS, Cooper LJN (1993) Cooperative binding by mouse IgG3 antibodies: implications for functional affinity, effector function, and isotype restriction. Springer Sem Immunopathol 15:275–291
Greenspan NS, Cooper LJN (1995) Complementarity, specificity, and the nature of epitopes and paratopes in multivalent interactions. Immunol Today 16:226–230
Greenspan NS, Di Cera E (1999) Defining epitopes: it's not as easy at it seems. Nat Biotechnol 17: 936–937
Greenspan NS, Dacek DA, Cooper LJN (1988) Fc region-dependence of IgG3 anti-streptococcal group A carbohydrate antibody functional affinity. I. The effect of temperature. J Immunol 141:4276–4282
Greenspan NS, Monafo WJ, Davie JM (1987) Interaction of IgG3 anti-streptococcal group A carbohydrate (GAC) antibody with streptococcal group A vaccine: enhancing and inhibiting effects of anti-GAC, anti-isotypic, and anti-idiotypic antibodies. J Immunol 138:285–292
Hawkins RE, Russell SJ, Baier M, Winter G (1993) The contribution of contact and non-contact residues of antibody in the affinity of binding to antigen: the interaction of mutant D1.3 antibodies with lysozyme. J Mol Biol 234:958–964
Heinz FX, Mandl C, Berger R, Tuma W, Kunz C (1984) Antibody-induced conformational changes result in enhanced avidity of antibodies to different antigenic sites on the tick-borne encephalitis virus glycoprotein. Virology 133:25–34
Holmes NJ, Parham P (1983) Enhancement of monoclonal antibodies against HLA-A2 is due to antibody bivalency. J Biol Chem 258:1580–1586
Horan N, Yan L, Isobe H, Whitesides GM, Kahne D (1999) Nonstatistical binding of a protein to clustered carbohydrates. Proc Natl Acad Sci USA 96:11782–11786
Hornick CL, Karush F (1972) Antibody affinity III. The role of multivalence. Immunochemistry 9: 325–340
Hua QX, Shoelson SE, Kochoyan M, Weiss MA (1991) Receptor binding redefined by a structural switch in a mutant human insulin. Nature 354:238–241
Jencks WP (1981) On the attribution and additivity of binding energies. Proc Natl Acad Sci USA 78:4046–4050
Jones S, Thornton JM (1996) Principles of protein-protein interactions. Proc Natl Acad Sci USA 93: 13–20
Karush F (1976) Multivalent binding and functional affinity. Contemp Topics Molec Immunol 5: 217–228
Karush F (1978) The affinity of antibody: range, variability, and the role of multivalence. In: Litman GW, Good RA (eds) Immunoglobulins. Plenum, New York, pp 85–116
Lavoie TB, Drohan WN, Smith-Gill SJ (1992) Experimental analysis by site-directed mutagenesis of somatic mutation effects on affinity and fine specificity in antibodies specific for lysozyme. J Immunol 148:503–513
Lawrence MC, Colman PM (1993) Shape complementarity at protein/protein interfaces. J Mol Biol 234: 946–950

Levine B, Hardwick JM, Trapp BD, Crawford TO, Bollinger RC, Griffin DE (1991) Antibody-mediated clearance of alphavirus infection from neurons. Science 254:856–860

Levitt M, Park BH (1993) Water: now you see it, now you don't. Structure 1:223–226

Lim WA, Sauer RT (1989) Alternative packing arrangements in the hydrophobic core of lambda repressor. Nature 339:31–36

McCoy AJ, Chandana Epa V, Colman PM (1997) Electrostatic complementarity at protein/protein interfaces. J Mol Biol 268:570–584

Metzger H (1992) Transmembrane signaling: the joy of aggregation. J Immunol 149:1477–1487

Minton AP (1992) Confinement as a determinant of macromolecular structure. Biophys J 63:1090–1100

Minton AP (1993) Macromolecular crowding and molecular recognition. J Mol Recog 6:211–214

Moyle WR, Lin C, Corson RL, Ehrlich PH (1983) Quantitative explanation for increased affinity shown by mixtures of monoclonal antibodies: importance of a circular complex. Mol Immunol 20:439–452

Oi VT, Herzenberg LA (1979) Localization of murine Ig-1b and Ig-1a ($IgG_{2a}$) allotypic determinants detected with monoclonal antibodies. Mol Immunol 16:1005–1017

Pack P, Muller K, Zahn R, Pluckthun A (1995) Tetravalent miniantibodies with high avidity assembling in *Escherichia coli*. J Mol Biol 246:28–34

Parham P, Antonelli P, Herzenberg LA, Kipps TJ, Fuller A, Ward FE (1986) Further studies on the epitopes of HLA-B7 defined by murine monoclonal antibodies. Human Immunol 15:44–67

Pauling L (1945, 1962) Molecular structure and intermolecular forces. In: Landsteiner K (ed) The specificity of serological reactions. The President and Fellows of Harvard College (1945) and Dover Publications (1962), New York

Pauling L, Delbrück M (1940) The nature of the intermolecular forces operative in biological processes. Science 92:77–79

Quiocho FA, Wilson DK, Vyas NK (1989) Substrate specificity and affinity of a protein modulated by bound water molecules. Nature 340:404–407

Rheinnecker M, Hardt C, Ilag LL, Kufer P, Gruber R, Hoess A, Lupas A, Rottenberger C, Pluckthun A, Pack P (1996) Multivalent antibody fragments with high functional affinity for a tumor-associated carbohydrate antigen. J Immunol 157:2989–2997

Rini JM, Schulze-Gahmen U, Wilson IA (1992) Structural evidence for induced fit as a mechanism for antibody-antigen recognition. Science 255:959–965

Steward MW (1981) The biological significance of antibody affinity. Immunol Today 2:134–139

Thompson RJ, Jackson AP (1984) Cyclic complexes and high avidity antibodies. Trends Biochem Sci 9:1–3

Tulip WR, Varghese JN, Webster RG, Laver WG, Colman PM (1992) Crystal structures of two mutant neuraminidase-antibody complexes with amino acid substitutions in the interface. J Mol Biol 227:149–159

Turner JR, Tartakoff AM, Greenspan NS (1990) Cytologic assessment of nuclear and cytoplasmic O-linked *N*-acetyl-glucosamine distribution by using anti-streptococcal monoclonal antibodies. Proc Natl Acad Sci USA 87:5608–5612

Ubol S, Levine B, Lee S-H, Greenspan NS, Griffin DE (1995) Role of immunoglobulin valency and heavy chain constant domain in antibody-mediated downregulation of Sindbis virus replication in persistently infected neurons. J Virol 69:1990–1993

Usinger WR, Lucas AH (1999) Avidity as a determinant of the protective efficacy of human antibodies to pneumococcal capsular polysaccharides. Infect Immun 67:2366–2370

Weber G (1975) Energetics of ligand binding to proteins. Adv Prot Chem 29:1–83

Webster DM, Henry AH, Rees AR (1994) Antibody-antigen interactions. Curr Opin Struct Biol 4: 123–129

Williams DH, Westwell MS (1998) Aspects of weak interactions. Chem Soc Rev 27:57–63

Wilson IA, Stanfield RL (1994) Antibody-antigen interactions: new structures and new conformational changes. Curr Opin Struct Biol 4:857–867

Zimmerman SB, Minton AP (1993) Macromolecular crowding: biochemical, biophysical, and physiological consequences. Annu Rev Biophys Biomol Struct 22:27–65

# Mechanisms of Virus Neutralization by Antibody

P.J. Klasse and Q.J. Sattentau



## 1 Introduction

Antibodies can interfere with virus infection both in vivo and in vitro. Their net effects and mechanisms of action in the two contexts may sometimes have little bearing on each other: examples of this possibility notwithstanding, we argue that the analysis of antibody-virus interactions in vitro can facilitate the design of strategies aimed at eliciting immune control of virus infections in vivo. In this chapter, while focusing on the virtues and pitfalls of in vitro systems for the dissection of some antiviral effects of antibodies, we discuss the relevance of these experimental systems to viral immunity in vivo.

Imperial College School of Medicine, Jefferiss Research Trust Laboratories, The Wright-Fleming Institute, St. Mary's Hospital, Norfolk Place, London W2 1PG, UK

We define neutralization of viruses as the reduction of viral infectivity in vitro by the binding of a ligand to the virion (virus particle). The ligand can be an antibody, a soluble receptor molecule, a peptide or a carbohydrate; but we mostly deal with the classical case of antibody-mediated neutralization. The definition thus does not require the antibody-binding site to be a viral antigen. It includes any abrogation of infectivity by the binding of antibodies to a cellular antigen on the virion surface; but it excludes reduction in viral infectivity by antibodies to the viral receptors on the cell surface. It also excludes inhibition of viral growth due to effects of antibody on progeny-virus release or maturation. Furthermore, the definition leaves the extent, reversibility and mechanism of viral inactivation as matters to be empirically investigated for each individual combination of antibody and virus. Aggregation of virions by antibody can reduce the number of infectious events in vitro; hence it fulfills the current criteria, although neutralization has sometimes been defined so as exclude it (Mandel 1978).

Whereas the binding of the antibody to the virion is a relatively simple physicochemical event, neutralization is a biological phenomenon of a complex assay system that involves living target cells. Indeed, there are antibodies that neutralize virus infection of some but not of other target cells. For example, one monoclonal antibody (mAb) to La Crosse virus, of the bunyaviridae family, neutralizes the infection of BHK cells but not of insect cells. Another mAb does the opposite (Grady and Kinch 1985).

Although antibody binding to the virion is, by definition, a requirement for neutralization, this does not mean that antibody has to stay bound: whether virus can still remain neutralized after the dissociation of the antibody is a matter for investigation. Likewise, the step in the viral life cycle that is blocked by neutralization needs to be experimentally determined: All viruses need to attach to their target cells. Enveloped viruses must fuse with host cell membranes at the cell surface or in an endosomal organelle, and naked viruses must penetrate membranes at one of these locations (for a recent review, see Klasse et al. 1998). Intuition may suggest that neutralization interferes primarily with these events of viral entry. However, only experiments can show whether effects on later steps in the replication cycle are instead responsible. The semantic demarcation might be drawn after interference with the "first viral, major biosynthetic event" (Dimmock 1984). Effects later than that should then be designated something other than neutralization (see Mazanec et al. 1995), even if they were caused by antibodies bound to the surface of the virion.

Many antibodies elicited by virus infections are directed to epitopes on proteins located in the interior of the virion; others recognize only monomeric or denatured forms of viral antigens (Parren et al. 1997). Accordingly, none of those will bind to intact virus particles. Other antibodies do recognize native epitopes on the surface of virions and may therefore mediate neutralization. However, not all antibodies that bind to infectious virions will neutralize. Those directed to the neuraminidase of influenza virus, for example, do not (Wiley and Skehel 1987), and the effect on infectivity of antibodies to the fiber antigen of adenovirus is controversial (Dimmock 1984). The key distinction may be whether the antigen mediates the entry of the virus or not. But is binding of antibodies to the functional form of a viral antigen that mediates entry the sole requirement for neutralization? Evidence for and against this simple relation between binding and neutralization is discussed elsewhere in this work (see chapter by Burton et al., this volume). The suggested existence of antibodies that do not by themselves neutralize but instead block neutralization by others could refute the simple rule that all antibody binding to functional surface proteins of the virus is neutralizing (Hashimoto and Prince 1963; Massey and Schochetman 1981b). The issue is linked to the question of mechanisms of neutralization, which we deal with here: does the antibody interfere with virus-cell contact in a broad manner, or with distinct functional aspects of the viral surface proteins?

A virus particle has many copies of its antigens. The question of how many of these sites must be occupied by antibody for neutralization to occur also concerns mechanism. A global effect on the whole virion by the binding of a few antibodies to surface structures would give different requirements from the knocking out of the function of each individual molecule on a one-to-one basis. The interpretations of the early kinetic analyses of the neutralization reaction still pervade theories of mechanism in this field. Here, we re-examine these interpretations and then discuss the evidence for various conceivable mechanisms of neutralization (Fig. 1).

## 2 Neutralization Assays

In experimental systems for assessing neutralization, three distinct processes can be distinguished, although they need not be separated in time (Klasse 1996): antibody-virus binding, attachment of virions to target cells, and subsequent viral entry and replication.

### 2.1 Antibody-Virion Binding

This reaction is determined by antibody affinity for viral epitopes and the concentration of the reactants. Normally, antibody is present in vast molar excess over virus and therefore the degree of antibody occupancy on the virus particles, as equilibrium is approached, approximately depends only on the affinity and concentration of the antibodies. The further from equilibrium that the degree of neutralization is measured, the greater the influence of the on-rate constant for antibody binding. Most classical antibody-antigen reactions have been measured by the use of rather simple antibody-hapten associations. In these cases, the off-rate constant tends to be the major factor determining binding. For more complex antigens expressed on the cell surface, however, including viral envelope glycoproteins, the on-rate may be dominant because equilibrium is not approximated

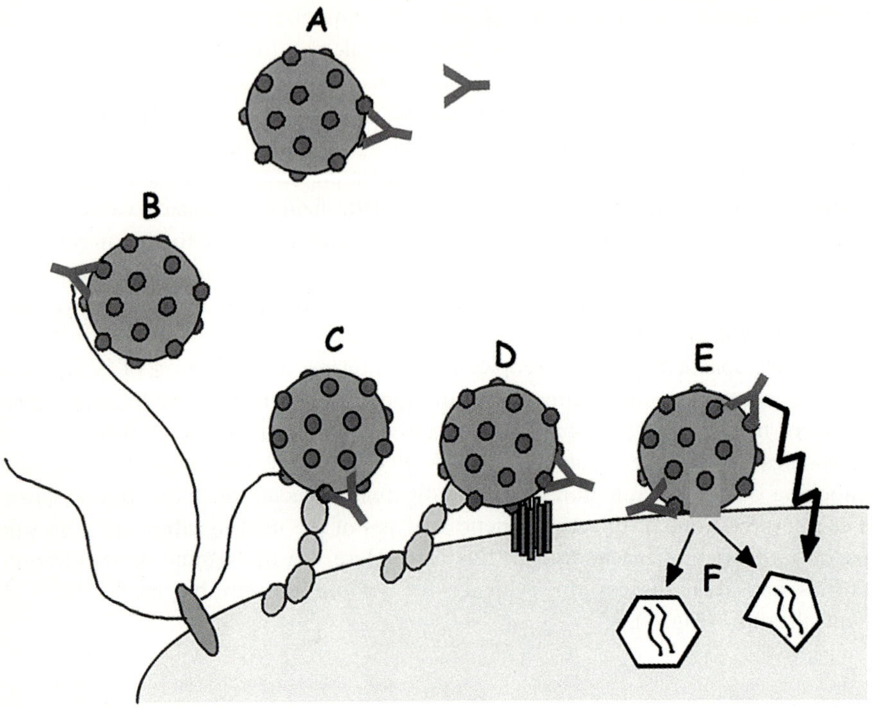


**Fig. 1.** Different possible steps in the first part of a replicative cycle of an enveloped virus (in this case based on HIV-1 and its receptors) that neutralizing antibodies may hypothetically interfere with. *A*, The virion has not yet attached to a target cell. Antibody has bound bivalently to envelope-glycoprotein spikes on the virion. *B*, The virion is attaching to the cell via a sulfated carbohydrate chain on a glycosaminoglycan. The antibody may interfere with the attachment process. *C*, The virion has approached its primary receptor, in this case CD4. Neutralizing antibodies may directly or indirectly block the interaction between the receptor and the viral envelope glycoprotein. *D*, The virus particle is brought closer to the cell membrane through the interaction with a coreceptor, in this case a seven-transmembrane-spanning chemokine receptor. The antibody may potentially block the interaction between the envelope glycoprotein and the coreceptor. *E*, The viral envelope has initiated fusion with the cellular membrane. In the illustrated example, this process takes place at the cell surface and is postulated to be mediated by the fusogenic form of the envelope glycoprotein that is elicited through interactions with both the primary receptor and the coreceptor. Alternatively, for many enveloped viruses the fusion can occur after internalization of the virus particle through endocytosis and be triggered by the low pH in the endosomes. The antibody may potentially interfere with the fusion process directly without having blocked receptor interactions. *F*, In some cases, mechanisms of neutralization of both enveloped (as illustrated) and naked viruses are hypothesized to act after the viral capsid has entered the cytoplasm of the target cell. For enveloped viruses in particular, this mechanism poses conceptual difficulties because it would require the transduction of a transmembrane signal from the envelope glycoprotein to the viral proteins in the interior of the virion

(Mason and Williams 1980; Sattentau and Moore 1995). The on-rate constant for binding to complex antigens may be influenced by epitope masking by carbohydrate moieties and protein quaternary structure, or by epitope induction through conformational changes.

### 2.2 Adsorption of Virions to Susceptible Cells

Virions bind to specific structures, receptors for the virus, on the surface of susceptible cells via viral attachment proteins. Although this is a vastly more complex process than antibody binding to virions, it can potentially be modeled as pursuant to the law of mass action (KLASSE and MOORE 1996). It should be noted that most viral preparations have a high ratio of non-infectious to infectious particles. Sometimes a great proportion of the virions that do not infect may be potentially infectious: they just do not get the chance to infect because adsorption is disfavored (KABAT et al. 1994). Phase 1, i.e. antibody binding to the virus, is normally allowed to proceed before phase 2, virus adsorption, begins. But it has been argued that this separation introduces more differences than necessary between the in vitro neutralization assay and the situation in vivo (LAYNE et al. 1991). If antibodies, virus, and cells are all mixed together, higher demands are generally placed on neutralization efficiency (LAYNE et al. 1991), which includes neutralization rate (SCHOFIELD et al. 1997). However, certain kinds of neutralization may actually require prior interactions between virus particles and cell-surface receptors (KJELLEN 1985; WYATT et al. 1998; WYATT and SODROSKI 1998).

### 2.3 Replication of Non-Neutralized Virus

Viral replication is measured to determine whether neutralization has taken place. Neutralization is often expressed as a percentage of decrease of the infectivity of the input virus. However, in the literature, "50% neutralization" does not necessarily mean that half of the infectious input virions have been inactivated: only under certain conditions of linearity between viral dose and assay signal are such quantitative statements meaningful (KLASSE 1996). Deviations from proportionality preclude valid determinations of neutralization kinetics and of the number of antibody molecules involved.

## 3 Physicochemical Aspects of Viral Neutralization

In this section, we examine some prevailing ideas of the last 70 years of research into the biochemical events underlying viral neutralization. We discuss the concepts of single- and multi-hit by antibody, and argue why we believe that multi-hit neutralization generally stands up best to scrutiny.

### 3.1 Kinetics and Molecularity

Although neutralization is a complex biological phenomenon, attempts were made early on to study the neutralization reaction physicochemically: the process of antibody binding to virions can be quenched, for example by dilution. The dissociation of divalently bound antibody may be negligible. Thus the duration of the neutralization reaction can be varied, which allows the determination of neutralization rate. In an influential paper, DULBECCO et al. (1956) obtained a linear relationship when they plotted the rate of neutralization of poliovirus and western equine encephalitis virus as a function of antibody concentration. This was interpreted as a reaction of first-order kinetics in antibody concentration. Furthermore, the linearity was said to imply that a single antibody molecule was enough to neutralize a virion, i.e. that the neutralization was a single-hit phenomenon. Both inferences can be criticized: firstly, a linear relationship of rate to the concentration of one reactant can be described as pseudo-first-order kinetics when that reactant is present in vast molar excess. Hence, the linear relationship does not establish the order of the reaction. Secondly, the order of the kinetics of a reaction does not imply the number of molecules in the reaction complex, i.e. the molecularity of the reaction does not follow from the kinetics (KLASSE and MOORE 1996). Single-hit neutralization cannot be deduced from the first-order kinetics, but the order of the reaction could potentially refute single-hit hypotheses.

Other evidence has been suggested to refute a single-hit molecularity. A lag in neutralization as a function of time has been observed in several virus systems (DELLA-PORTA and WESTAWAY 1978). This could be interpreted as a requirement for antibody binding to exceed a certain level before neutralization can begin; the single-hit complexes, formed first, would still be infectious. However, there are other interpretations of such a lag in neutralization, or lack of it, such as heterogeneities among virions, as shown below. Moreover, there are more direct methods for determining the number of antibodies bound per virion.

## 3.2 Stoichiometry and Occupancy

When the number of epitopes on the virion is known, the molecularity of neutralization, i.e. the number of hits required per virion, can be translated into a relative occupancy. For example, a single hit on a virion with 100 epitopes would constitute a relative occupancy of 2%, if both paratopes of a bivalent antibody were bound. However, at low antibody occupancies it can be argued that the number of antibody molecules bound will be Poisson-distributed around an average number. Stoichiometric data, obtained for example by the use of radiolabeled antibody, can only provide this average number for the whole population of virions. For inferences about the number of hits required for neutralization, the assumption has to be made that infectious and non-infectious particles bind antibody equally well and that each virion has the same number of epitopes. The molecularity of neutralization can then be determined from a plot of the natural logarithm of the relative infectivity as a function of the average number of antibodies per virion. The actual curve needs to be compared with those representing various numbers of hits (see KLASSE and MOORE 1996). The single-hit curve alone is a straight line. Hence, only when testing the hypothesis of single-hit neutralization is it relevant to read the antibody number that gives a relative infectivity of 1/e (i.e. with the natural logarithm −1, for a derivation see KLASSE and MOORE 1996): this value should then be 1 on the $x$ axis. The reading of any other value on the $x$ axis at $y = -1$, corresponding to a relative infectivity of 1/e, hence lacks theoretical justification (see ICENOGLE et al. 1983). Obviously the entire curve obtained should be compared with the theoretical curves for different molecularities in order to find the best fit.

Stoichiometric data have indicated multi-hit molecularities of neutralization for poliovirus (ICENOGLE et al. 1983) and influenza virus (TAYLOR et al. 1987), while both were neutralized with close to first-order kinetics. This has been considered a discrepancy, and in order to explain it the hypothesis of "neutralization-relevant" epitopes was devised. As mentioned above, first-order kinetics, and particularly pseudo-first-order kinetics, are compatible with multi-hit neutralization. Furthermore, the concentration dependence of antibody binding would not relate to only a subset of the binding antibodies but the total. So had there been a discrepancy between theory and data, the hypothesis of "neutralization-relevant" epitopes would not have accounted for it.

An elegant way of refuting a hypothesis of single-hit neutralization would be to use phenotypically mixed virus carrying two distinct neutralization-sensitive antigens and antisera recognizing each antigen specifically. If the two sera together neutralize but neither does so alone, then that indicates that more than one antibody per virion is required (MANDEL 1978). It can also be argued that reduction in infectivity by addition of a secondary antibody recognizing a primary, weakly neutralizing one indicates the existence of infectious antibody-virus complexes (IORIO and BRATT 1984; TRAUTMAN and HARRIS 1977). Or, put another way, some virions with at least one antibody bound to them are not neutralized. However, alternative interpretations are that the secondary antibody aggregates the complexes or strengthens the binding of the primary antibody, thereby diminishing the fraction of virions without antibody. A special case that is not consistent with single-hit hypotheses is provided by neutralization epitopes that are not exposed on the native state of the surface antigens on the virion. It has been proposed that the binding of a first antibody to an external epitope of the adenovirus hexon protein induces a conformational change that allows the binding of a second one to an epitope that is internal in the native structure (KJELLEN and PEREIRA 1968). Neither antibody neutralizes on its own, and only the first would bind on its own. A related example is synergy between two or more antibodies binding to the same molecule: added separately, antibodies to distinct epitopes on the same protein neutralize very weakly, whereas when added together they neutralize more strongly than what would be predicted if effects were merely additive (ALLAWAY et al. 1993; LI et al. 1998; TILLEY et al. 1992). But none of this proves that more than one antibody of each specificity is needed per virion, or that more than one viral spike or antigen need be knocked out per virion.

Regardless of how many bound antibody molecules are required per virion for neutralization to occur, the binding can be expressed as an occupancy, i.e. the fraction or percentage of epitopes that are complexed on each virion. There appear to be examples of threshold occupancies that give neutralization (Icenogle et al. 1983; Layne et al. 1990; Parren et al. 1998; Taylor et al. 1987; Wohlfart 1988). The postulate of such a critical threshold of occupancy is complemented by, but does not presuppose, the hypothesis that a certain minimal number of antigens on the virion surface are required for infectivity (Klasse and Moore 1996; Trautman and Harris 1977). If a single unoccupied antigen on the virion surface is sufficient for infectivity, this would merely be the special case of a requirement of 100% occupancy for neutralization. However, it has been pointed out that steric restraints may preclude simultaneous occupancy of all epitopes, and the torsional flexibility of the antibody will determine the distance between epitopes that can be bridged. Cross-linking of separate polypeptide chains may confer different neutralization mechanisms from occupancy per se (Thouvenin et al. 1997).

It is also possible that neutralization is not an all-or-nothing event but a mere statistical decrease in a virion's propensity to infect (Icenogle et al. 1983). A critical number of viral surface proteins could conceivably be equal to, or larger than, the number necessary to make up a fusion or entry complex. HIV-1 pseudotypes of antigenic and non-antigenic, but functional, forms of envelope glycoprotein have been used to study minimal occupancy (Schonning et al. 1999). The data refuted a requirement of complete occupancy for neutralization, but left undecided whether a threshold exists or whether there is incremental neutralization running the whole gamut from one hit to complete occupancy. The difficulty in resolving this issue may be due to the spontaneous shedding of the outer envelope glycoprotein, gp120, from the virions of some strains of HIV-1, particularly those that are adapted to growth in T-cell lines (reviewed in Moore and Ho 1995). A certain number of antibody hits would constitute different levels of occupancy depending on how many gp120 molecules the virions retain (Klasse and Moore 1996). Since gp120 retention would vary from virion to virion in a population, it is conceivable that discrete threshold effects in neutralization would be blurred (Klasse and Moore 1996). The complication would be exacerbated if the propensity of the individual virion to infect also varied with the number of gp120 molecules per virion. But the increased sensitivity to neutralizing antibodies and soluble receptor molecules after treatment of virions under conditions that promote shedding (Layne et al. 1990) suggests that the critical occupancy does indeed depend on the total average number of intact envelope molecules available.

The stronger neutralization of HIV-1 by IgG3 than IgG1, in spite of identical paratopes on the two antibodies (Cavacini et al. 1995), may mean that there can be IgG1 bound that does not neutralize, while IgG3 similarly bound does. Thus IgG3 would require a lower occupancy than IgG1, which suggests that the number of envelope spikes available for infection is not identical to the number of unoccupied spikes. Steric hindrance by the larger Fc portion of IgG3 could be responsible for impeding the function of a greater number of spikes (Parren et al. 1998). Alternatively, greater hinge flexibility of IgG3 (Roux et al. 1997) could lead to enhanced cross-linking of antigen (see Thouvenin et al. 1997), which would affect functional affinity and hence occupancy. Clearly, some occupancies of antibody on HIV-1 particles are sub-neutralizing. There may exist a distinct neutralizing occupancy for every antibody that can bind to functional envelope spikes on the virus surface (PARREN et al. 1998), but this issue requires further attention. Indeed, it was suggested that one antibody used at different concentrations neutralized by different mechanisms (ARMSTRONG and DIMMOCK 1992): it is conceivable that the critical level of occupancy would be mechanism-specific.

## 3.3 Reversibility of Antibody Binding and Neutralization

Another consideration of antibody occupancy on virions in neutralization concerns the reversibility of the reaction. Dilution of antibody complexes with poliovirus and western equine encephalitis were shown not to reactivate viral infectivity (DULBECCO et al. 1956). Therefore dilution was used to stop further neutralization. Exposure of neutralized poliovirus to acid in order to strip off the antibody restored infectivity. Some antibodies that neutralize HIV-1 have been shown to induce the dissocation of gp120 from the virions (POIGNARD et al. 1996). It is conceivable that, after such an event, antibodies would dissociate from the antigen when the latter is no longer presented in a polyvalent form, but such dissociation would be irrelevant to the neutralizing effect. Antigen dissociation from the virions is not necessary for neutralization, but could be sufficient and would render neutralization irreversible. Moreover, it could decrease the required degree of occupancy on the gp120 molecules that remain on the virions. However, neutralization of HIV-1 by antibodies that did not induce gp120 shedding has also been found to be irreversible (McDOUGAL et al. 1996). Examples from poliovirus neutralization indicate that antibody dissociation is neither sufficient nor necessary for restoring infectivity: at low ionic strength a mAb acted by converting the virions to non-infectious, empty capsids from which the antibody subsequently dissociated (MANDEL 1976). In contrast, acid treatment may sometimes reactivate viral infectivity without leading to antibody dissociation (MANDEL 1976, 1978).

## 3.4 The Persistent Fraction

Reversibility of antibody binding was one possible explanation for the phenomenon of the persistent fraction that has dogged much of neutralization research. DULBECCO et al. (1956) described neutralization as biphasic: only the first phase was kinetically of approximately the first order. After that, no further neutralization occurred, but a constant infectious fraction remained. It has been pointed out that it is a non-neutralized but not necessarily non-neutralizable fraction, since virus in such fractions has been expanded and then neutralized. Before the era of mAbs, heterogeneities in the neutralizing serum were implicated as a cause of the persistent infectivity. Moreover, the addition of a secondary antibody directed to the neutralizing ones can reduce the size of the persistent fraction (HASHIMOTO and PRINCE 1963; MANDEL 1976). It has also been suggested that some non-neutralizing antibodies can block the binding of the neutralizing ones

(Hashimoto and Prince 1963; Massey and Schochetman 1981b). If there is more to neutralization than the mere binding of antibody to the right site, one possibility is that events that lead from antibody binding to the neutralized state, such as conformational changes in the ligated viral antigen or globally in the virion, are sometimes modified or arrested. The neutralizing changes taking place within the virus could conceivably be induced by the antibody with variable efficiency, allowing the presence of a persistent fraction (Mandel 1978). Another mechanism by which virus could bind antibody but still retain infectivity is through pre-existing virus aggregates or the formation of antibody-virion aggregates. The reduction in infectivity of each individual particle in the aggregate may not be sufficient to abrogate the capacity of the aggregated virions to infect, because the aggregates may adsorb to cells more efficiently than single virions (Mandel 1978). It is clear, however, that aggregation of virions cannot account for all viral persistence observed (Iorio and Bratt 1984): such fractions also prevail in monodispersed virus preparations and with monovalent antibody fragments. In summary, therefore, the wide variety of hypotheses concerning the cause of the persistent fraction has been divided into two distinct categories: (1) dissociation of antibody, and (2) retention of, or reversion to, the infectious state in spite of a degree of antibody binding that neutralizes the majority of virions (Mandel 1978). A more straightforward possibility would be stochastically predicted lower occupancies at one extremity of the spectrum of degrees of antibody binding to the population of virions in the neutralization reaction mixture. These lower occupancies may be insufficient to cause neutralization. In our view, this may be the most plausible explanation in many cases of neutralization persistence.

## 4 Replicative Steps Blocked by Neutralization

So far neutralization by antibodies alone has been considered. Such neutralization can sometimes be reversible. A disruption of the surrounding membrane, or virolysis, would be a plausible irreversible mechanism of neutralization of enveloped viruses. Their neutralization can indeed be enhanced by complement (Spear et al. 1991). For example, antibodies to the retroviral transmembrane protein were shown to mediate lysis of virions in the presence of complement, whereas those to the outer envelope glycoprotein did not (Oroszlan and Nowinski 1980). However, HIV-1 is to a great extent resistant to lysis because of the incorporation of cellular complement resistance factors CD46, CD55, and CD59 within the viral envelope during budding (Montefiori et al. 1994; Saifuddin et al. 1997; Stoiber et al. 1996). Moreover, complement-mediated enhancement of HIV-1 infection with and without antibodies is probably attributable to coating of virions by complement factors that do not neutralize, thereby facilitating viral adsorption onto the cell surface via complement receptors (for a review see Speth et al. 1997).

### 4.1 Virus Attachment

The earliest neutralization studies showed that under some conditions neutralized virus could not adsorb to target cells. Since attachment or adsorption to susceptible cells is a necessary early step in replication, it was conceptually straightforward that inhibition of this process was identified as one mechanism of neutralization (Rubin and Franklin 1957). This appears to be a major neutralization mechanism of both infection of vero cells by dengue virus (He et al. 1995) and of T-cell lines by HIV-1 (Ugolini et al. 1997). However, equally well-established for a range of viruses is that neutralized virus can sometimes adsorb to cells; indeed, virus that has adsorbed but which has been prevented from penetrating can be neutralized (Lu et al. 1992, Highlander et al. 1987). In contrast, mouse mammary tumor virus could not be neutralized after adsorption to cells, which was interpreted as support for interference with receptor interactions as a major neutralization mechanism (Massey and Schochetman 1981a). But post-attachment neutralization can actually inhibit receptor interactions (Lu et al. 1992) and even attachment. Thus rabies virus that was neutralized after it had attached was released from the cells by the antibody (Dietzschold et al. 1987). It is interesting that increasing the concentration of antibody to above that which is necessary for polio-, influenza and herpes simplex virus neutralization impaired viral attachment (Mandel 1978; Armstrong and Dimmock 1992; Highlander et al. 1987). Either another mechanism of neutralization, such as inhibition of attachment, thereby came into play, or the same mechanism of interference with virus-cell interactions was simply reflected in more drastic effects on virus binding. The efficiency of attachment might be reflected in how much adsorbed virus can be eluted from the cells, and the relationship to neutralization can be complex. Thus, poliovirus can be more difficult to elute after neutralization, whereas others such as Newcastle disease virus and rabbit poxvirus are more easily so (Mandel 1978).

Similar molecules can apparently neutralize by different mechanisms: A short, soluble fragment representing only two domains of the rhinovirus receptor ICAM-1 neutralizes by blocking attachment in competition with the cellular ICAM-1 molecules. But the soluble molecule corresponding to the entire extracellular part of the receptor neutralizes at lower concentrations and inhibits a post-attachment entry step (Greve et al. 1991). More intriguing still, a monospecific IgG (IgG1b12) directed to the CD4-binding region of HIV-1 gp120 apparently neutralized by a mechanism different from that of the Fab derived from it. It was suggested that both the IgG and the Fab inhibit HIV-1 infection at a post-attachment stage of the virus life cycle, the IgG at the fusion event and the Fab at a post-fusion step (McInerney et al. 1997).

A neutralizing antibody to the transmembrane protein of HIV-1 was shown not to block virus attachment to any cells tested (Mondor et al. 1998; Ugolini et al. 1997). The mechanism of neutralization would have to act post-attachment. Some potently neutralizing antibodies to HIV-1 even enhance the binding of virions to some cell types while they block attachment to others (Mondor et al. 1998; Ugolini et al. 1997). HIV-1 neutralization has also been described as occurring before or independently of virus attachment (McDougal et al. 1996). In contrast, there are neutralizing antibodies to the enterovirus 71, a picornavirus, that can only act, indeed only bind to virions, after adsorption of the virus to RD cells. Hence they are unable to neutralize infection of green monkey kidney cells (Kjellen 1985). The explanation may be that certain receptor interactions induce the exposure or antigenic conformation of epitopes, and that the virus uses different receptors on the two cell types. But the situation can be more complicated: viruses may use auxillary molecules in addition to the main attachment receptors for binding to and entry into cells. Such usage by the virus could differ between cell types. The use of auxillary molecules could potentially affect the exposure of neutralization epitopes and hence the sensitivity to neutralization. However, different strains of HIV-1 that can use either or both of the chemokine receptors CCR5 or CXCR4 for entry have not shown any general differences in neutralization sensitivity due to such coreceptor usage (Cecilia et al. 1998; LaCasse et al. 1998; Montefiori et al. 1998; Trkola et al. 1998). One epitope on the viral envelope glycoprotein that is induced by binding to the primary receptor for the virus, CD4, has since been strongly implicated in interactions with the chemokine receptors (Rizzuto et al. 1998). It could thus be investigated whether pre-incubation of receptor-positive cells with the virus, in order to induce the antigenicity of the epitope, leads to differential neutralization via this epitope depending on coreceptor usage. One prediction is that such neutralization cannot take place because of steric restrictions (Wyatt and Sodroski 1998).

Blocking virus-receptor interactions is an important neutralization mechanism (Massey and Schochetman 1981a; Rubin and Franklin 1957; Ugolini et al. 1997) but clearly not the only one (Fig. 1). Furthermore, such interference can affect different viral functions. There may be a degree of interference with virus-receptor interactions that allows virus attachment, even in the absence of ancillary attachment receptors, but prevents the main receptors from performing other functions in entry. Such functions could be activation of the fusion potential of viral envelope glycoproteins or mediation of internalization through the endocytic pathway. A virus can also use yet other receptors with little or no role in attachment of virions to target cells. Blocking the interactions with these would not be expected to impede virus attachment. The interactions of the HIV-1 envelope glycoprotein with chemokine receptors may be a case in point. These interactions are thought to bring the virus and cell membranes into close apposition and to participate in the activation of the fusion potential of the envelope glycoprotein after virus attachment. Nevertheless, antibodies and some polyanionic ligands that bind to chemokine-receptor-interactive surfaces on the envelope glycoproteins block virus attachment to some cell types (Mondor et al. 1998; Moulard et al. 2000; Ugolini et al. 1997).

Changing the isotype of a mAb directed to the CD4-binding site on HIV-1 gp120 from IgG1 to IgG3, i.e., substituting a bulkier Fc portion, vastly increased its neutralizing capacity without affecting the immunoreactivity (Cavacini et al. 1995). If the engineered isotype switch has no effect on the functional affinity of the mAb for virus particles, this is indirect evidence that the neutralization mechanism is steric hindrance of virion-cell contacts. Since IgG3 directed against envelope glycoproteins is rarely detected in natural infection (Khalife et al. 1988; Klasse and Blomberg 1987), there may be potential gains in vivo in eliciting or supplying neutralizing antibodies of this isotype. Additional evidence for steric hindrance as a mechanism may come from neutralization by antibodies to cellular antigens incorporated into viral envelopes. Thus the simian immunodeficiency virus (SIV) and HIV-1 can be neutralized by antisera to antigens of the MHC of both classes I and II (Arthur et al. 1992).

In conclusion, therefore, it should be noted that there can be numerous distinct mechanisms for how the antibodies block interactions with virus receptors: direct competition, allosteric conformational changes, dissociation of the antigen from virions, and steric hindrance by binding to epitopes that do not overlap with the receptor-interactive surfaces. Antibody interference with virus-receptor interactions may well manifest itself differently depending on the assay conditions. Blocking of virus attachment may represent a dominant mechanism of neutralization in many systems; but virus receptors are involved in both virus attachment and other entry functions. Thus neutralization that interferes with virus-receptor interactions comprises a qualitative miscellany of mechanisms at the level of molecular events.

## 4.2 Post-Attachment Events

As we have discussed above, neutralizing antibody interference with virus-receptor interactions does not necessarily inhibit virion attachment. Blocking of virus-receptor interactions can affect other steps in virus entry than attachment. Moreover, neutralization sometimes interferes with steps other than receptor interactions. For enveloped virus there is the possibility that fusion of the viral and cell membranes can be specifically blocked after virus attachment, independently of interference with receptor interactions (Imai et al. 1998; Kida et al. 1985). Antibodies that allow attachment of the flavivirus West Nile virus to macrophages block the pH-dependent fusion between the viral envelope and the endosomal membrane after internalization of the antibody-virion complex (Gollins and Porterfield 1986). Studies based on recombinant receptor and envelope glycoprotein interactions suggested that neutralization of bovine leukemia virus, a retrovirus, occurs via two different mechanisms: a block of envelope glycoprotein interactions with the receptor for the virus and interference with a later step in entry which could be membrane fusion (Orlik et al. 1997). For visna virus, another retrovirus, it is unclear whether a specific block of membrane fusion can act as a neutralization mechanism. Antibodies that block cell-to-cell fusion induced by virus particles (fusion from without) neutralize infection very poorly. In contrast, antibodies to a distinct epitope neutralize well but are poor inhibitors of fusion from without (Crane et al. 1988). However, it is uncertain whether cell-to-cell fusion from without mimics that involved in viral entry.

The case of HIV-1 is complicated. Firstly, antibodies that do not block binding to the primary receptor for the virus, CD4, can nevertheless, under conditions of high occupancy, block viral attachment to certain cells (Bahraoui et al. 1988; Ugolini et al. 1997; Valenzuela et al. 1997), possibly through steric hindrance (see Barbas et al. 1992; Parren et al. 1998). Secondly, both CD4 and the coreceptors, CXCR4 and CCR5, are believed to have distinct consecutive functions in activating the fusogenicity of the envelope protein. Therefore, antibodies interfering with these interactions can prevent fusion more or less directly. Nevertheless, the neutralization by an antibody specific for the transmembrane glycoprotein (TM) neither blocks attachment nor is believed to interfere with any receptor interactions. TM has the capacity to form coiled coils and these are thought to be related the fusogenic state of the envelope (Chan et al. 1997; Weissenhorn et al. 1997). Peptides representing the part of TM involved in coiled-coil formation can specifically block fusion, presumably without interfering with receptor interactions. The anti-TM antibody recognizes a peptide that is contiguous in sequence to the coiled-coil region (Gorny and Zolla-Pazner 2000; Muster et al. 1993; Sattentau et al. 1995). It can therefore be hypothesized that this antibody, like the peptides, prevents coiled-coil formation.

The discussion so far has dealt with neutralization as inhibition of distinct steps involved in virus entry. There is considerable support for such inhibition. In the case of enveloped viruses, hypotheses of such mechanisms are also conceptually simple in that they posit effects of antibodies only on the functions of those viral proteins to which they bind. Yet, later replicative steps have been proposed to be blocked by neutralization.

## 4.3 Post-Entry Events

The processes dealt with so far comprise virus entry (Klasse et al. 1998), but the definition of neutralization encompasses potential inhibition of some later events in the replicative cycle. As mentioned above, neutralized poliovirus can sometimes attach more strongly to target cells than infectious virus. In one study, even penetration was not uniformly prevented by neutralization, but the virus that infected was uncoated to a greater extent than neutralized virus that had penetrated. Moreover, a greater proportion of the RNA in the uncoated virus was degraded (Mandel 1976, 1978). Such late neutralizing effects have also been described for the enveloped virus of avian infectious bronchitis (Stinski and Cunningham 1970). Other post-entry mechanisms have been proposed: inhibition of the activation of the ribonucleoprotein (Armstrong and Dimmock 1992) and transcriptase actvity (Possee et al. 1982) of influenza virus and effects on the viral core HIV-1 in the cytoplasm of the target cell (McInerney et al. 1997).

How can antibodies interfere with uncoating or enhance RNA degradation? One example of rearrangement in a capsid by an antibody has been given above: antibody binding to an external epitope on the hexon protein of adenovirus induces a conformational change that exposes an otherwise internal epitope. Only antibody binding to the latter yields neutralization (Kjellen and Pereira 1968). Similarly, the VP4 protein of the poliovirus capsid becomes exposed upon binding of neutralizing antibody to the capsid surface; VP4 may also bind certain antibodies after the initial rearrangement of the capsid (MANDEL 1978). The poliovirus capsid is metastable and oscillates between two isoelectric states. The binding of neutralizing antibody, regardless of how many molecules are bound per virion, can lock the capsid in one of the conformations. Furthermore, it was shown in vitro that the binding of neutralizing antibody blocked uncoating. How should RNA degradation then be interpreted? It may be the result of an aberrant uncoating process that occurs both for neutralized and non-neutralized virus. The absence of intact RNA from uncoated capsids may thus hypothetically represent the neutralization (MANDEL 1976, 1978). However, a later study (BRIOEN et al. 1985) failed to find a correlation between the amount of poliovirus that was pI-shifted and the neutralized fraction. Several neutralizing antibodies failed to induce the previously described shift to pI 4. Nevertheless, as mentioned in connection with reversibility, antibodies can induce a conformational change in the whole virion, which makes it extrude the RNA before the encounter with cells. Since this conformational change abolishes the antigenicity for the same antibody, the mechanism has been called "hit-and-run neutralization". Caution should be exercised in the interpretation of this result, since the reaction requires unphysiologically low pH: at physiological pH the same antibody is reported to inhibit infectivity through the aggregation of virions (BRIOEN et al. 1985). Soluble forms of ICAM-1 can neutralize rhinovirus by inducing an uncoating reaction outside the target cells. However, the majority of the neutralization caused by the soluble forms of the virus receptor acts through a different, fully reversible mechanism (GREVE et al. 1991). Neutralization of another picorna virus, foot-and-mouth-disease virus, by the induction of an uncoating-like reaction that extrudes the viral genome has also been described (MCCULLOUGH et al. 1987). It should be noted that uncoating-triggering neutralization can occur extracellularly and prior to attachment, although it perverts a late step in the entry process.

Post-entry effects of adenovirus neutralization were also observed in one extensive study: antibodies to distinct antigens on the surface of the adenovirus particle neutralized by different mechanisms, while none of them blocked attachment. Nor did they substantially reduce virus internalization through endocytosis. Fiber antibodies aggregated the virus, which ended up in multivesicular bodies and lysosomes. After hexon-antibody neutralization, the majority of the virus was detected in the endosomal and lysosomal compartments of the cells. Penton-base-antibody neutralization let a substantial proportion of virus enter the cytoplasm (WOHLFART et al. 1985).

It has been suggested that the lentivirus causing visna in sheep is neutralized by different mechanisms in fibroblasts derived from the choroid plexus and in macrophages (KENNEDY-STOSKOPF and NARAYAN 1986). In fibroblasts, the mechanism appeared to be inhibition of attachment to the cell surface, whereas in macrophages the neutralizing serum increased internalization of the virus, whilst seemingly blocking a step after uncoating but before reverse transcription. However, one could speculate that this enhancement of internalization is unrelated to the infectious route of the virus. It was later shown that while this enhancement of virus internalization depended on the Fc portion of the antibodies, $F(ab)_2$ fragments delayed infection of the macrophages (JOLLY et al. 1989). It may represent a germane parallel that HIV-1, like visna virus a member of the lentivirus family, enters HeLa cells expressing CD4 via two different routes. The predominant pathway is by endocytosis, which is largely non-infectious, whereas fusion at the plasma membrane (MARECHAL et al. 1998) appears to be proportionately more productive (PELCHEN-MATTHEWS et al. 1995).

In one study of two mAbs to the V3 loop of gp120 of HIV-1, neither appeared to affect virus attachment. One was reported to block fusion and virus internalization, while the other one did neither. It was therefore hypothesized that the latter sent a transmembrane signal through the envelope and abrogated the functionality of the viral core (ARMSTRONG et al. 1996). There are other examples of antibodies that neutralize HIV-1 without blocking cell-to-cell fusion, but at which stage of the entry or early infection process they act has not been decided (BARBAS et al. 1992). We note that quantitative and qualitative differences between inhibition of cell-to-cell fusion and virion fusion with the target cell may be considerable. Thus lack of effect on cell-to-cell fusion by an antibody does not directly refute inhibition of virion-to-cell fusion as its neutralization mechanism.

Particularly intriguing examples of putative post-fusion neutralization are provided by influenza virus and antibodies to its hemagglutinin protein. Although high concentrations of antibody can block virus attachment, lower but still neutralizing concentrations seemingly only change the distribution of the viral nucleoprotein from its normal concentration in the nucleus to a perinuclear, cytoplasmic localization. Furthermore, although neutralization was proportional to inhibition of hemagglutinin-induced fusion, restoring fusion gave the normal nucleoprotein distribution but did not restore infectivity (ARMSTRONG and DIMMOCK 1992).

For conceptual clarity, we should like to distinguish between, on the one hand, the stages in the viral replicative cycle at which an antibody can bind in order to neutralize, and, on the other, the replicative steps that it blocks or perturbs. Sophisticated future research may give precise resolution of these distinct issues for individual combinations of virus and antibody.

## 5 Artifacts and the Biological Relevance of Neutralization

In the preceding section, we describe various hypotheses of mechanisms of virus neutralization by antibodies, and appraise their corroboration. Antibodies prevent viral entry through different mechanisms that are not covered by any one simple theory of neutralization; furthermore, post-entry mechanisms remain a theoretical and empirical challenge. Hence, each virus-antibody and virus-cell combination must be examined in its own right. Many controversies about the reproducibility of data and their interpretation need to be resolved. Here, we comment on the distinct problem of the biological relevance of neutralization phenomena. In vitro systems are riddled with artifacts, many of which could be avoided. Whilst it is tempting to rely on highly artificial systems that have the benefits of high sensitivity or rapid throughput for the dissection of neutralization mechanisms, the loss in biological relevance can be severe. One striking example of this is the use of cell-line adaptations of HIV-1. The common practice of propagating HIV-1 in immortalized $CD4^+$ T cell lines had the effect of artificially, but dramatically, increasing viral sensitivity to neutralizing antibodies and soluble receptor mimics as well as selecting for variants of unusual tropism (Moore and Ho 1995). The adaptation may lead to a higher functional affinity of antibodies for the oligomeric envelope glycoproteins of the virions, an increased propensity of the outer envelope glycoprotein to dissociate from its transmembrane-protein anchorage, and a greater capacity of the protein to induce membrane fusion (for reviews see Moore and Ho 1995; Poignard et al. 1996; Sattentau et al. 1999). All of these changes can be predicted to affect neutralization sensitivity in quantitatively different ways (Klasse and Moore 1996). Furthermore, the adapted variants have more positively charged residues in the outer envelope glycoprotein than clinical isolates do, and therefore bind more avidly to heparan sulfate proteoglycans expressed extensively on the surface of immortalized T-cell lines used frequently for analysis of neutralization mechanisms. A given neutralizing antibody can therefore have distinct effects on virus attachment depending both on the degree of adaptation of the virus and on the cell type used in the assay (Mondor et al. 1998; Ugolini et al. 1997). Since attachment is a sine qua non for infection, the mechanism of neutralization could thus change by viral adaptation. However, there is no evidence that viral adaptation could mediate a drastic change in how the virus is neutralized, for example from pre- to post-entry neutralization. Nevertheless, if we aim to relate different mechanisms of neutralization to what might take place in vivo, we must shun artifacts. Practically speaking, this would mean, in the case of HIV-1, the use of virus cloned or isolated directly from infected tissues, and propagated only in primary-cell culture and to a minimal extent. The mechanisms of neutralization of such virus would then be dissected on cell types as close as possible to those most important in vivo: primary cultures of $CD4^+$ T cells, macrophages and dendritic cells. Neutralization assays can never be used to elucidate net effects of antiviral antibodies in vivo; but they can be more or less relevant to the dissection of some mechanisms of immunity.

## 6 Conclusions

From the above, four major areas of challenges in current virus-neutralization research emerge: Firstly, the study of antibody binding to virus suffers from dogged theoretical confusion, uncertainty of equilibrium assumptions and the use of artifactual model antigens. Secondly, the biochemical detection of viral entry can be flawed by the inability to distinguish infectious from inert particles and productive from non-productive routes of internalization; recombinant fluorescent virions and quantitation of marker-gene expression may help overcome these conundra. Future research may thus resolve controversies surrounding effects on virus attachment, viral entry, and subsequent biochemical events. Thirdly, uncertainty regarding the precise events taking place during viral entry may be dispelled by the increasingly sophisticated information about the structure of the viral surface proteins that mediate attachment and entry, and by insights into how these interact with the cell-surface receptors. This knowledge will inspire detailed molecular hypotheses about the mechanisms of neutralization. Moreover, improved understanding of the fusion process between the viral envelope and cellular membranes, as well as of membrane penetration by naked viruses, will suggest well-defined steps at which antibodies can interfere. Fourthly, neutralization as an in vitro model of one aspect of antibody immunity may be brought into closer proximity with the in vivo situation; and it may be complemented in the future by new models that simulate other important effects of antibody-virus interactions in the organism.

## References

Allaway GP, Ryder AM, Beaudry GA, Maddon PJ (1993) Synergistic inhibition of HIV-1 envelope-mediated cell fusion by CD4- based molecules in combination with antibodies to gp120 or gp41. AIDS Res Hum Retroviruses 9:581–587

Armstrong SJ, Dimmock NJ (1992) Neutralization of influenza virus by low concentrations of hemagglutinin-specific polymeric immunoglobulin A inhibits viral fusion activity, but activation of the ribonucleoprotein is also inhibited. J Virol 66:3823–3832

Armstrong SJ, McInerney TL, McLain L, Wahren B, Hinkula J, Levi M, Dimmock NJ (1996) Two neutralizing anti-V3 monoclonal antibodies act by affecting different functions of human immunodeficiency virus type 1. J Gen Virol 77:2931–2941

Arthur LO, Bess JW Jr, Sowder RCD, Benveniste RE, Mann DL, Chermann JC, Henderson LE (1992) Cellular proteins bound to immunodeficiency viruses: implications for pathogenesis and vaccines. Science 258:1935–1938

Bahraoui E, Clerget-Raslain B, Chapuis F, Olivier R, Parravicini C, Yagello M, Montagnier L, Gluckman JC (1988) A molecular mechanism of inhibition of HIV-1 binding to CD4$^+$ cells by monoclonal antibodies to gp110. Aids 2:165–169

Barbas CFD, Bjorling E, Chiodi F, Dunlop N, Cababa D, Jones TM, Zebedee SL, Persson MA, Nara PL, Norrby E, et al. (1992) Recombinant human Fab fragments neutralize human type 1 immunodeficiency virus in vitro. Proc Natl Acad Sci USA 89:9339–9343

Brioen P, Rombaut B, Boeye A (1985) Hit-and-run neutralization of poliovirus. J Gen Virol 66:2495–2499

Brioen P, Thomas AA, Boeye A (1985) Lack of quantitative correlation between the neutralization of poliovirus and the antibody-mediated pI shift of the virions. J Gen Virol 66:609–613

Cavacini LA, Emes CL, Power J, Desharnais FD, Duval M, Montefiori D, Posner MR (1995) Influence of heavy chain constant regions on antigen binding and HIV-1 neutralization by a human monoclonal antibody. J Immunol 155:3638–3644

Cecilia D, KewalRamani VN, O'Leary J, Volsky B, Nyambi P, Burda S, Xu S, Littman DR, Zolla-Pazner S (1998) Neutralization profiles of primary human immunodeficiency virus type 1 isolates in the context of coreceptor usage. J Virol 72:6988–6996

Chan DC, Fass D, Berger JM, Kim PS (1997) Core structure of gp41 from the HIV envelope glycoprotein. Cell 89:263–273

Crane SE, Clements JE, Narayan O (1988) Separate epitopes in the envelope of visna virus are responsible for fusion and neutralization: biological implications for anti-fusion antibodies in limiting virus replication. J Virol 62:2680–2685

Della-Porta AJ, Westaway EG (1978) A multi-hit model for the neutralization of animal viruses. J Gen Virol 38:1–19
Dietzschold B, Tollis M, Lafon M, Wunner WH, Koprowski H (1987) Mechanisms of rabies virus neutralization by glycoprotein-specific monoclonal antibodies. Virology 161:29–36
Dimmock NJ (1984) Mechanisms of neutralization of animal viruses. J Gen Virol 656:1015–1022
Dulbecco R, Vogt M, Strickland AGR (1956) A study of the basic aspects of neutralization. Two animal viruses. Western equine encephalitis virus and poliomyelitis virus. Virology 2:162–205
Gollins SW, Porterfield JS (1986) A new mechanism for the neutralization of enveloped viruses by antiviral antibody. Nature 321:244–246
Gorny MK, Zolla-Pazner S (2000) Recognition by human monoclonal antibodies of free and complexed peptides representing the prefusogenic and fusogenic forms of human immunodeficiency virus type 1 gp41. J Virol 74:6186–6192
Grady LJ, Kinch W (1985) Two monoclonal antibodies against La Crosse virus show host-dependent neutralizing activity. J Gen Virol 66:2773–2776
Greve JM, Forte CP, Marlor CW, Meyer AM, Hoover-Litty H, Wunderlich D, McClelland A (1991) Mechanisms of receptor mediated rhinovirus neutralization defined by two soluble forms of ICAM-1. J Virol 65:6015–6023
Hashimoto N, Prince AM (1963) Virology 19:261–272
He RT, Innis BL, Nisalak A, Usawattanakul W, Wang S, Kalayanarooj S, Anderson R (1995) Antibodies that block virus attachment to Vero cells are a major component of the human neutralizing antibody response against dengue virus type 2. J Med Virol 45:451–461
Highlander SL, Sutherland SL, Gage PJ, Johnson DC, Levine M, Glorioso JC (1987) Neutralizing monoclonal antibodies specific for herpes simplex virus glycoprotein D inhibit virus penetration. J Virol 61:3356–3364
Icenogle J, Shiwen H, Duke G, Gilbert S, Rueckert R, Anderegg J (1983) Neutralization of poliovirus by a monoclonal antibody: kinetics and stoichiometry. Virology 127:412–425
Imai M, Sugimoto K, Okazaki K, Kida H (1998) Fusion of influenza virus with the endosomal membrane is inhibited by monoclonal antibodies to defined epitopes on the hemagglutinin. Virus Res 53:129–139
Iorio RM, Bratt MA (1984) Neutralization of Newcastle disease virus by monoclonal antibodies to the hemagglutinin-neuraminidase glycoprotein: requirement for antibodies to four sites for complete neutralization. J Virol 51:445–451
Jolly PE, Huso DL, Sheffer D, Narayan O (1989) Modulation of lentivirus replication by antibodies: Fc portion of immunoglobulin molecule is essential for enhancement of binding, internalization, neutralization of visna virus in macrophages. J Virol 63:1811–1813
Kabat D, Kozak SL, Wehrly K, Chesebro B (1994) Differences in CD4 dependence for infectivity of laboratory-adapted and primary patient isolates of human immunodeficiency virus type 1. J Virol 68:2570–2577
Kennedy-Stoskopf S, Narayan O (1986) Neutralizing antibodies to visna lentivirus: mechanism of action and possible role in virus persistence. J Virol 59:37–44
Khalife J, Guy B, Capron M, Kieny MP, Ameisen JC, Montagnier L, Lecocq JP, Capron A (1988) Isotypic restriction of the antibody response to human immunodeficiency virus. AIDS Res Hum Retroviruses 4:3–9
Kida H, Yoden S, Kuwabara M, Yanagawa R (1985) Interference with a conformational change in the haemagglutinin molecule of influenza virus by antibodies as a possible neutralization mechanism. Vaccine 3:219–222
Kjellen L (1985) A hypothesis accounting for the effect of the host cell on neutralization-resistant virus. J Gen Virol 66:2279–2283
Kjellen L, Pereira HG (1968) Role of adenovirus antigens in the induction of neutralizing antibody. J Gen Virol 2:177–185
Klasse J, Blomberg J (1987) Patterns of antibodies to human immunodeficiency virus proteins in different subclasses of IgG. J Infect Dis 156:1026–1030
Klasse PJ (1996) Physico-chemical analysis of the humoral immune response to HIV-1: quantification of antibodies, their binding to viral antigens and neutralization of viral infectivity. In: Korber B, Walker B, Koup R, et al. (eds) HIV molecular immunology database, vol 2. Los Alamos National Laboratory, Theoretical Biology and Biophysics, Los Alamos, New Mexico, pp IV: 22–52
Klasse PJ, Bron R, Marsh M (1998) Mechanisms of enveloped virus entry into animal cells. Adv Drug Deliv Rev 34:65–91

Klasse PJ, Moore JP (1996) Quantitative model of antibody- and soluble CD4-mediated neutralization of primary isolates and T-cell line-adapted strains of human immunodeficiency virus type 1. J Virol 70:3668–3677

LaCasse RA, Follis KE, Moudgil T, Trahey M, Binley JM, Planelles V, Zolla-Pazner S, Nunberg JH (1998) Coreceptor utilization by human immunodeficiency virus type 1 is not a primary determinant of neutralization sensitivity. J Virol 72:2491–2495

Layne SP, Merges MJ, Dembo M, Spouge JL, Nara PL (1990) HIV requires multiple gp120 molecules for CD4-mediated infection. Nature 346:277–279

Layne SP, Merges MJ, Spouge JL, Dembo M, Nara PL (1991) Blocking of human immunodeficiency virus infection depends on cell density and viral stock age. J Virol 65:3293–3300

Li A, Katinger H, Posner MR, Cavacini L, Zolla-Pazner S, Gorny MK, Sodroski J, Chou TC, Baba TW, Ruprecht RM (1998) Synergistic neutralization of simian-human immunodeficiency virus SHIV-vpu$^+$ by triple and quadruple combinations of human monoclonal antibodies and high-titer anti-human immunodeficiency virus type 1 immunoglobulins. J Virol 72:3235–3240

Lu S, Putney SD, Robinson HL (1992) Human immunodeficiency virus type 1 entry into T cells: more-rapid escape from an anti-V3 loop than from an antireceptor antibody. J Virol 66:2547–2550

Mandel B (1976) Neutralization of poliovirus: a hypothesis to explain the mechanism and the one-hit character of the neutralization reaction. Virology 69:500–510

Mandel B (1978) Neutralization of animal viruses. Adv Virus Res 23:205–268

Marechal V, Clavel F, Heard JM, Schwartz O (1998) Cytosolic Gag p24 as an index of productive entry of human immunodeficiency virus type 1. J Virol 72:2208–2212

Mason DW, Williams AF (1980) The kinetics of antibody binding to membrane antigens in solution and at the cell surface. Biochem J 187:1–20

Massey RJ, Schochetman G (1981a) Topographical analysis of viral epitopes using monoclonal antibodies: mechanism of virus neutralization. Virology 115:20–32

Massey RJ, Schochetman G (1981b) Viral epitopes and monoclonal antibodies: isolation of blocking antibodies that inhibit virus neutralization. Science 213:447–449

Mazanec MB, Coudret CL, Fletcher DR (1995) Intracellular neutralization of influenza virus by immunoglobulin A anti-hemagglutinin monoclonal antibodies. J Virol 69:1339–1343

McCullough KC, Smale CJ, Carpenter WC, Crowther JR, Brocchi E, De Simone F (1987) Conformational alteration in foot-and-mouth disease virus virion capsid structure after complexing with monospecific antibody. Immunology 60:75–82

McDougal JS, Kennedy MS, Orloff SL, Nicholson JK, Spira TJ (1996) Mechanisms of human immunodeficiency virus Type 1 (HIV-1) neutralization: irreversible inactivation of infectivity by anti-HIV-1 antibody. J Virol 70:5236–5245

McInerney TL, McLain L, Armstrong SJ, Dimmock NJ (1997) A human IgG1 (b12) specific for the CD4 binding site of HIV-1 neutralizes by inhibiting the virus fusion entry process, but b12 Fab neutralizes by inhibiting a postfusion event. Virology 233:313–326

Mondor I, Ugolini S, Sattentau QJ (1998) Human immunodeficiency virus type 1 attachment to HeLa CD4 cells is CD4 independent and gp120 dependent and requires cell surface heparans. J Virol 72:3623–3634

Montefiori DC, Collman RG, Fouts TR, Zhou JY, Bilska M, Hoxie JA, Moore JP, Bolognesi DP (1998) Evidence that antibody-mediated neutralization of human immunodeficiency virus type 1 by sera from infected individuals is independent of coreceptor usage. J Virol 72:1886–1893

Montefiori DC, Cornell RJ, Zhou JY, Zhou JT, Hirsch VM, Johnson PR (1994) Complement control proteins, CD46, CD55, CD59, as common surface constituents of human and simian immunodeficiency viruses and possible targets for vaccine protection. Virology 205:82–92

Moore JP, Ho DD (1995) HIV-1 neutralization: the consequences of viral adaptation to growth on transformed T cells. Aids 9:S117–S136

Moulard M, Lortat-Jacob H, Mondor I, Roca G, Wyatt R, Sodroski J, Zhao L, Olson W, Kwong PD, Sattentau QJ (2000) Selective interactions of polyanions with basic surfaces on human immunodeficiency virus type 1 gp120. J Virol 74:1948–1960

Muster T, Steindl F, Purtscher M, Trkola A, Klima A, Himmler G, Ruker F, Katinger H (1993) A conserved neutralizing epitope on gp41 of human immunodeficiency virus type 1. J Virol 67:6642–6647

Orlik O, Ban J, Hlavaty J, Altaner C, Kettmann R, Portetelle D, Splitter GA (1997) Polyclonal bovine sera but not virus-neutralizing monoclonal antibodies block bovine leukemia virus (BLV) gp51 binding to recombinant BLV receptor BLVRcp1. J Virol 71:3263–3267

Oroszlan S, Nowinski RC (1980) Lysis of retroviruses with monoclonal antibodies against viral envelope proteins. Virology 101:296–299

Parren PW, Burton DR, Sattentau QJ (1997) HIV-1 antibody–debris or virion? Nat Med 3:366–367

Parren PW, Mondor I, Naniche D, Ditzel HJ, Klasse PJ, Burton DR, Sattentau QJ (1998) Neutralization of human immunodeficiency virus type 1 by antibody to gp120 is determined primarily by occupancy of sites on the virion irrespective of epitope specificity. J Virol 72:3512–3519

Pelchen-Matthews A, Clapham P, Marsh M (1995) Role of CD4 endocytosis in human immunodeficiency virus infection. J Virol 69:8164–8168

Poignard P, Fouts T, Naniche D, Moore JP, Sattentau QJ (1996) Neutralizing antibodies to human immunodeficiency virus type-1 gp120 induce envelope glycoprotein subunit dissociation. J Exp Med 183:473–484

Poignard P, Klasse PJ, Sattentau QJ (1996) Antibody neutralization of HIV-1. Immunol Today 17:239–246

Possee RD, Schild GC, Dimmock NJ (1982) Studies on the mechanism of neutralization of influenza virus by antibody: evidence that neutralizing antibody (anti-haemagglutinin) inactivates influenza virus in vivo by inhibiting virion transcriptase activity. J Gen Virol 58?:373–386

Rizzuto CD, Wyatt R, Hernandez-Ramos N, Sun Y, Kwong PD, Hendrickson WA, Sodroski J (1998) A conserved HIV gp120 glycoprotein structure involved in chemokine receptor binding. Science 280:1949–1953

Roux KH, Strelets L, Michaelsen TE (1997) Flexibility of human IgG subclasses. J Immunol 159:3372–3382

Rubin H, Franklin RM (1957) On the mechanism of Newcastle disease virus neutralization by immune serum. Virology 3:84–89

Saifuddin M, Hedayati T, Atkinson JP, Holguin MH, Parker CJ, Spear GT (1997) Human immunodeficiency virus type 1 incorporates both glycosyl phosphatidylinositol-anchored CD55 and CD59 and integral membrane CD46 at levels that protect from complement-mediated destruction. J Gen Virol 78:1907–1911

Sattentau QJ, Moore JP (1995) Human immunodeficiency virus type 1 neutralization is determined by epitope exposure on the gp120 oligomer. J Exp Med 182:185–196

Sattentau QJ, Moulard M, Brivet B, Botto F, Guillemot JC, Mondor I, Poignard P, Ugolini S (1999) Antibody neutralization of HIV-1 and the potential for vaccine design. Immunol Lett 66:143–149

Sattentau QJ, Zolla-Pazner S, Poignard P (1995) Epitope exposure on functional, oligomeric HIV-1 gp41 molecules. Virology 206:713–717

Schofield DJ, Stephenson JR, Dimmock NJ (1997) High and low efficiency neutralization epitopes on the haemagglutinin of type A influenza virus. J Gen Virol 78:2441–2446

Schonning K, Lund O, Lund OS, Hansen JE (1999) Stoichiometry of monoclonal antibody neutralization of T-cell line- adapted human immunodeficiency virus type 1. J Virol 73:8364–8370

Spear GT, Sullivan BL, Takefman DM, Landay AL, Lint TF (1991) Human immunodeficiency virus (HIV)-infected cells and free virus directly activate the classical complement pathway in rabbit, mouse and guinea-pig sera; activation results in virus neutralization by virolysis. Immunology 73: 377–382

Speth C, Kacani L, Dierich MP (1997) Complement receptors in HIV infection. Immunol Rev 159:49–67

Stinski MF, Cunningham CH (1970) Antibody-neutralized avian infectious bronchitis virus in chicken embryo kidney cells: entry and degradation. J Gen Virol 8:173–186

Stoiber H, Pinter C, Siccardi AG, Clivio A, Dierich MP (1996) Efficient destruction of human immunodeficiency virus in human serum by inhibiting the protective action of complement factor H and decay accelerating factor (DAF, CD55) J Exp Med 183:307–310

Taylor HP, Armstrong SJ, Dimmock NJ (1987) Quantitative relationships between an influenza virus and neutralizing antibody. Virology 159:288–298

Thouvenin E, Laurent S, Madelaine MF, Rasschaert D, Vautherot JF, Hewat EA (1997) Bivalent binding of a neutralising antibody to a calicivirus involves the torsional flexibility of the antibody hinge. J Mol Biol 270:238–246

Tilley SA, Honnen WJ, Racho ME, Chou TC, Pinter A (1992) Synergistic neutralization of HIV-1 by human monoclonal antibodies against the V3 loop and the CD4-binding site of gp120. AIDS Res Hum Retroviruses 8:461–467

Trautman R, Harris WF (1977) Modeling and computer stimulation approach to the mechanism of foot-and-mouth disease virus neutralization assays. Scand J Immunol 6:831–841

Trkola A, Ketas T, Kewalramani VN, Endorf F, Binley JM, Katinger H, Robinson J, Littman DR, Moore JP (1998) Neutralization sensitivity of human immunodeficiency virus type 1 primary isolates to antibodies and CD4-based reagents is independent of coreceptor usage. J Virol 72:1876–1885

Ugolini S, Mondor I, Parren PW, Burton DR, Tilley SA, Klasse PJ, Sattentau QJ (1997) Inhibition of virus attachment to CD4+ target cells is a major mechanism of T cell line-adapted HIV-1 neutralization. J Exp Med 186:1287–1298

Valenzuela A, Blanco J, Krust B, Franco R, Hovanessian AG (1997) Neutralizing antibodies against the V3 loop of human immunodeficiency virus type 1 gp120 block the CD4-dependent and -independent binding of virus to cells. J Virol 71:8289–8298

Weissenhorn W, Dessen A, Harrison SC, Skehel JJ, Wiley DC (1997) Atomic structure of the ectodomain from HIV-1 gp41. Nature 387:426–430

Wiley DC, Skehel JJ (1987) The structure and function of the hemagglutinin membrane glycoprotein of influenza virus. Annu Rev Biochem 56:365–394

Wohlfart C (1988) Neutralization of adenoviruses: kinetics, stoichiometry, mechanisms. J Virol 62:2321–2328

Wohlfart CE, Svensson UK, Everitt E (1985) Interaction between HeLa cells and adenovirus type 2 virions neutralized by different antisera. J Virol 56:896–903

Wyatt R, Kwong PD, Desjardins E, Sweet RW, Robinson J, Hendrickson WA, Sodroski JG (1998) The antigenic structure of the HIV gp120 envelope glycoprotein. Nature 393:705–711

Wyatt R, Sodroski J (1998) The HIV-1 envelope glycoproteins: fusogens, antigens, immunogens. Science 280:1884–1888

# A Model for Neutralization of Viruses Based on Antibody Coating of the Virion Surface

D.R. Burton, E.O. Saphire, and P.W.H.I. Parren



## 1 Introduction

Historically, a school of thought in virology has argued that viruses are often neutralized by the binding of one or a few antibody molecules to a critical site, resulting in conformational changes in envelope or capsid proteins that lead to inactivation of the virion. Such a mechanism, however, is difficult to reconcile with the pathway of antibody generation and maturation. Antibodies are not molecules

Departments of Immunology and Molecular Biology, The Scripps Research Institute, 10550 N. Torrey Pines Road, La Jolla, CA 92037, USA like enzymes that have evolved over long time periods for narrow, highly specialized functions. Rather, they are products of mutation and selection that are generated afresh in each individual animal as a result of antigen challenge. Each antibody, even directed against the same epitope, has a different sequence and different detailed mode of binding to antigen, and we therefore consider it unlikely that such molecules would trigger common conformational changes in the antigen. We therefore primarily see antibodies as superb binding molecules selected to bind with the highest affinity. It is with this in mind that we present this review of the neutralization literature. Neutralization, we argue, is a function of occupancy of binding sites on the virion and may be mostly mediated by the bulk of the antibody molecule. Antibody bound to virions may induce the activation of effector functions aiding the inactivation of virus in vivo. A more extensive review on this subject has been published elsewhere (Parren and Burton 2001).

## 2 Mechanisms of Neutralization

At the end of the nineteeth century, Willem Beijerinck made an important conceptional advance with the recognition that viruses, which he termed the "contagium vivum fluidum", represent infectious entities distinct from bacteria (Beijerinck 1898; Van Kammen 1999). Great advances in understanding the inactivation of viruses by antibody were then made in the early twentieth century, as reviewed by Burnet, Keogh and Lush (Burnet et al. 1937). They assumed that neutralization studies of bacteriophage and plant and animal viruses showed that, generally: (1) neutralization was reversible, and (2) more than one antibody was required for neutralization. Based on studies with a number of animal viruses, including vaccinia and influenza, they summarized: "virus and antibody combined reversibly, and that the infectivity or otherwise of a given virus particle was determined by the amount of antibody united to it at the time of effective contact with the susceptible cell". Reversibility indicated that the "union of antibody has no intrinsic inactivating effect on the virus". The conclusions that neutralization is reversible and requires the binding of multiple antibody molecules to the virion have since been disputed (and rediscovered) in many later publications. In particular, the observation of single-hit kinetics in a number of experimental systems has been interpreted as evidence for neutralization by the binding of a single antibody to a critical site on the virion.

In 1956, Jerne and Avegno showed that inactivation of T4 bacteriophage followed first-order kinetics, and inactivation may be the result of a single event. This was plausible as neutralizing antibodies bind to the tail-plate of the phage, an essential component in the infective process, and the width of the tail-plate is similar to the length of an antibody molecule. An influential study which sharply contradicted the conclusions by Burnet et al. (1937) was published by Dulbecco, Vogt and Strickland, also in 1956, who analyzed neutralization of western equine encephalitis virus (WEE) and poliovirus by specific antisera (Dulbecco et al. 1956). It was found that curves describing the logarithm of virus survival as a function of time generally did not display an initial shoulder or lag-time, which was interpreted as neutralization following first-order kinetics. After an initial time period, however, the slope of the curves decreased, eventually becoming horizontal, which indicated the existence of a persistent and apparently neutralization-resistant fraction of virus. Dissociation of poliovirus-antibody complexes could not be observed, and there was only a very slight dissociation of WEE-antibody complexes. It was concluded that the attachment of a single neutralizing antibody molecule to a critical site is sufficient to inactivate a virus particle and that neutralization is irreversible (Dulbecco et al. 1956).

However, not all the results reported by Dulbecco et al. (1956) could be explained by the neutralization model they developed. First, a clear lag time of WEE neutralization was observed when kinetic neutralization experiments were performed at temperatures lower than 37°C. Even at 37°C, Dulbecco and coworkers noted that a small lag time may exist, although the experimental conditions used may have made detection difficult. Furthermore, multiplicity curves, which describe the logarithm of virus survival as a function of increasing antibody concentration, did not follow theoretical curves based on the model; experimental curves bent upward at high antibody concentration rather than downward. This deviation could not be completely explained by the persistent virus fraction. The authors concluded that different fractions may be present in the challenge stock consisting of virions with a different number of critical sites or unequal distribution of critical sites (Dulbecco et al. 1956).

Krummel and Uhr (1969) recognized that the critical site neutralization models developed often did not fit the data at later time points during the course of neutralization in kinetic experiments. They argued that, for a highly symmetrical virus with a surface array of repeated coat-protein subunits, such as the bacteriophage ϕX174 studied, it was likely that multiple antibody molecules were required for neutralization. Kinetic neutralization experiments, in which residual infectivity was assessed using two different methods, indeed indicated that bacteriophage covered with antibody may retain its infectivity. Significantly, Krummel and Uhr (1969) observed an apparent faster neutralization rate if residual infectivity was assessed using a "decision tube" method rather than the conventional direct-plating method. In the direct plating kinetic neutralization experiment, antibody and phage are mixed and samples, drawn at multiple time points, incubated with bacteria, and plated to assess residual infectivity. In the decision tube method, an excess of additional neutralizing antiserum is added a short time after adding the reaction sample to the bacteria. Whereas infectious titers determined with both methods were similar prior to the addition of neutralizing antibody, the decision tube method measured markedly decreased infectious titers compared to the conventional plating method at time points after the addition of antibody. As neutralization of ϕX174 phage was effectively irreversible under the conditions used, the difference observed could not be explained by the prevention of reactivation of neutralized phage (e.g., by dissociation of antibody-virus complexes). A plausible explanation is that some phage are partially coated with antibody molecules but not neutralized, and these phage infect bacteria at a slower rate than naked phage. A similar delay in the initiation of bacteriophage-induced lysis in neutralization experiments has previously been interpreted to indicate that partially antibody-coated but infectious phage particles exist (reviewed in BURNET et al. 1937). By entering an infectivity function for such phage-antibody complexes into the mathematical model developed by KRUMMEL and UHR (1969), the model matched their data better. This clearly indicated that the observed reaction rates of neutralizing antibody with bacteriophage were not first-order, even under conditions of antibody excess.

## 2.1 Neutralization of Picornaviruses

Several mechanisms have been described for the neutralization of picornaviruses including aggregation, virion stabilization, inhibition of virus attachment, and induction of conformational changes in the viral capsid (see Smith, this volume for a detailed review). The induction of conformational changes (resulting in a large change in isoelectric point) upon antibody binding is the most prominent and has prompted the most detailed investigations. Most studies have been carried out with poliovirus and human rhinovirus (HRV).

Neutralization of poliovirus has been studied extensively. DULBECCO et al. (1956) argued that neutralization of poliovirus is irreversible and follows first-order kinetics, as discussed. Studies performed by MANDEL (1961) confirmed that reactivation of virus infectivity after dilution of antibody-virus complexes could not be demonstrated in measurements taken over several months. The apparent irreversibility of the reaction, however, appeared to be due to a very tight binding of antibody to poliovirus rather than the induction of permanent changes in the viral particles, as it was demonstrated that neutralizing rabbit antibody could be eluted from poliovirions at low pH accompanied by quantitative reactivation of virus activity (MANDEL 1961). MANDEL (1976) found that the capsid protein of poliovirus may exist in two conformational states characterized by distinct isoelectric points: state A with an isoelectric point of about 7, and state B with an isoelectric point of about 4.5. Neutralizing antibody was found to stabilize the virion in state B. MANDEL (1976) proposed that binding of neutralizing antibody locked the viral capsid in a non-infectious conformation by rendering it resistant to uncoating following its adsorption to the target cell. Single-hit kinetics in poliovirus neutralization could be explained by the induction of a generalized molecular rearrangement in the viral capsid after binding of a single antibody to a critical site on the capsid molecule. The mechanism is feasible, as the poliovirus capsid is a closely packed structure, making conformational cooperativity among capsid subunits a possibility (MANDEL 1976). However, the relationship between the conformational state and infectivity was not complete, as infectious virus in the B state could be detected, particularly at low antibody multiplicity (even after subtracting the infectivity assigned to the small fraction of virus naturally occurring in the B state) (MANDEL 1976). A correlation between binding of neutralizing antibody to poliovirus and a shift in isoelectric point has generally held up in studies with monoclonal antibodies (EMINI et al. 1983a), although exceptions have been noted (EMINI et al. 1983a; BRIOEN et al. 1985).

Changes in isoelectric point following the binding of neutralizing antibody have also been described for another picornavirus, namely, rhinovirus (HRV) (COLONNO et al. 1989). However, in this case it is particularly clear that the induced conformational changes do not play a major role in picornavirus neutralization (reviewed in Hewat and Blaas, this volume; Smith, this volume). The most convincing data come from cryo-electron microscopy and X-ray crystallography of virus-antibody complexes, which show that rhinovirus can be neutralized without the induction of large conformational changes in the virion capsid (SMITH et al. 1993, 1996; Smith, this volume; HEWAT and BLAAS 1996). Furthermore, as the change in isoelectric point has been observed following binding of a large number of different antibodies against a range of epitopes on rhinovirus, it appears unlikely that all these antibodies would induce a common conformational change required for neutralization. The most likely hypothesis is that the changes in isoelectric point are the result of changes occurring in exposed, highly flexible portions of the capsid proteins rather than structural rearrangement of the viral capsid (CHE et al. 1998; Smith, this volume). Although these changes are the result of the binding of neutralizing antibody, they are probably mostly irrelevant to the mechanism of neutralization itself.

The antibody activity most likely to affect HRV infectivity appears to be inhibition of virus attachment. The strongly neutralizing mAb 17-Ia binds to an epitope located on a loop between the B and C strands of the capsid protein VP1 β-barrel and situated close to the canyon that harbors the binding site for the cellular HRV receptor ICAM-1 (reviewed in Smith, this volume). mAb 17-Ia neutralizes by inhibition of attachment, and it has been demonstrated that attachment of six to seven antibody molecules per HRV particle is required to reduce attachment by 63% (SMITH et al. 1993). Significantly, COLONNO et al. (1989) have shown that Fab fragments of antibodies against all four neutralizing epitopes on HRV inhibit attachment, some of which are located more-distal from the HRV receptor binding site. A linear correlation was found when inhibition of attachment was plotted as a function of antibody concentration. Furthermore, the rank order of neutralization potency followed the rank order of inhibition of attachment. The studies on HRV clearly demonstrate that the binding of more than one antibody molecule is required for neutralization and suggest a role for antibody occupancy of binding sites on the virus, as will be discussed in the following sections.

Further studies on the neutralization of poliovirus suggested that neutralization of this picornavirus also requires the attachment of multiple antibody molecules. Particular insight into the relationship between antibody bound to poliovirus and neutralization was provided by PHILIPSON et al. (1966), who developed an aqueous polymer phase system to study the interaction between neutralizing antibody and poliovirus. It was found that virions complexed with antibody could be separated from free virions by adding a polymer to the mixture followed by a simple phase-separation technique. The phase system was more sensitive than conventional neutralization tests in detecting antibody binding to the virion, clearly indicating that neutralizing antibody was bound to the virions before neutralization could be detected. Interestingly, binding titers in the phase system method were two to four times greater than the serum neutralization titers against poliovirus (Philipson et al. 1966). The significance of this number will become clear below. Additional studies by Philipson (1966) on the interaction of antibody with poliovirus using countercurrent distribution furthermore showed that, at low antibody input, virion-antibody complexes separated from free virus retained infectivity.

The neutralizing antibody dose is defined as the antibody concentration at which virus infectivity is reduced to 37% (1/e) of the infectivity of the non-neutralized virus stock. Icenogle et al. (1983) determined the neutralizing dose for poliovirus in a study using radiolabeled virus and antibody; they demonstrated that the binding of an average of four to five antibody molecules to a poliovirus particle resulted in neutralization. Two neutralization models were proposed to explain these results in the context of first-order kinetics: a critical-site hypothesis in which binding to one out of four sites on the virus leads to neutralization, and a stepwise model in which binding to each site reduces infectivity by three-fourths. Both models, however, are in conflict with the studies by Philipson et al. (1966), discussed above, which demonstrated that antibody binding to virions can be detected prior to neutralization. Significantly, the observation that an occupancy of four to five antibody molecules per virion is required for neutralization is consistent with the observed differences in binding and neutralization titers in the Philipson studies. The most straightforward conclusion is that the single-hit hypothesis of poliovirus (picornavirus) neutralization is not correct and that the mechanism of neutralization of poliovirus may be similar to that of HRV.

The overall architecture of poliovirus and HRV is similar, and both viruses have similar proportions, consisting of a 30-nm diameter capsid surrounding the positive-strand RNA genome. The three-dimensional structure of poliovirus in complex with a soluble form of its receptor (CD155) has recently been solved (Belnap et al. 2000; He et al. 2000). The poliovirus receptor, CD155, has a footprint similar to that observed for the rhinovirus receptor, ICAM-1, bound to HRV. Both bind to the canyon, a narrow surface depression on the viral surface. It may therefore not be completely surprising that the stoichiometric analyses discussed above yield similar results for HRV and poliovirus. The data suggest that the attachment of four to seven antibody molecules is required for the neutralization of picornaviruses of poliovirus/HRV dimensions.

Studies by Horaud and colleagues (Crainic et al. 1983; Diamond et al. 1985; Blondel et al. 1986) suggested that antibodies may bind to poliovirus without neutralization. This conclusion was made on the basis of studies using neutralizing antibodies that were analyzed for their activity (neutralization and binding) against a range of antigenic poliovirus variants (distinct serotypes and escape mutants); e.g., some serotypes appeared to be bound by antibody in an immunoprecipitation assay but were not neutralized. The studies, however, made no distinction between strong and weak binding. Mutations on poliovirus envelope in or near the antibody binding site may result in decreased antibody affinity for the poliovirus mutants in question, resulting in a level of occupancy too low to neutralize the virus (for example, as shown by the studies discussed above). This interpretation is supported by the observation that addition of anti-antibody in the studies above did result in neutralization of the mutant viruses (BLONDEL et al. 1986), presumably because anti-antibody effectively increases affinity (avidity) and raises occupancy to the level required for neutralization (see below).

## 2.2 Evidence for Multi-Hit Models of Neutralization

Single-hit models are based on the observation that infectivity starts to decrease immediately following addition of antibody to virus. The absence of a detectable lag phase, i.e., a period in which infectivity remains unaffected following antibody addition, has often been taken as supporting evidence for the single-hit model (DULBECCO et al. 1956; MANDEL 1976; ICENOGLE et al. 1983; TAYLOR et al. 1987; MCLAIN and DIMMOCK 1994). It may be argued, however, that reactions of antibodies of nanomolar affinities for their cognate antigen occur rapidly (e.g., SCICLUNA and MCCULLOUGH 1999), and, as samples are being drawn in kinetic studies at intervals of minutes or greater, it may not be possible to measure the surviving virus fraction fast enough to observe the presence or absence of a lag phase under standard conditions (DELLA-PORTE and WESTAWAY 1977). Furthermore, lag times in neutralization kinetics have indeed been observed in studies with most viruses when the experiments were performed at lower incubation temperature, more dilute antibody concentration, or with antibodies of lower affinity (BURNET et al. 1937; DULBECCO et al. 1956; LAFFERTY 1963a; PHILIPSON 1966; TAYLOR et al. 1987). This is difficult to reconcile with the single-hit model, as it requires the assumption that the mechanism of neutralization changes as temperature, antibody concentration, or affinity are varied.

KLASSE and MOORE (1996, and references therein) have argued that a first-order neutralization reaction does not give direct proof for neutralization by single-hit kinetics. One antibody molecule neutralizing one virion implies first-order kinetics, but the reverse is not true. Stoichiometric analyses (see below) suggest that the binding of more than one antibody molecule to a virus particle is usually required for neutralization. An alternative explanation for the absence of a lag phase to the one discussed above, at least for enveloped viruses, may be found in a heterogeneity of spike density. In virus stocks, virions may exist with a low number of active spikes that may only require one or a few hits for neutralization (KLASSE and MOORE 1996). HIV-1 is an example where this hypothesis may be applicable, as shedding of gp120 from the virion results in the formation of inactive spikes (SATTENTAU et al. 1995; PARREN et al. 1999).

A plethora of evidence indicates that virus neutralization in general is the result of the attachment of a number of antibody molecules to the infective particle. Many of these arguments have been cogently reviewed by DELLA-PORTE and WESTAWAY

(1977). A summary of these arguments supplemented with evidence for multi-hit neutralization from more recent studies is presented here. These arguments include: the occurrence of a persistent fraction of infectious virus, the neutralization of infectious virus-antibody complexes (sensitized virus) by addition of anti-antibody, neutralization of sensitized virus by addition of complement, retardation of infection of sensitized virus, the phenomenon of antibody-mediated enhancement of infection, and additive and synergistic effects of antibodies in neutralization.

### 2.2.1 The Persistent Fraction of Infectious Virus

The presence of a persistent fraction of non-neutralizable virus in most virus preparations has been the source of much debate. The non-neutralizable virus fraction has been attributed to the presence of low-affinity antibodies in antibody preparations (particularly early immune sera), virus heterogeneity, neutralization-interference of non-neutralizing antibodies, virus association with lipids and lipoproteins, and virus aggregation. In some studies it has been shown that the infectious titer of the persistent fraction can be further reduced by the addition of anti-antibody, implying that this fraction contains virus particles with some level of antibody coating.

An early explanation of the non-neutralizable virus fraction attributed the phenomenon to association-dissociation equilibria between virus and antibody (Burnet et al. 1937). Indeed, dissociation may play a role when low-affinity antibody preparations are being used and dissociation of the virus-antibody complexes occurs before or during the addition of the complexes to the target cell (Jerne and Avegno 1956; Lafferty 1963b). With high-affinity antibody preparations, however, this does not seem to be an important factor (Dulbecco et al. 1956). It has also been suggested that a low-affinity antibody or non-neutralizing antibody may bind to the virion, preventing the binding of neutralizing antibody and giving rise to a non-neutralizable fraction. In some cases, however, such a fraction is more easily explained in terms of heterogeneity of the virus preparation. Vaccinia virus and rabbitpox virus preparations, for example, contain a small fraction of extracellular enveloped virus (EEV) in an excess of intracellular mature virus (IMV). The EEV contains an additional wrapping membrane and is relatively resistant to neutralization (Ichihashi 1996; Vanderplasschen et al. 1997). The wrapping membrane, however, is easily damaged during manipulations, thereby exposing the more neutralization-sensitive IMV particle (Ichihashi 1996; Vanderplasschen et al. 1997). This mechanism, rather than dissociation of low-affinity or non-neutralizing antibody, may explain experiments that show that the persistent fraction in rabbitpox virus can be neutralized following dilution and washing (Lafferty 1963a).

In a study on neutralizing antibodies against mouse mammary tumor virus (MMTV), it was suggested that non-neutralizing mAbs against MMTV gp52 could interfere with the binding and neutralizing activity of a neutralizing mAb (Massey and Schochetman 1981). This would be one of the only examples of such an activity for a mAb. The inhibition described, however, is extremely weak and only apparent at high non-neutralizing mAb:neutralizing mAb ratios. On closer examination, the observed blocking is well within the range of the experimental error of the MMTV pseudotype neutralization assay used (Massey and Schochetman 1981) and is therefore unlikely to be of any significance.

Aggregation of virus may be responsible for many observations of a non-neutralizable virus fraction. Aggregates of virus particles already present in the preparation or formed by aggregation with antibody may inhibit access of antibody to virus particles at the inside of a small aggregate, protecting it from neutralization. Filtration to remove the aggregate or treatment with anti-antibody to coat the aggregate more extensively can indeed remove the persistent fraction in some instances (Wallis and Melnick 1967, 1970).

### 2.2.2 Neutralization of Sensitized Virus with Anti-Antibody or Complement

Neutralization of sensitized virus by anti-antibody or complement components has been cited as an argument against single-hit kinetics, as it indicates that infectious virus-antibody complexes exist (Della-Porte and Westaway 1977).

### 2.2.3 Retardation of Infection by Antibody

Burnet et al. (1937) noted in their review that infectious antibody-bacteriophage complexes exist, as evidenced by observations in the literature that plaque sizes were reduced in the presence of antibody, which presumably occurred through delayed infection of phages with a low level of antibody attached. Krummel and Uhr (1969; also see above) demonstrated reduced infectivity rates for bacteriophages complexed with neutralizing antibody in kinetic neutralization experiments. Westaway (1965) further noted that the rate of penetration of West Nile virus pre-incubated with neutralizing antibody was delayed, as the virus was susceptible to post-attachment neutralization for a longer period of time after adsorption to target cells than non-sensitized virus. These experiments indicate that antibody may combine with virus at an occupancy too low to neutralize the virus but high enough to decrease the rate of entry into the target cell.

### 2.2.4 Antibody-Mediated Enhancement of Infection

Antibody-mediated enhancement of infection is a phenomenon that appears to occur for neutralizing antibodies at sub-neutralizing concentrations (Hawkes and Lafferty 1967; Halstead 1982; Morens et al. 1987). Enhancement is very sensitive to the target cell and is dependent on the types and expression levels of Fc and/or complement receptors on the target-cell surface. A classical example is enhancement of dengue virus infection, which is dependent on the interaction between virion-bound antibody and Fc receptors expressed on the target cell. In typical assays, neutralization is observed at relatively high concentrations whereas enhancement is observed at lower concentrations. Using neutralizing antibodies against dengue virus, Morens et al. (1987) indeed showed that the mAb concentration maximizing enhancement was predicted by the neutralization titer. HAWKES and LAFFERTY (1967) noted that the balance between enhancement and neutralization is dependent on the incubation time of virus and antibody, i.e., they observed a shift of the dose-response curve as time progressed, and enhancement as well as neutralization occurred at increasingly lower concentrations. Presumably, at early time points, attachment of a small number of antibodies leads to enhancement, and as the reaction progresses more antibodies attach eventually leading to neutralization of the virion. This clearly indicates that enhancement and neutralization are two different biological outcomes of the interaction of an antibody with virus at different levels of occupancy.

For HIV-1, both Fc-receptor-mediated and complement-mediated antibody-dependent enhancement of infection have been described (TAKEDA et al. 1988; ROBINSON et al. 1988; HOMSY et al. 1989; MASCOLA et al. 1993; SCHUTTEN et al. 1995; LUND et al. 1995). Fc-receptor-mediated endocytosis of virion-antibody complexes may lead to the internalization of virus and infection. A plausible alternative is that binding to Fc receptors stabilizes the interaction of virion and target cell permitting interaction of the envelope spike and virus receptor at low levels of antibody coating of the virion (CONNOR et al. 1991). At higher levels of coating, this interaction may be inhibited. Complement receptor 2 (CD21) may mediate enhancement of HIV-1 infection via two proposed mechanisms, either by acting as a receptor for HIV-1 opsonized with antibody and complement, or by increasing virus binding to the cell due to an interaction of CD21 with opsonized virus (BOYER et al. 1992; LUND et al. 1995). In addition, virion-bound C1q may directly interact with receptors on the host cell leading to enhancement of infection (PROHASZKA et al. 1997). In some cases, such as that of HIV-1, the enhancement does not require the Fc part of the antibody molecule (SULLIVAN et al. 1998b and this volume). It is suggested that low-level coating may trigger conformational changes in the envelope that, for example, favor fusion between the virion and target cell, or such coating may nonspecifically reduce repulsion between the virion and target cell surface. A similar mechanism of antibody-mediated enhancement of infection has been described for Sindbis virus (FLYNN et al. 1988).

The phenomenon of enhancement indicates that antibodies bind to virions at subneutralizing concentrations and is therefore a strong argument for the importance of antibody occupancy in virus neutralization and neutralization following multi-hit kinetics. High levels of occupancy lead to neutralization, whereas in the presence of permissive target cells low levels of occupancy may lead to enhancement of infection.

### 2.2.5 Additive and Synergistic Effects in Neutralization

The presence of additive and synergistic effects in neutralization are indicative of a role of antibody occupancy in neutralization of viruses and provide an argument for multi-hit neutralization. Phenotypic mixing experiments in particular are informative in this respect. It has been shown that cells dually infected with certain viruses may yield phenotypically mixed virions. These are relatively poorly neutralized by single-specificity antisera against each of the parent viruses but are more effectively neutralized by mixed neutralizing antisera. Cells dually infected with the myxovirus fowl plaque virus (FPV) and the rhabdovirus vesicular stomatitis virus (VSV) produced a virus progeny of which about 20% was composed of virus particles with mixed envelopes. Neutralization of virions carrying mixed envelopes was poor to non-existent in neutralization assays with anti-VSV and anti-FPV alone. However, strong neutralization occurred when the assay was carried out with a mixture of the two sera (Zavada and Rosenbergova 1972). Similar observations were made with phenotypically mixed virions of VSV and Sendai virus (Kimura 1973) and VSV and measles virus (Wild et al. 1975–1976).

Additive effects are also observed for antibodies binding to the paramyxovirus Newcastle disease virus (NDV). Binding of single mAbs against the NDV hemagglutinin-neuramidase (HN) glycoprotein leaves a large persistent fraction of non-neutralized virus (Iorio and Bratt 1983). The persistent fraction appeared to be bound by neutralizing mAb as it was almost completely neutralized by the addition of anti-antibody (Iorio and Bratt 1984). It did not seem to contain large aggregates of virus or genotypically distinct viruses (Iorio and Bratt 1984). Interestingly, the persistent fraction could be neutralized when antibodies against several epitopes on the HN glycoprotein were combined and, in addition, a level of neutralization comparable to that achieved with a polyclonal immune serum could only be obtained by combining antibodies against four distinct epitopes on NDV (Iorio and Bratt 1983, 1984). These observations are consistent with some heterogeneity of the NDV HN glycoprotein which may exist in different conformations on the virion surface. Neutralization of La Crosse virus to the level achieved with a polyclonal antibody could similarly only be obtained with combinations of two mAbs against the G1 glycoprotein (Kingsford 1984). These studies suggest a role for antibody occupancy of viral-surface binding sites in neutralization of NDV and La Crosse virus.

## 2.3 Neutralization of Adenovirus

Adenovirus is an icosahedral capsid with characteristic long, thin fibers protruding from the 12 vertices. A knob at the end of each fiber serves as the primary receptor attachment site. The penton-base protein, located at the base of the fiber, binds to αv integrins via an RGD motif (reviewed by Stewart and Nemerow 1997). An unusual observation is that an antibody directed against the RGD motif neutralizes adenovirus infection as a Fab fragment but not as a whole IgG molecule (Stewart et al. 1997). Cryo-electron microscopy of adenovirus complexed with the Fab indicated that the Fab may occupy all five RGD epitopes on the penton-base complex. The whole IgG molecule was shown to bind soluble penton base with a stoichiometry of 2.8, indicating that, per penton base, two IgG molecules may bind bivalently and one monovalently. In the intact virus particle, however, the central fiber presumably provides a source of steric hindrance preventing the IgG molecules but not the Fab fragments from binding to all five sites on the penton base

(Stewart et al. 1997). Neutralization of adenovirus by an antibody against the penton-base protein therefore requires occupancy of all five integrin binding sites.

## 2.4 Neutralization of Rabies Virus

A number of elegant and detailed studies have been carried out by Flamand and colleagues on antibody neutralization of rabies virus. These studies support the general notion that neutralization is the result of coating the virion with antibody molecules (Flamand et al. 1993; Raux et al. 1995). The number of antibody molecules bound per rabies virus particle was determined carefully and correlated with the magnitude of neutralization achieved. It was found that infectivity was completely preserved with fewer than 130 IgG or 30 IgM molecules bound per virion (Flamand et al. 1993). The neutralizing dose, the dose at which infectivity is decreased to 37% (1/e), was between 130 and 320 IgG (40–50 IgM) molecules bound per virion depending on the antibody tested. There is a marked threshold effect, with a two- to three-fold increase in the number of antibody molecules close to the threshold leading to a drop in infectious titer of about 1–2 logs (Flamand et al. 1993).

Three mAbs were identified that bound to the virus without neutralization. Two of these were subsequently shown to bind to a minor population of envelope spikes in an altered acidic conformation of envelope (Raux et al. 1995). Coating of the virion via this altered spike was too incomplete to permit neutralization. If the virion was maintained at lower pH to convert most spikes to the acidic form, then more antibody bound and the virus was neutralized. One mAb, which effectively neutralized wild-type virus, appeared to saturate the envelope spikes of a neutralization escape variant (1,080 molecules of IgG per virion) but did not neutralize it. The antibody-coated virus was still able to attach to target cells. The authors considered it unlikely that this attachment was occurring through envelope because of the thickness of the antibody shell coating the virus (Flamand et al. 1993). This case appears to constitute a bona fide example of virion coating by antibody without neutralization. Although it is the rabies mutant virus rather than the antibody that seems be an exception to the rule here, it would be very interesting to elucidate how this mutant virus can escape neutralization while being completely coated with antibody.

Overall, the studies by Flamand et al. (1993) strongly suggest that there is a correlation between antibody occupancy and neutralization of rabies virus. The experiments show that neutralization occurs at a mean critical coating density of about 225 IgG or 45 IgM molecules bound per virion.

## 2.5 Neutralization of HIV-1

We have investigated the relative importance of binding site occupancy and epitope specificity in antibody neutralization of HIV-1 using an extensive panel of neutralizing mAbs against HIV-1 (Parren et al. 1998). Neutralization was studied using T-cell-line-adapted (TCLA) HIV-1 isolates. The strategy was to compare the binding of a number of antibodies to different gp120 epitopes, presented in the form of functional oligomeric gp120 on infected cells, with their capacity to neutralize the corresponding virus. A concentration of mAb yielding half-maximal binding ($K_{50}$ ) and a neutralization titer ($ID_{50}$) of similar magnitude would be consistent with antibody occupancy of virion binding sites playing a major role in HIV-1 neutralization. Similar $K_{50}/ID_{50}$ ratios for different epitopes would suggest that neutralization is independent of the epitope recognized. Indeed, an excellent correlation ($r = 0.882$; $P < 0.0001$) between neutralization ($ID_{50}$) and binding to oligomeric gp120 ($K_{50}$) was observed. Furthermore, the $K_{50}/ID_{50}$ ratios fell, in general, within a relatively narrow range for antibodies to different neutralization epitopes, suggesting that occupancy of binding sites on HIV-1 virions is the major factor in determining neutralization, irrespective of epitope recognized (Parren et al. 1998). This is illustrated in Fig. 1, in which the $K_{50}/ID_{50}$ ratios for a panel of mAbs against five distinct neutralizing epitopes on TCLA HIV-1 isolates are plotted.

We proposed a model in which neutralization is effected by coating of the virion with antibody. For attachment of the virus to the target cell to occur and infection to initiate, it is presumed that multiple contacts in a localized area between gp120 and the primary HIV-1 receptor CD4 must be established. Coating of the virus above a critical density would interrupt this process and thereby inhibit infection (Parren et al. 1998; reviewed in Parren et al. 1999).

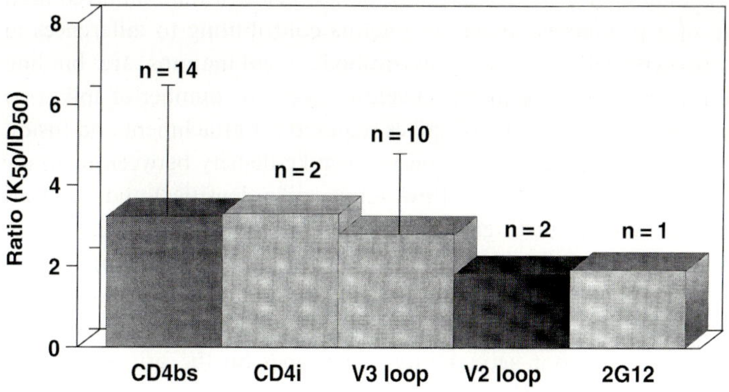


**Fig. 1.** Neutralization of HIV-1 is determined by antibody occupancy irrespective of epitope recognized. The antibody concentration yielding half-maximal binding ($K_{50}$) and neutralization titer ($ID_{50}$) and $K_{50}/ID_{50}$ ratios were determined for a number of neutralizing monoclonal antibodies (mAbs) against the CD4 binding site, CD4-induced epitope, V2 loop, V3 loop, and 2G12 epitopes on gp120 in binding and neutralization assays with HIV-$1_{LAI}$ and HIV-$1_{MN}$ (Parren et al. 1998). Similar $K_{50}/ID_{50}$ ratios for different epitopes suggest that neutralization is independent of the epitope recognized. This suggests that occupancy of binding sites rather than epitope recognized is the major determining factor in neutralization. The ratios shown indicate that the $K_{50}$ and $ID_{50}$ values roughly are of similar magnitude, indicating that neutralization occurs at antibody coating densities below the maximal level achievable

Schønning et al. (1999), in an elegant study, analyzed the stoichiometry of HIV-1 neutralization by use of virions carrying mixed envelope spikes on their surface. Cells were transfected with an HIV-1 clone (A308), which is potently neutralized by a mAb against the V3 loop on gp120, and a mutant of this clone, T321, which resists neutralization by this antibody by a loss-of-binding mutation. The proportion of antigen binding sites on the virion could now be titrated by transfecting plasmids encoding the two clones in different ratios. Neutralization of the mixed envelope virions was performed with a high concentration of neutralizing antibody (500× the $ID_{50}$ of the neutralization-sensitive clone), so that all available binding sites would be occupied. Neutralization by virion aggregation did not appear to play a significant role in inactivation. The experiment showed that, as the number of binding sites (and therefore the number of bound antibody molecules) increased, viruses become susceptible to neutralization. Neutralization of 63% of the virions occurred at a level of about 35% occupancy; a level which is consistent with our studies discussed above. The increase in neutralization was incremental, indicating a multi-hit rather a single-hit model (Schønning et al. 1999).

It has been shown that average, freshly budded HIV-1 virions carry approximately 72 envelope spikes (216 gp120 molecules) on their surface (Gelderblom 1991). From the analyses above, we then predict that the attachment of approximately 70 IgG molecules is required for the neutralization of such an HIV-1 virion, which would be equivalent to about one IgG molecule/spike.

Klasse and Moore (1996) developed a quantitative model of HIV-1 neutralization to explain the large difference in neutralization sensitivity between TCLA strains and primary isolates of HIV-1 (see Moore and Ho 1995 for reviews of HIV-1 neutralization). The model predicts that neutralization of HIV-1 is due to the reduction of the number of active envelope spikes below a minimum required for the induction of a productive infection. Factors contributing to differences in neutralization sensitivity of different virus/antibody combinations are binding constants of neutralizing antibodies to the envelope spike, the number of spikes on the surface, and the minimum number of spikes required for attachment and fusion (Klasse and Moore 1996). Possible differences in spike density between neutralization-sensitive TCLA strains of HIV-1 (low active spike densities) and neutralization-resistant primary isolates (high active spike densities) are quoted as an important factor in neutralization resistance (Klasse and Moore 1996). Differences in spike densities between TCLA strains and primary isolates, however, are controversial (Sullivan et al. 1995; Karlsson et al. 1996). Alternatively, the resistance may be due to the poor accessibility of epitopes on the primary isolate envelope spike (Parren et al. 1999).

## 2.6 The Stoichiometry of Antibody Binding to Virus at Neutralization

The stoichiometry of antibody binding at neutralization has been determined for a number of viruses. We briefly summarize the relevant studies, focusing on those in which stoichiometry was determined by a direct and quantitative method. Methods that determined the amount of antibody bound by use of purified radiolabeled virion-antibody complexes (e.g., TAYLOR et al. 1987), or quantitative SDS-PAGE (e.g., FLAMAND et al. 1993) may give reliable data, whereas methods in which the stoichiometry was extrapolated from kinetic analysis may be inaccurate because of uncertainties with respect to the mathematical models used, particularly in the early studies.

RAPPAPORT (1970) determined that an average of four antibody molecules bound per neutralized MS2 bacteriophage represented a neutralizing dose. Studies by Collono and Smith and their colleagues indicated that HRV is neutralized by inhibition of attachment and that the binding of six to seven antibody molecules per HRV virion is required to reduce HRV attachment by 63% (COLONNO et al. 1989; SMITH et al. 1993). ICENOGLE et al. (1983) determined that neutralization of poliovirus requires the attachment of four to five antibody molecules, as discussed above. The picornaviruses poliovirus and HRV therefore require the attachment of an average of five to six antibody molecules for neutralization. A study that assessed the number of IgG molecules necessary to neutralize influenza virus showed for two different mAbs (HC2 and HC10), that approximately 70 molecules were required (TAYLOR et al. 1987). Studies by Flamand and colleagues on rabies virus have shown that an average occupancy of 225 IgG molecules per virion is necessary for neutralization (see above; FLAMAND et al. 1993; RAUX et al. 1995). Studies on bovine papillomavirus (BPV) determined that the number of antibody molecules required for neutralization was approximately 14, 30, 72, and 36 molecules, respectively, for the four mAbs, 5B6, 3, 6, and 9, studied (RODEN et al. 1994). An average of about 38 bound IgG molecules is therefore required to neutralize a BPV virion. Complexes of mAb 5B6 and 9 with the major capsid BPV protein L1 have been analyzed by cryo-electron microscopy. It was shown that mAb 5B6, which apparently requires a slightly lower occupancy for neutralization, binds only to hexavalent capsomers, whereas mAb 9 also binds to pentameric capsomers (BOOY et al. 1998). mAb 9 binds monovalently at the tip of the capsomers and is highly contorted, projecting the Fc domain at an 120° angle into the space between the capsomers while the free Fab domain projects outwards. mAb 9 provides a dense, extended antibody coat, which is consistent with its ability to inhibit attachment of BPV to the target cell. mAb 5B6 binds deeply between the hexavalent capsomers and adopts a more linear conformation. mAb 5B6 does not inhibit virion attachment efficiently and likely interferes with a post-attachment step (BOOY et al. 1998). It is possible that mAb 5B6 may coat the virus more extensively by its lower molecular profile and may therefore be more effective at a somewhat lower level of occupancy.

The relatively low number of antibodies required to neutralize some viruses compared to others has generated much discussion. Critical-site hypotheses have been widely used to explain neutralization. However, how critical "neutralization-relevant" spikes differ from other "neutralization-irrelevant" spikes has not been defined, and although molecular rearrangements in the capsid on non-enveloped viruses may provide a possible mechanism, it is unclear how signals would be transmitted between loosely attached spikes in the membrane of an enveloped virus. Some studies have suggested that binding of a neutralizing antibody, for example to the hemagglutinin (HA) of influenza virus, transmits a signal across the envelope effecting neutralization (POSSEE et al. 1982; TAYLOR et al. 1987). A molecular mechanism for this is unknown.

A much simpler hypothesis favored by us is that neutralizing antibodies generally act by coating of the virus surface, thereby interfering with critical steps in the infectious pathway such as obstructing interactions with cell surface receptors or interfering with molecular rearrangements required for virus-cell fusion. Figure 2 provides a schematic representation of the size of the viruses discussed above relative to antibody. The average number of antibody molecules required to neutralize the viruses depicted by increasing size from left to right is 4, 5–6, 38, 70 and 225. A plot of the number of antibodies required for neutralization as a function of the average surface area for each of the viruses is shown in Fig. 3a. The straight line demonstrates that a roughly equivalent fraction of the viron surface, independent of the size of the virion, must be bound by antibody for neutralization to occur.

To calculate the average surface area on a virion associated with each antibody molecule at neutralization, we determined the slope of the line in Fig. 3. The line in Fig. 3a can be described by the formula: $N = 0.0033 \times A$, in which $N$ is the number of IgG molecules required for neutralization and $A$ is the total virion surface area. It follows that each antibody molecule is associated with an average of $300\text{nm}^2$ virion surface area at neutralization. It is not unreasonable to assume that an antibody molecule can provide steric hindrance for an area this size. A $300\text{nm}^2$ area is covered by a circle with a radius of about 9.7nm. The length of an antibody molecule is about 12.5nm, of which approximately 7nm are provided by the Fc tail, which has a high degree of translational and rotational flexibility relative to the Fab domains. This indicates that it is likely that an antibody, even when bound bivalently whereby movement of the Fab domains is restricted, could hinder the interaction of virion surface proteins with critical sites on the target cell in an area surrounding the binding site or, alternatively, hinder the formation of intermediates in the viral entry process, such as fusion complexes.

We believe that the available data, in agreement with the occupancy model, indicate that antibody-covered virions are either neutralized or not, and that there is no convincing evidence for any intermediate states. Virions covered with antibodies at low occupancy may infect their target cells more slowly, as discussed above. Such particles, however, are still infectious and therefore do not represent a neutralization intermediate in our view. Once antibody coating proceeds beyond a critical density, which appears to equal a relatively constant number of antibody molecules per unit of virion surface area, the virion is rendered non-infectious. Neutralization is expected to be reversible by dissociation of the antibody-virion complexes in most cases. Neutralization of a virus population can thus be understood in similar terms. The partial neutralization of a homogeneous virus stock at a certain concentration of antibody (e.g., 50%) simply represents a Poisson distribution of antibody molecules attached to virions at a given point in time. The fraction of the viruses coated with antibody molecules below the neutralization threshold are fully infectious, whereas the complementary fraction is fully neu-

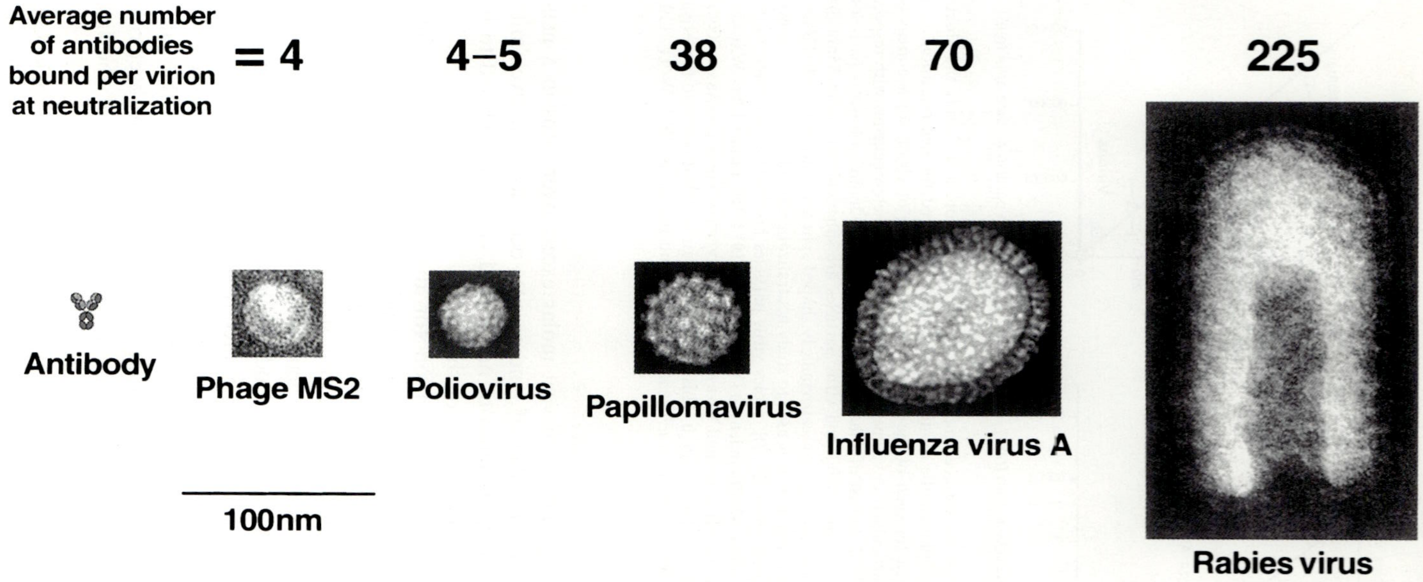


**Fig. 2.** Schematic representation of virion size relative to antibody. Electron microscopic images of each of the viruses considered are depicted on a scale relative to the size of an antibody molecule. Viruses with a larger average virion size appear to require the attachment of more antibody molecules to be neutralized. Clearly, however, viruses are generally neutralized well before all available mAb binding sites have been occupied. Saturation of binding sites, for example, only occurred with 30 IgG molecules bound per poliovirus and 1,080 IgG molecules bound per rabies virus particle (ICENOGLE et al. 1983, FLAMAND et al 1993). Electron microscopic images were provided by: Dr. Linda M. Stannard, University of Cape Town, South Africa (poliovirus, influenza virus, and polyoma virus); Dr. H.W. Ackermann, Université Laval, Quebec, Canada (bacteriophage MS2); Dr. S.K. Vernon, Wyeth Laboratories, Philadelphia, PA [rabies virus; reprinted from (VERNON et al. 1972) with permission granted by Academic Press, Orlando, FL]

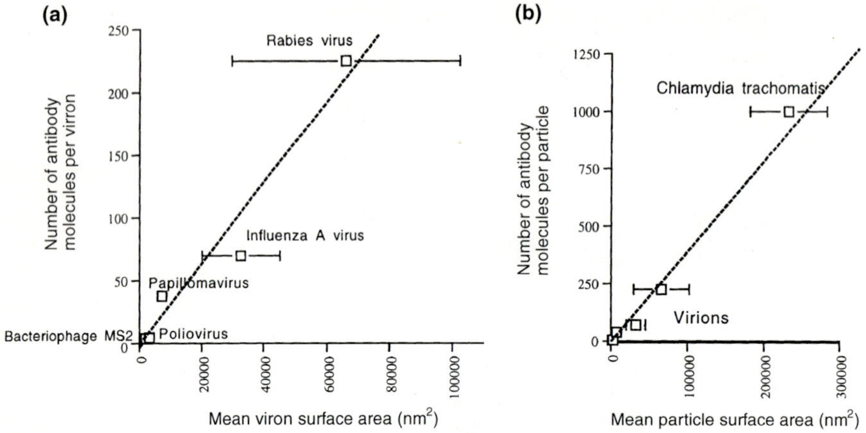


**Fig. 3a,b.** Antibody occupancy of virion surface binding sites at neutralization. **a** To study the observation in Fig. 2 in more detail, we plotted the number of antibodies required for neutralization as a function of the average surface area for each of the viruses shown. A range of virion surface areas is indicated, as virus preparations, particularly enveloped viruses, are typically heterogeneous with respect to size and shape. Bacteriophage MS2 consist of spherical particles of about 26nm; poliovirus particles are spherical with a diameter of 27–30nm; papilloma virus particles are spherical with a 40–55nm diameter; influenza A virus particles typically appear as irregularly shaped spherical particles of 80–120nm diameter when derived from infected eggs [the virus used to study neutralization stoichiometry was grown in embryonated chicken eggs (TAYLOR et al. 1987)]; rabies virus particles are bullet-shaped and range from 45 to 100nm in diameter and 100 to 430nm in length (VERNON et al. 1972; RUECKERT 1996; WAGNER and ROSE 1996; LAMB and KRUG 1996; MURPHY and WEBSTER 1996; UNIVERSAL VIRUS DATABASE 2000). The straight line demonstrates a linear relationship. **b** This relationship can be extrapolated to the larger pathogen, the intracellular bacterium *Chlamydia trachomatis* (PEELING and BRUNHAM 1991; WOLF et al. 2000)

tralized. Because antibody typically is in vast molar excess over virus in a neutralization reaction, the average number of binding sites occupied at a given antibody concentration, and thus the fraction of neutralized virus, is predicted to be primarily determined by antibody affinity for the virion.

The antibody occupancy at neutralization for the intracellular bacterium *Chlamydia trachomatis* has been determined in a study that was much less quantitative than the studies on viruses discussed above. We include it in our discussion here as it is the only study of which we are aware that provides an estimate of the number of antibody molecules required to neutralize a non-viral pathogen. The study found that approximately 1,000 antibody molecules of the strongly neutralizing mAb UM-4 were required to reduce infectivity of infectious elementary bodies by 50% (PEELING and BRUNHAM 1991). (A higher number was given for a weakly neutralizing antibody, but we consider this less reliable since the study assumed that all the antibody in an equilibrium mixture was bound to bacteria.) Interestingly, the *C. trachomatis* data can be fit by extrapolation of the line described for viruses in Fig. 3a (Fig. 3b).

The occupancy model suggests that one mechanism by which virus could seek to escape neutralization by antibody would be to decrease the number of neutralizing epitopes on the viral surface. Such a mechanism has indeed been described for human cytomegalovirus (HCMV) (Li et al. 1995). A neutralization escape mutant was isolated from a patient who was being treated with a neutralizing human antibody against the HCMV surface glycoprotein gH. The escape mutant was unstable when cultured in the absence of anti-gH neutralizing antibody and displayed a resistant phenotype against a number of distinct neutralizing antibodies and a polyclonal serum against gH. Biochemical analysis revealed that the escape mutant displayed a marked decrease in the amount of gH expressed on the surface. Interestingly, this mechanism of resistance would be consistent with the large differences in susceptibility to neutralization of clinical HCMV isolates and might be a mechanism by which the virus persists in the presence of neutralizing antibody in the serum (Li et al. 1995). If antibody occupancy indeed plays a major role in neutralization as we suggest, then down-regulation of surface molecules on the virus may be a more common mechanism of neutralization resistance than currently thought. However, as in the case of HCMV, it may be necessary to culture fresh clinical isolates in the presence of neutralizing antibody to observe these effects, since passaging would allow rapid reversion to wild type.

The occupancy model suggests that Fab fragments may neutralize a viral particle less efficiently than the corresponding whole antibody, as steric hindrance by the Fab is provided at a significantly reduced level. Generally, a ~1- to 50-fold difference is found in neutralization potency between whole IgG and Fab, which has been explained by differences in avidity (reviewed in (Parren and Burton 1997)). Some examples from the literature, however, indicate that a loss of steric hindrance may also play a role in this decrease. Icenogle et al. (1983) showed that treatment of neutralized poliovirus particles at low occupancy (an occupancy of the order of six antibody molecules per poliovirus particle, reducing infectivity of the virus stock to 28%) with papain to remove the IgG Fc tail reactivated the infectivity. It was shown that the Fab fragments were still bound to the virion, however, as addition of an anti-antibody again reduced infectivity to the initial neutralized level (Icenogle et al. 1983). Similar observations were made by Emini et al. (1983b). For the paramyxovirus SV5, Merz et al. (1981) have shown that anti-HN antibody completely neutralized the virus, whereas Fab fragments could only cause partial neutralization and a relatively large fraction of non-neutralizable virus remained. The neutralization mechanism of this antibody did not appear to involve aggregration or cross-linking, excluding this as the basis for the difference.

In agreement with our model, it has been shown that the neutralization potency of mAbs against the arterivirus porcine reproductive and respiratory syndrome virus (PRRSV) correlated with occupancy, as potent neutralizing mAbs stained virions in immuno-electron microscopy more strongly than weakly neutralizing mAbs (Weiland et al. 1999).

The occupancy model may finally explain the reduction in neutralization rate as a function of time in kinetic experiments. As viruses are more densely decorated with antibody molecules, the reaction rate will start to decrease, as incoming antibody will be sterically hindered by antibody already bound to the virus. Such steric hindrance has been shown to exist by kinetic neutralization experiments on herpes simplex virus (HSV), performed by Ashe and Notkins (1967). These authors demonstrated a drop in the neutralization rate constant as neutralization progressed, the magnitude of which correlated with the level of sensitization. Kinetic neutralization studies using HSV pre-sensitized with a neutralizing antibody furthermore showed that neutralization rate constants were strongly reduced if the subsequent neutralization reaction was performed with the whole antibody, but were only mildly reduced when performed with the Fab fragment (Ashe et al. 1969). Therefore, under the conditions used, the antibody already bound to the virion reduced the binding rate of whole antibody relative to that of the smaller Fab fragment presumably by a steric hindrance mechanism.

## 2.7 Neutralization by Antibodies Against Host Cell-Derived Proteins Incorporated into Virions

Many instances of viruses, particularly retroviruses, that incorporate non-viral host-cell-derived proteins into their viral envelopes have been described. HIV-1 has been shown to incorporate HLA class I and class II molecules, ICAM-1, LFA-1, and CD44 (Hoxie et al. 1987; Gelderblom et al. 1987; Arthur et al. 1992, 1995; Orentas and Hildreth 1993; Capobianchi et al. 1994; Saarlos et al. 1997). SIV and FIV furthermore may also incorporate HLA antigens (Lee et al. 1982; Arthur et al. 1992). This may have important consequences, as macaques immunized with human cells or HLA-DR, for example, can be protected from challenge with SIV grown in human host cells (Stott 1994; Arthur et al. 1995). It has been suggested that cellular adhesion molecules associated with HIV-1 may have a function in increasing adherence to the target cell and thus enhance virus entry; a particularly well-studied example of which is ICAM-1 (Rizzuto and Sodroski 1997). Antibodies against such host-cell-derived molecules can have neutralizing activity, e.g., ICAM-1-expressing HIV-1 virions can be completely neutralized by an antibody against ICAM-1. Interestingly, the neutralizing antibody activity observed is comparable in magnitude to that of "classic" neutralization. The neutralizing effect therefore is far greater than can be explained by the loss of adhesion provided by virion ICAM-1 in contact with its cognate receptor, LFA-1, on the target cell (Rizzuto and Sodroski 1997).

Antibodies against HLA class I and II have been shown to neutralize SIV, in which neutralization susceptibility appears to be determined by the number of HLA molecules present on the virion (Vzorov and Compans 2000). Incorporation of influenza virus HA molecules on phenotypically mixed SIV virions furthermore renders the virions susceptible to neutralization by antibodies against influenza HA (Vzorov and Compans 2000). This was also observed for phenotypically mixed virions of VSV with FPV, Sendai and measles virus (Zavada and Rosenbergova 1972; Kimura 1973; Wild et al. 1975–1976), as discussed above.

Neutralization of viruses by antibodies to foreign proteins on the viral surface supports the occupancy model. Clearly, viruses can be neutralized by the binding of antibodies to sites on the viral surface that do not directly serve a function in the infectious process (e.g., influenza HA on SIV). Coating of the viral surface with antibodies against such molecules, and subsequent steric interference with the virion-host cell interaction required to initiate infection, is an attractive hypothesis to explain these observations.

## 2.8 A Molecular Model for Picornavirus Neutralization

It has been shown that the attachment of an average of five to six IgG molecules can lead to the neutralization of a picornavirus particle. Neutralization of HRV in particular has been correlated with the inhibition of attachment to the target cell (Colonno et al. 1989; Smith et al. 1993).

To visualize how the binding of a small number of IgG molecules to the surface of a 30-nm picornavirus particle can impact the virus-receptor interaction at the cell surface, we have recently prepared a molecular model of a picornavirus, with a low number of antibodies bound (Parren and Burton 2001). In the diagram in Fig. 4, three IgG molecules are shown interacting with the visible half of the surface area of a picornavirus particle. This level of coating is close to that predicted to neutralize the virus. The considerable rotational flexibility of the Fc region (with dimensions of $7 \times 5.7 \times 2.4$nm) is demonstrated for one IgG molecule (Fig. 4a). Taking the flexibility of Fc and the translational flexibility of the antigen-bound Fab fragment into account, the antibody could be thought to hinder the attachment of virus to cellular receptors in a significant area surrounding the antigen binding site, as shown in Fig. 4b. As the initiation of productive infection requires the interaction of virus with a number of receptor molecules in a localized area, it can be envisioned that the binding of a relatively small number of antibody molecules to a picornavirus particle can effectively interrupt attachment and initiation of infection via steric hindrance and geometric constraints.

## 2.9 Binding but Non-Neutralizing Antibodies

Neutralization of a virus by antibody requires that the antibody binds to the coat of the virus. Therefore antibody binding to virions may be a useful indicator of in vivo activity. In reality, the measurement of antibody binding to virions is often technically demanding. Furthermore for many virus preparations, infectious particle: total particle ratios are low, introducing an element of uncertainty with regard to envelope homogeneity. Therefore measurement of binding to envelope at the surface of infected cells has often been preferred. It should be noted though that some viruses express non-structural or modified viral proteins on the infected-cell surface but not on virions. Such (non-neutralizing) antibodies are ineffective against free virus but may be able to protect against viral infection in vivo by mediating the destruction of infected cells. This has been discussed in more detail elsewhere (Parren and Burton 2001).

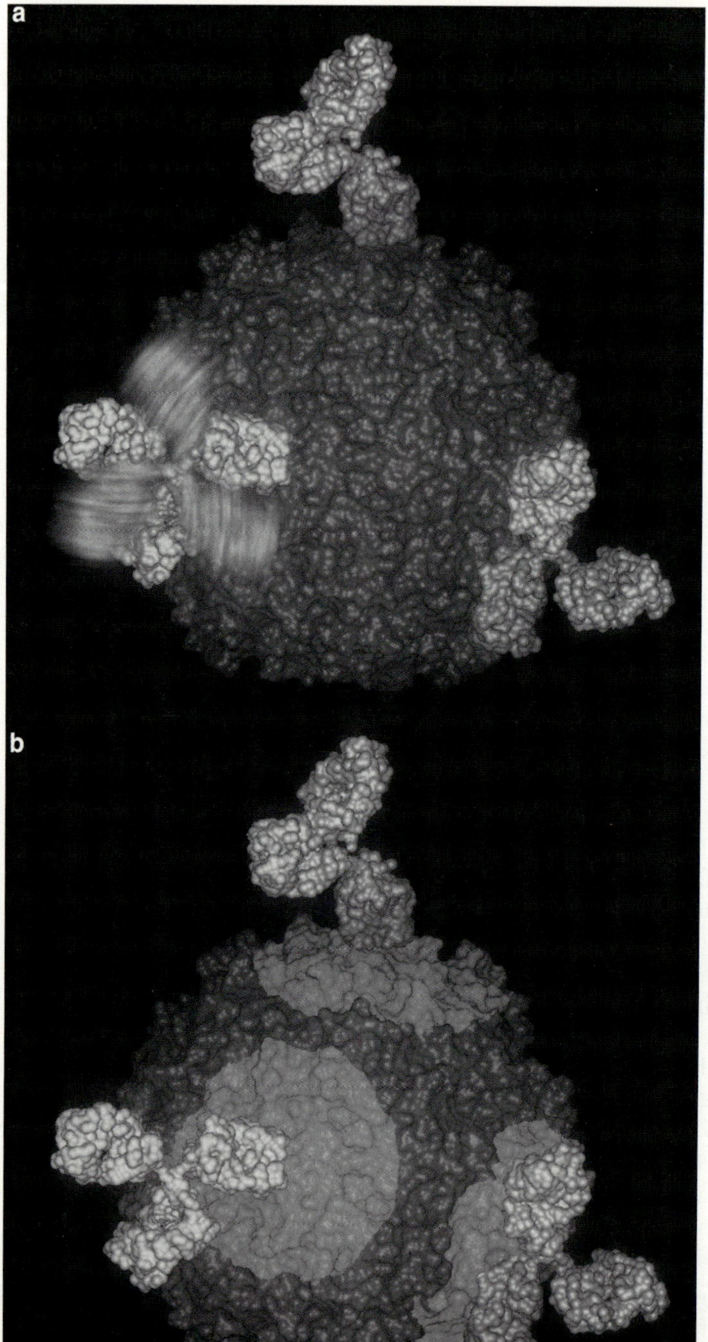
a
b

For HIV-1 a very good correlation has been described between neutralization and binding to envelope spikes on the surface of infected cells (Roben et al. 1994; Sattentau and Moore 1995; Parren et al. 1998). However, problems can arise with the measurement of binding to envelope on cell surfaces. It is important to minimize these problems as much as possible by proper experimental design. First, envelope spikes such as those of HIV-1 can shed protein, providing opportunities for observation of antibodies that apparently bind well to the infected cell surface but do not bind to functional envelope spikes. Some antibodies against the HIV-1 transmembrane (TM) unit gp41 probably fall into this category, as they may bind to inactivated envelope spikes in which gp41 is exposed following shedding of the surface unit (SU) gp120. A second type of artifact in HIV-1 is provided by antibody binding to shed gp120 in interaction with the primary HIV-1 receptor, CD4; antibodies apparently bind to the cells but this does not reflect binding to envelope spikes. This is significant, as many non-neutralizing epitopes that are not accessible on the oligomeric HIV-1 envelope spike are exposed on monomeric gp120 (Moore et al. 1995; Parren et al. 1999). A third type of artifact may arise if envelope is overexpressed, for example by expression of recombinant envelope glycoprotein under the control of a strong promoter. Thus it seems clear that the unprocessed HIV-1 envelope precursor glycoprotein gp160 can be expressed at cell surfaces under certain conditions (e.g., Karlsson et al. 1996; Dubay et al. 1995). Since many antibodies that do not bind to envelope spikes bind well to unprocessed gp160, there is again an opportunity for erroneous conclusions (Parren et al. 1997).

Whilst there is general agreement that only anti-envelope antibodies that bind to the envelope spike on the virion will be neutralizing, there is considerably less agreement about whether all antibodies that bind to envelope spike will neutralize virus, i.e., are there antibodies that bind well to envelope spikes but do not neutralize virus? Binding of antibodies to virions at high levels of occupancy would be in conflict with the neutralization model discussed in this review.

Our opinion on the issue of non-neutralizing but binding antibodies has been presented recently (Burton et al. 2000). We summarize here. We argue that overwhelmingly there is a good correlation between neutralization and virion binding with few exceptions. Our own experience in studies on human mAbs to HIV-1, RSV, and Ebola virus has been that there is an excellent correlation between binding to envelope spikes and neutralization, which in these studies was measured as binding to infected cells. We have generally seen a close correlation between half-maximal antibody binding and antibody concentration required to give 50% neutralization, suggesting that neutralization is directly related to occupancy of sites on the virion. For HIV-1, as discussed above, neutralization is

**Fig. 4a,b.** Picornavirus neutralization. **a** Picornavirus particle (*gray*) with three IgG molecules attached to approximately half of the virion surface area. **b** Region of virus particle surface (*light area*) subject to direct obstruction by a bound IgG molecule. Rotational flexibility of the Fc region is demonstrated for one IgG molecule. Binding of an average of about 5–6 IgG molecules/picornavirus particle results in neutralization. This number of antibody molecules bound leads to inhibition of HRV binding to its target cells. Surfaces were calculated as described. (Sanner et al. 1996) incremental, with increasing antibody occupancy, irrespective of the epitope recognized, leading to increased inhibition of infectivity (PARREN et al. 1998; SCHØNNING et al. 1999).

Irrespective of our observations, there is a strong tradition in virology of "binding (i.e., to the virion) but non-neutralizing" antibodies. As discussed above, early studies described a non-neutralizable fraction of virus that persisted even at high antibody concentrations. Addition of anti-antibody could reduce infectivity of this fraction. It seems that virus aggregation, at least in part, may have been responsible for this phenomenon (WALLIS and MELNICK 1967). We also note that non-neutralizable fractions have been described for hepatitis A and hepatitis C as a result of virus association with lipids or lipoproteins (LEMON and BINN 1985; PRINCE 1994).

Some descriptions of binding, non-neutralizing antibodies arose because of a failure to appreciate that antibodies that bind to isolated envelope molecules do not necessarily, and very often do not, bind to envelope spikes. A classic example is HIV-1, where many antibodies have been described that bind with high affinity to monomeric gp120 or unprocessed gp160; very few of these, however, show substantial affinity for envelope spikes (reviewed in PARREN et al. 1999).

Antibody-mediated enhancement of infection is a phenomenon that may support the existence of non-neutralizing but binding antibodies since the antibodies involved must bind to virions. However, as discussed above, a key observation here is that enhancement, when described, appears to occur for neutralizing antibodies at sub-neutralizing concentrations (e.g., HALSTEAD 1982; MORENS et al. 1987). We favor the view that antibody enhancement is an effect that can arise under conditions of low-level antibody coating of virions rather than an indication of antibodies with the ability to bind well to virions but not neutralize. The phenomenon does suggest, however, a potential problem for some vaccines that induce low levels of neutralizing antibodies.

Antibodies against influenza virus neuraminidase and matrix proteins 2 have been used as an example of binding but non-neutralizing antibodies. The surface of influenza viruses contain three types of surface proteins, namely, hemagglutinin (HA), neuraminidase (NA) and matrix (M2) protein, respectively, inserted in a host-cell derived membrane (MURPHY and WEBSTER 1996). Whereas antibodies against HA neutralize influenza, antibodies against NA and M2 do not (WEBSTER and LAVER 1967; KILBOURNE et al. 1968; ZEBEDEE and LAMB 1988). In the presence of antibodies against NA and M2, however, virus growth in vitro and in vivo is reduced, indicating that these antibodies do recognize these proteins on the surface of the virus or infected cells. The activities observed are most likely the result of reduced virus release and antibody-mediated destruction of infected cells (WEBSTER and LAVER 1967; MURPHY et al. 1972; ZEBEDEE and LAMB 1988). The absence of neutralizing activity of antibodies against NA and M2 can be explained by their abundance and distribution on the viral surface and is in direct agreement with the occupancy hypothesis. The NA and M2 protein are far less abundant on the viral surface than the HA protein: generally there are four to five times more HA than NA spikes, and NA spikes furthermore seem to be organized in patches. Each virus usually contains only a few copies of the M2 protein (Murphy and Webster 1996; Lamb and Krug 1996). Therefore, although antibodies against NA and M2 may bind to these proteins on the viral surface, the maximum level of occupancy achievable is relatively low and likely to remain below the level required to inhibit infectivity. This is reminiscent of the observations with the minor acid-inducible form or the rabies spike protein described above (Raux et al. 1995).

Structural studies on influenza HA furthermore are fully consistant with the occupancy model described in this review. The structure of HA in complex with a neutralizing antibody which binds at a site distal from the receptor binding site was solved and compared to the complex of HA with a receptor-binding site antibody. It was concluded that no significant conformational changes occurred outside of the antibody binding sites, and that neutralization correlated only with antibody avidity. It was suggested that the relative bulk of the antibody molecule was sufficient to interrupt virus-cell binding irrespective of the exact location of its binding site on HA (Fleury et al. 1999; Bizebard, this volume).

## 2.10 Inhibition of Infection by Anti-Receptor Antibodies

If the interruption of viral infection is primarily mediated by an occupancy and steric hindrance mechanism, as we propose, then binding of antibodies to the viral receptor on the target cell should also inhibit the initiation of infection. We would predict that the ability of anti-receptor antibodies to inhibit infection would then also be a function of occupancy and mostly independent of the epitope recognized. One of the best studied examples is probably the inhibition of HIV-1 infection by antibodies against its primary receptor, CD4. CD4 is a molecule of four Ig-like domains. The first two domains project from the cell surface and are connected by a hinge to the last two domains, which have been suggested to lie parallel to the cellular membrane (Wu et al. 1997). The gp120 binding site has been mapped to the first, most membrane-distal domain of CD4 (Arthos et al. 1989).

A large array of mAbs against distinct epitopes on CD4 have been studied for their ability to inhibit gp120 binding to $CD4^+$ cells and HIV-1 infection; most antibodies in fact were found to inhibit infection (Sattentau et al. 1986; Healey et al. 1990). It is interesting that even antibodies mapped to the third domain, which do not affect the interaction between CD4 and gp120, effectively inhibit infection. A comparison of mAbs against the first and third domains indicated that these antibodies inhibited HIV-1-induced syncytium formation for a range of isolates with similar potency. Inhibition of HIV-1 infectivity required that binding saturation of all available binding sites on the target cell surface was achieved. This inhibition was not correlated with a down-regulation of CD4 during the experiment (Healey et al. 1990). The potency of mAbs to the third domain of CD4 is noteworthy given that this domain is not required for HIV-1 infection, as shown by replacement of the last two domains of CD4 with those of CD8 with retention of receptor activity (Bedinger et al. 1988). One antibody, which probably binds to domain 3 (OKT4), did not inhibit HIV-1 infection. However, this may reflect the relatively poor affinity of this antibody for CD4 (Healey et al. 1990; Truneh et al. 1991). One antibody against the fourth domain of CD4 did not appear to block HIV-1 infection (Healey et al. 1990; Truneh et al. 1991). Even though antibodies against the third domain of CD4 inhibit HIV-1 infection, they do not appear to inhibit virion binding to $CD4^+$ T cells (Healey et al. 1990; Ugolini et al. 1997). Interestingly, it has been shown that most anti-CD4 antibodies, irrespective of the epitope recognized, are able to interrupt MHC class II-dependent T cell activation. Similar to their HIV-1 inhibitory activity, significant inhibition only occurs at near-saturating antibody concentrations (Merkenschlager et al. 1990).

The studies described above appear to show that inhibition of HIV-1 infection by anti-CD4 antibodies is dependent on the saturation of all available binding sites and is mostly independent of epitope recognized. However, a difference exists between antibodies against the virus and CD4 in their ability to inhibit virus-cell binding. Whereas all neutralizing antibodies against gp120 inhibit virus attachment, only antibodies that bind in the direct vicinity of the high-affinity gp120 binding site on CD4, and not antibodies to more-distant epitopes, including those in the first domain, inhibit attachment (Truneh et al. 1991; Ugolini et al. 1997). It may be that antibody affinity plays a role in the observed difference, as the inhibitory CD4 mAbs tested were of higher affinity than the non-inhibitory CD4 mAbs (Healey et al. 1990; Truneh et al. 1991). Alternatively, it may be a consequence of differences in antibody coating densities. The spike density on the virus and therefore its antibody coating density is several orders of magnitude greater than the receptor and resulting antibody coating density of the target cell. It is therefore conceivable that a virus surrounded by a dense antibody cannot approach the target cell sufficiently to initiate binding, whereas viruses are not inhibited in their approach to separated clusters of CD4 molecules with antibodies attached to epitopes distant from the gp120 binding site. Presumably, however, the antibodies attached to CD4 then perturb the necessary spatial organization of the multiple envelope-receptor contacts and conformational changes required to initiate a productive infection event. Notably, it has been suggested that these same antibodies can interrupt the precise spatial organization of CD4-T cell receptor and accessory molecules required for triggering T cell proliferation (Merkenschlager et al. 1990).

Following CD4 binding, HIV-1 requires the interaction with a chemokine receptor, belonging to the seven-transmembrane G-coupled receptor family, as an obligatory coreceptor for HIV-1 entry into the target cell. While a large number of chemokine receptors can serve as coreceptors for HIV-1 entry, the majority of isolates use either CXCR4 or CCR5 as its major coreceptor (see Berger 1997 for a review). A recent study investigated the inhibition of HIV-1 infection by antibodies against a number of epitopes on CCR5 (Lee et al. 2000). An antibody against the second extracellular loop (ECL2-A epitope), mAb 2D7, effectively inhibited HIV-1 infection and gp120 and chemokine binding to receptor. All other antibodies studied and directed against a different epitope on ECL2 (ECL2-B epitope), the N-terminus and epitopes involving multiple loops, blocked infection more poorly or not at all, and varied in their ability to inhibit gp120 and chemokine binding to receptor. MAbs against the N-terminus blocked gp120 binding better than chemokine binding, while the reverse was found for antibodies against the ECL2-B epitope. The affinities of these antibodies for CCR5 varied but were equal or greater than that of mAb 2D7 in some cases (Lee et al. 2000). Interestingly, it was found that CCR5 may exist in multiple conformational forms on the cell surface. MAb 2D7, the antibody with the strongest neutralizing activity, recognizes an epitope that is expressed on a much greater proportion of the CCR5 molecules than all the other mAbs tested, which appear to bind to only a fraction of the CCR5 molecules expressed (Lee et al. 2000). A correlation between inhibition of HIV-1 infection and occupancy of binding sites therefore exists, similar to the studies discussed above. We suggest that most antibodies against CCR5 are not effective in inhibition of HIV-1 infection, as they only recognize a subset of CCR5 molecules on the cell surface and therefore cannot saturate potential HIV-1 coreceptor binding molecules.

## 3 Inactivation of Virus In Vivo

Antibody activity in vivo can arise through binding to free virions or infected cells (Parren and Burton 2001). Neutralization, as we argue above, is generally mediated by antibodies that bind to capsid or envelope molecules exposed on the mature virus surface and which effectively coat the virion preventing viral attachment and/or entry. Neutralization is the antiviral activity we usually measure in vitro, and probably plays a major role in inhibiting free virus infectivity in vivo. Virus neutralization, particularly at low antibody coating densities, however, can be enhanced by effector mechanisms mediated by the Fc moiety of antibody in complex with virus.

First, the binding of antibodies to the viral surface may lead to the activation of complement, which is dependent on the isotype of the antibody and the spacing of the epitopes recognized. Complement activation and virolysis may lead to virus inactivation at sub-neutralizing concentrations of antibody. Coating virions by early components of the complement system (C1–4), however, may also enhance neutralization without inducing virolysis (reviewed by Oldstone 1975). For HIV-1, it has been shown that the infectivity of plasma virions may be reduced by an in vivo deposition of C3 on the virion surface (Sullivan et al. 1998a). Presumably, the complement coat interferes with virus interaction with the target cell thereby contributing to neutralization.

Second, virions coated by antibody may be eliminated by phagocytosis followed by inactivation in an intracellular compartment within the phagocyte. This process has been described for the picornavirus foot-and-mouth disease virus (FMDV). FMDV sensitized with antibody at sub-neutralizing concentrations was actively phagocytosed in vitro by monocytes and macrophages (McCullough et al. 1988).

Third, in mucosal transmission, antibodies may bind intracellularly to transcytosing viruses, leading to neutralization or redirection of the transcytosing virus. Polymeric IgA and IgM are actively transported over the mucosal epithelium after binding to the polymeric Ig receptor and may, during transport, contact and neutralize transcytosing viruses. Sendai virus added to the apical side of an epithelial cell monolayer was shown to interact intracellularly with polymeric IgA added basolaterally, and viral titers were reduced in cellular supernatant and lysate (MANZANEC et al. 1992; FUJIOKA et al. 1998). MANZANEC et al. (1995; reviewed in Kato et al., this volume) demonstrated neutralization of influenza virus via a similar mechanism. Studies by BOMSEL et al. (1998) have shown that that transcytosis of primary HIV-1 isolates over tight epithelium can be blocked by dimeric immune IgA and IgM. It was shown that basolaterally internalized antibodies met transcytosing virions in an intracellular compartment, thereby redirecting the virions back to the mucosal compartment (BOMSEL et al. 1998). Although the affinity for virions of the polyclonal dimeric IgA and IgM, purified from serum from infected individuals, is likely to be low, efficient binding may be occurring by a multivalent interaction at relatively high concentrations within the intracellular compartment. Polymeric IgA and IgM therefore may bind to virus in an intracellular compartment neutralizing infectivity or redirecting transcytosis.

## 4 Conclusions

There has been much discussion of the mechanisms of neutralization of viruses in vitro. It has been proposed that antibodies can act at many different stages of the infectious process, including post-viral entry to the target cell, that several different mechanisms may operate in concert, and that critical sites on the virion surface must be occupied for neutralization (DIMMOCK 1993). We do not agree with this interpretation of the data. In the majority of cases, we believe that the data are consistent with a simple occupancy model, essentially as initially proposed by Macfarlane Burnet in 1937. According to this model, neutralization occurs when a sizable fraction of available sites on the virion are occupied by antibody, leading to inhibition of viral attachment or interference with the entry (fusion) process. The relatively large bulk of the antibody molecule, very roughly similar to that of a typical envelope spike for an enveloped virus, is suggested to be critical.

The model is consistent with a number of observations. First, we have shown that there is a roughly linear relationship between the surface area of a virus and the number of antibody molecules bound at neutralization. This number is approximately that predicted to effectively coat the virion particle, given the size of the antibody molecule. Second, any antibody that binds well to and coats the virion surface should neutralize the virus – as is almost universally found. Non-neutralizing antibodies should not coat the virus, although they may bind at lower levels of occupancy. Heterogeneous envelope spikes provide an opportunity for binding to virions without neutralization because of low occupancy, as discussed above for rabies virus and the FPV/VSV mixed virions. Third, the model predicts that the precise epitope recognized by the antibody on the virion surface should not be crucial. Rather the number of antibody molecules bound/unit area, which will be determined by the affinity of antibody for virion-expressed antigen, will be most important. This is precisely what is found for HIV-1, as discussed above.

Antibody enhancement of viral infection in vitro has received much attention, although its significance in vivo has only been demonstrated for dengue virus infection. It appears to occur in the presence of sub-neutralizing concentrations of neutralizing antibodies.

The question of whether some antibodies bind well to the virus without neutralization is often raised. In particular, the concern has been expressed that non-neutralizing antibodies might compete with the binding of neutralizing antibodies to virus, thereby interfering with neutralization. First, we strongly question the evidence for effective binding but non-neutralizing antibodies. Generally, there is a very strong correlation between occupancy and neutralization, and most examples of such non-neutralizing antibodies can be explained by poor virus binding, or even a failure to appreciate that antibodies which bind well to isolated envelope or capsid molecules do not necessarily bind well to the virus particle. Some convincing but isolated examples exist, such as the rabies mutant virus discussed above, which merit further investigation. Second, to our knowledge there is no convincing and confirmed evidence for a non-neutralizing antibody interfering with neutralization. Virion coating but non-neutralizing antibodies therefore do not appear to play a significant role in the humoral response against viruses, if any.

The model described above asserts that neutralization is determined by the extent of coating of virus by antibody. At a given antibody concentration, this is in turn determined by the affinity of antibody for the antigen on the virion surface. Hence the model predicts that a humoral component of a vaccine should simply aim to elicit antibodies of the highest affinity for the virion surface antigen. This is most directly achieved by immunization with molecules identical to or as similar to the viral surface antigen as possible. This may not be easy. We have argued that a number of enveloped viruses have evolved surface proteins of low immunogenicity (Burton and Parren 2001). Subunit envelope proteins appear, in a number of instances, to elicit antibody responses that show little reactivity with the form of the envelope (usually oligomeric) expressed on the virion surface (Roben et al. 1994; Sattentau and Moore 1995; Parren et al. 1998; Sakurai et al. 1999).

*Acknowledgements.* E.O.S. thanks Dr. I.A. Wilson for his generous support and continued encouragement. We thank Dr. Michel Sanner for assistance in the preparation of the picornavirus models. We are grateful to Drs. L.M. Stannard and H.W. Ackermann for providing electron microscopic images. We acknowledge the financial support of the National Institutes of Health under grant numbers AI33292, HL59727 (to D.R.B.); AI40377, AI48494 (to P.WH.I.P.); and GM46192 (to I.A.W.).

## References

Arthos J, Deen KC, Chaikin MA, Fornwald JA, Sathe G, Sattentau QJ, Clapham PR, Weiss RA, McDougal JS, Pietropaolo C, Axel R, Truneh A, Maddon PJ, Sweet RW (1989) Identification of the residues in human CD4 critical for the binding of HIV. Cell 57:469–481

Arthur LO, Bess JW Jr, Sowder RCI, Benveniste RE, Mann DL, Chermann JC, Henderson LE (1992) Cellular proteins bound to immunodeficiency viruses: implications for pathogenesis and vaccines. Science 258:1935–1938

Arthur LO, Bess JW Jr, Urban RG, Strominger JL, Morton WR, Mann DL, Henderson LE, Benveniste RE (1995) Macaques immunized with HLA-DR are protected from challenge with simian immunodeficiency virus. J Virol 69:3117–3124

Ashe WK, Mage M, Notkins AL (1969) Kinetics of neutralization of sensitized herpes simplex virus with antibody fragments. Virology 37:290–293

Ashe WK, Notkins AL (1967) Kinetics of sensitization of herpes simplex virus and its relationship to the reduction in the neutralization rate constant. Virology 33:613–617

Bedinger P, Moriarty A, von Borstel RC, Donovan NJ, Steimer KS, Littman DR (1988) Internalization of the human immunodeficiency virus does not require the cytoplasmic domain of CD4. Nature 334:162–165

Beijerinck MW (1898) Over een Contagium vivum fluidum als oorzaak van de vlekziekte der tabaksbladen. Versl. Gew. Verg. Wis en Natuurk. Afd., Kon. Acad. Wetensch. Amsterdam VII:229–235

Belnap DM, McDermott BMJr, Filman DJ, Cheng N, Trus BL, Zuccola HJ, Racaniello VR, Hogle JM, Steven AC (2000) Three-dimensional structure of poliovirus receptor bound to poliovirus. Proc Natl Acad Sci USA 97:73–78

Berger EA (1997) HIV entry and tropism: the chemokine receptor connection. AIDS 11 (Suppl A):S3–S16

Bizebard T, Barbey-Martin C, Fleury O, GiGant B, Barrère B, Skehel JJ, Knossow (2001) Structural studies on viral escape from antibody neutralization. Current Topics Microbiol Immunol 260: 55–64

Blondel B, Crainic R, Fichot O, Dufraisse G, Candrea A, Diamond D, Girard M, Horaud F (1986) Mutations conferring resistance to neutralization with monoclonal antibodies in type 1 poliovirus can be located outside or inside the antibody-binding site. J Virol 57:81–90

Bomsel M, Heyman M, Hocini H, Lagaye S, Belec L, Dupont C, Desgranges C (1998) Intracellular neutralization of HIV transcytosis across tight epithelial barriers by anti-HIV envelope protein dlgA or IgM. Immunity 9:277–287

Booy FP, Roden RBS, Greenstone HL, Schiller JT, Trus BL (1998) Two antibodies that neutralize papillomavirus by different mechanisms show distinct binding patterns at 13Å resolution. J Mol Biol 281:95–106

Boyer V, Delibrias C, Noraz N, Fischer E, Kazatchkine MD, Desgranges C (1992) Complement receptor type 2 mediates infection of the human CD4-negative Raji B-cell line with opsonized HIV. Scand J Immunol 36:879–883

Brioen P, Thomas AAM, Boeye A (1985) Lack of quantitative correlation between the neutralization of poliovirus and the antibody-mediated pI shift of the virions. J Gen Virol 66:609–613

Burnet FM, Keogh EV, Lush D (1937) The immunological reactions of the filterable viruses. Austral J Exp Biol Med Sci 15:227–368

Burton DR, Parren PWHI (2000) Vaccines and the induction of functional antibodies: time to look beyond the molecules of natural infection? Nat Med 6:123–125

Burton DR, Williamson RA, Parren PWHI (2000) Antibody and virus: binding and neutralization. Virology 270:1–3

Capobianchi MR, Fais S, Castilletti C, Gentile M, Ameglio F, Dianzani F (1994) A simple and reliable method to detect cell membrane proteins on infectious human immunodeficiency virus type 1 particles. J Infect Dis 169:886–889

Che Z, Olson NH, Leippe D, Lee WM, Mosser AG, Rueckert RR, Baker TS, Smith TJ (1998) Antibody-mediated neutralization of human rhinovirus 14 explored by means of cryoelectron microscopy and X-ray crystallography of virus-Fab complexes. J Virol 72:4610–4622

Colonno RJ, Callahan PL, Leippe DM, Rueckert RR, Tomassini JE (1989) Inhibition of rhinovirus attachment by neutralizing monoclonal antibodies and their Fab fragments. J Virol 63:36–42

Connor RI, Dinces NB, Howell AL, Romet-Lemonne JL, Pasquali JL, Fanger MW (1991) Fc receptors for IgG (FcγRs) on human monocytes and macrophages are not infectivity receptors for human immunodeficiency virus type 1 (HIV-1): studies using bispecific antibodies to target HIV-1 to various myeloid cell surface molecules, including the FcγR. Proc Natl Acad Sci USA 88:9593–9597

Crainic R, Couillin P, Blondel B, Cabau N, Boue A, Horodniceanu F (1983) Natural variation of poliovirus neutralization epitopes. Infect Immun 41:1217–1225

Della-Porte AJ,Westaway EG (1977) A multi-hit model for the neutralization of animal viruses. J Gen Virol 38:1–19

Diamond DC, Jameson BA, Bonin J, Kohara M, Abe S, Itoh H, Komatsu T, Arita M, Kuge S, Nomoto A, Osterhaus ADME, Crainic R, Wimmer E (1985) Antigenic variation and resistance to neutralization in poliovirus type 1. Science 229:1090–1093
Dimmock NJ (1993) Neutralization of animal viruses. Current Topics Microbiol Immunol 183:1–149
Dubay JW, Dubay SR, Shin HJ, Hunter E (1995) Analysis of the cleavage site of the human immunodeficiency virus type 1 glycoprotein: requirement of precursor cleavage for glycoprotein incorporation. J Virol 69:4675–4682
Dulbecco R, Vogt M, Strickland AGR (1956) A study of the basic aspects of neutralization of two animal viruses, western equine encephalitis virus and poliomyelitis virus. Virology 2:162–205
Emini EA, Kao SY, Lewis AJ, Crainic R, Wimmer E (1983a) Functional basis of poliovirus neutralization determined with monospecific neutralizing antibodies. J Virol 46:466–474
Emini EA, Ostapchuk P, Wimmer E (1983b) Bivalent attachment of antibody onto poliovirus leads to conformational alteration and neutralization. J Virol 48:547–550
Flamand A, Raux H, Gaudin Y, Ruigrok RW (1993) Mechanisms of rabies virus neutralization. Virology 194:302–313
Fleury D, Barrère B, Bizebard T, Daniels RS, Skehel JJ, Knossow M (1999) A complex of influenza hemagglutinin with a neutralizing antibody that binds outside the virus receptor binding site. Nat Struct Biol 6:530–534
Flynn DC, Olmsted RA, Mackenzie JMJr, Johnston RE (1988) Antibody-mediated activation of Sindbis virus. Virology 166:82–90
Fujioka H, Emancipator SN, Aikawa M, Huang DS, Blatnik F, Karban T, DeFife K, Mazanec MB (1998) Immunocytochemical colocalization of specific immunoglobulin A with sendai virus protein in infected polarized epithelium. J Exp Med 188:1223–1229
Gelderblom H, Reupke H, Winkel T, Kunze R, Pauli G (1987) MHC-antigens: constituents of the envelopes of human and simian immunodeficiency viruses. Z Naturforsch 42:1328–1334
Gelderblom HR (1991) Assembly and morphology of HIV: potential effect of structure on viral function. AIDS 5:617–637
Halstead SB (1982) Immune enhancement of viral infection. Prog Allergy 31:301–364
Hawkes RA, Lafferty KJ (1967) The enhancement of virus infectivity by antibody. Virology 33:250–261
He Y, Bowman VD, Mueller S, Bator CM, Bella J, Peng X, Baker TS, Wimmer E, Kuhn RJ, Rossmann MG (2000) Interaction of the poliovirus receptor with poliovirus. Proc Natl Acad Sci USA 97: 79–84
Healey D, Dianda L, Moore JP, McDougal JS, Moore MJ, Estess P, Buck D, Kwong PD, Beverley PC, Sattentau QJ (1990) Novel anti-CD4 monoclonal antibodies separate human immunodeficiency virus infection and fusion of $CD4^+$ cells from virus binding. J Exp Med 172:1233–1242
Hewat E, Blaas D (2001) Structural studies on antibody interacting with viruses. Current Topics Microbiol Immunol 260:29–44
Hewat EA, Blaas D (1996) Structure of a neutralizing antibody bound bivalently to human rhinovirus 2. EMBO J 15:1515–1523
Homsy J, Meyer M, Tateno M, Clarkson S, Levy J (1989) The Fc and not CD4 receptor mediates antibody enhancement of HIV infection in human cells. Science 244:1357–1360
Hoxie JA, Fitzharris TP, Youngbar PR, Matthews DM, Rackowski JL, Radka SF (1987) Nonrandom association of cellular antigens with HTLV-III virions. Hum Immunol 18:39–52
Icenogle J, Shiwen H, Duke G, Gilbert S, Rueckert R, Anderegg J (1983) Neutralization of poliovirus by a monoclonal antibody: kinetics and stoichiometry. Virology 127:412–425
Ichihashi Y (1996) Extracellular enveloped vaccinia virus escapes neutralization. Virology 217:478–485
Iorio RM, Bratt MA (1983) Monoclonal antibodies to Newcastle disease virus: delineation of four epitopes on the HN glycoprotein. J Virol 48:440–450
Iorio RM, Bratt MA (1984) Neutralization of Newcastle disease virus by monoclonal antibodies to the hemagglutinin-neuraminidase glycoprotein: requirement for antibodies to four sites for complete neutralization. J Virol 51:445–451
Jerne NK, Avegno P (1956) The development of the phage-inactivating properties of serum during the course of specific immunization of an animal: reversible and irreversible inactivation. J Immunol 76:200–208
Karlsson GB, Gao F, Robinson J, Hahn B, Sodroski J (1996) Increased envelope spike density and stability are not required for the neutralization resistance of primary human immunodeficiency viruses. J Virol 70:6136–6142
Kato H, Kato R, Fujihashi K, McGhee JR (2001) Role of mucosal antibodies in viral infections. Current Topics Microbiol Immunol 260:201–228

Kilbourne ED, Laver WG, Schulman JL, Webster RG (1968) Antiviral activity of antiserum specific for an influeza virus neuraminidase. J Virol 2:281–288

Kimura Y (1973) Phenotypic mixing of vesicular stomatitis virus with HVJ (Sendai virus). Jpn J Microbiol 17:373–381

Kingsford L (1984) Enhanced neutralization of La Crosse virus by the binding of specific pairs of monoclonal antibodies to the G1 glycoprotein. Virology 136:265–273

Klasse PJ, Moore JP (1996) Quantitative model of antibody-and soluble CD4-mediated neutralization of primary isolates and T-cell line-adapted strains of human immunodeficiency virus type 1. J Virol 70:3668–3677

Krummel WM, Uhr JW (1969) A mathematical and experimental study of the kinetics of neutralization of bacteriophage ϕX 174 by antibodies. J Immunol 102:772–785

Lafferty KJ (1963a) The interaction between virus and antibody. Virology 21:61–75

Lafferty KJ (1963b) The interaction between virus and antibody II. Mechanism of the reaction. Virology 21:76–90

Lamb RA, Krug RM (1996) Orthomyxoviridia: The viruses and their replication. In: Fields BN, Knipe DM, Howley PM, et al. (eds) Fields virology. Lippincott Williams and Wilkins, Philadelphia, pp 1353–1395

Lee B, Sharron M, Blanpain C, Doranz BJ, Vakili J, Setoh P, Berg E, Liu G, Guy HR, Durell SR, Parmentier M, Chang CN, Price K, Tsang M, Doms RW (2000) Epitope mapping of CCR5 reveals multiple conformational states and distinct but overlapping structures involved in chemokine and coreceptor function. J Biol Chem 274:9617–9626

Lee TH, Essex M, de Noronha F, Azocar J (1982) Neutralization of feline leukemia virus with feline antisera to leukocyte alloantigens. Cancer Res 42:3995–3999

Lemon SM, Binn LN (1985) Incomplete neutralization of hepatitis A virus in vitro due to lipid-associated virions. J Gen Virol 66:2501–2505

Li L, Coelingh KL, Britt WJ (1995) Human cytomegalovirus neutralizing antibody-resistant phenotype is associated with reduced expression of glycoprotein H. J Virol 69:6047–6053

Lund O, Hansen J, Sorensen AM, Mosekilde E, Nielsen JO, Hansen JE (1995) Increased adhesion as a mechanism of antibody-dependent and antibody-independent complement-mediated enhancement of human immunodeficiency virus infection. J Virol 69:2393–2400

Mandel B (1961) Reversibility of the reaction between poliovirus and neutralizing antibody of rabbit origin. Virology 14:316–328

Mandel B (1976) Neutralization of poliovirus: a hypothesis to explain the mechanism and the one-hit character of the neutralization reaction. Virology 69:500–510

Manzanec MB, Coudret CL, Fletcher DR (1995) Intracellular neutralization of influenza virus by immunoglobulin A anti-hemagglutinin monoclonal antibodies. J Virol 69:1339–1343

Manzanec MB, Lamm ME, Lyn D, Porter A, Bedrud JG (1992) Comparison of IgA versus IgG monoclonal antibodies for passive immunization of the murine respiratory tract. Virus Res 23:1–12

Mascola JR, Mathieson BJ, Zack PM, Walker MC, Halstead SB, Burke DS (1993) Summary report: workshop on the potential risks of antibody-dependent enhancement in human HIV vaccine trials. AIDS Res Hum Retroviruses 9:1175–1184

Massey RJ, Schochetman G (1981) Viral epitopes and monoclonal antibodies: isolation of blocking antibodies that inhibit virus neutralization. Science 213:447–449

McCullough KC, Parkinson D, Crowther JR (1988) Opsonization-enhanced phagocytosis of foot-and-mouth disease virus. Immunology 65:187–191

McLain L, Dimmock NJ (1994) Single- and multi-hit kinetics of immunoglobulin G neutralization of human immunodeficiency virus type 1 by monoclonal antibodies. J Gen Virol 75:1457–1460

Merkenschlager M, Buck D, Beverley PC, Sattentau QJ (1990) Functional epitope analysis of the human CD4 molecule. The MHC class II-dependent activation of resting T cells is inhibited by monoclonal antibodies to CD4 regardless whether or not they recognize epitopes involved in the binding of MHC class II or HIV gp120. J Immunol 145:2839–2845

Merz DC, Scheid A, Choppin PW (1981) Immunological studies of the functions of paramyxovirus glycoproteins. Virology 109:94–105

Moore JP, Cao Y, Qing L, Sattentau QJ, Pyati J, Koduri R, Robinson J, Barbas CFI, Burton DR, Ho DD (1995) Primary isolates of human immunodeficiency virus type 1 are relatively resistant to neutralization by monoclonal antibodies to gp120, and their neutralization is not predicted by studies with monomeric gp120. J Virol 69:101–109

Moore JP, Ho DD (1995) HIV-1 neutralization: the consequences of viral adaptation to growth on transformed T cells. AIDS 9 (suppl A):S117–S136

Morens DM, Halstead SB, Marchette NJ (1987) Profiles of antibody-dependent enhancement of dengue virus type 2 infection. Microb Pathog 3:231–237
Murphy BR, Kasel JA, Chanock RM (1972) Association of serum anti-neuraminidase antibody with resistance to influenza in man. N Eng J Med 286:1329–1332
Murphy BR, Webster RG (1996) Orthomyxoviruses. In: Fields BN, Knipe DM, Howley PM, et al. (eds) Fields virology. Lippincott Williams and Wilkins, Philadelphia, pp 1397–1446
Oldstone MBA (1975) Virus neutralization and virus-induced immune complex disease. Virus-antibody union resulting in immunoprotection or immunology injury–two sides of the same coin. Prog Med Virol 19:84–119
Orentas RJ, Hildreth JEK (1993) Association of host cell surface adhesion receptors and other membrane proteins with HIV and SIV. AIDS Res Hum Retroviruses 9:1157–1165
Parren PWHI, Burton DR (1997) Antibodies against HIV-1 from phage display libraries: mapping of an immune response and progress towards antiviral immunotherapy. Chem Immunol 65:18–56
Parren PWHI, Burton DR (2001) The anti-viral activity of antibodies in vitro and in vivo. Adv Immunol 77:195–262
Parren PWHI, Burton DR, Sattentau QJ (1997) HIV-1 antibody – debris or virion? Nat Med 3: 366–367
Parren PWHI, Mondor I, Naniche D, Ditzel HJ, Klasse PJ, Burton DR, Sattentau QJ (1998) Neutralization of HIV-1 by antibody to gp120 is determined primarily by occupancy of sites on the virion irrespective of epitope specificity. J Virol 72:3512–3519
Parren PWHI, Moore JP, Burton DR, Sattentau QJ (1999) The neutralizing antibody response to HIV-1: viral evasion and escape from humoral immunity. AIDS 13 (Suppl A):S137–S162
Peeling RW, Brunham RC (1991) Neutralization of Chlamydia trachomatis: kinetics and stoichiometry. Infect Immun 59:2624–2630
Philipson L (1966) Interaction between poliovirus and immunoglobulins. II. Basic aspects of virus-antibody interaction. Virology 28:35–46
Philipson L, Killander J, Albertsson PA (1966) Interaction between poliovirus and immunoglobulins. I. Detection of virus antibodies by partition in aqueous polymer phase systems. Virology 28:22–34
Possee RD, Schild GC, Dimmock NJ (1982) Studies on the mechanism of neutralization of influenza virus by antibody: evidence that neutralizing antibody (anti-haemagglutinin) inactivates influenza virus in vivo by inhibiting virion transcriptase activity. J Gen Virol 58:373–386
Prince AM (1994) Challenges for development of hepatitis C virus vaccines. FEMS Microbiol Rev 14:273–277
Prohaszka Z, Nemes J, Hidvegi T, Toth FC, Kerekes K, Erdei A, Szabo J, Ujhelyi E, Thielens N, Dierich MR, Spath P, Ghebrehiwet B, Hampl H, Kiss J, Arlaud G, Fust G (1997) Two parallel routes of the complement-mediated antibody-dependent enhancement of HIV-1 infection. AIDS 11:949–958
Rappaport I (1970) An analysis of the inactivation of MS2 bacteriophage with antiserum. J Gen Virol 6:25–32
Raux H, Coulon P, Lafay F, Flamand A (1995) Monoclonal antibodies which recognize the acidic configuration of the rabies glycoprotein at the surface of the virion can be neutralizing. Virology 210:400–408
Rizzuto CD, Sodroski JG (1997) Contribution of virion ICAM-1 to human immunodeficiency virus infectivity and sensitivity to neutralization. J Virol 71:4847–4851
Roben P, Moore JP, Thali M, Sodroski J, Barbas CF, Burton DR (1994) Recognition properties of a panel of human recombinant Fab fragments to the CD4 binding site of gp120 that show differing abilities to neutralize human immunodeficiency virus type 1. J Virol 68:4821–4828
Robinson WEJr, Montefiori DC, Mitchell WM (1988) Antibody-dependent enhancement of human immunodeficiency virus type 1 infection. Lancet 1:790–794
Roden RS, Weissinger EM, Henderson DW, Booy F, Kirnbauer R, Mushinski JF, Lowy DR, Schiller JT (1994) Neutralization of bovine papillomavirus by antibodies to L1 and L2 capsid proteins. J Virol 68:7570–7574
Rueckert RR (1996) Picornaviridae: the viruses and their replication. In: Fields BN, Knipe DM, Howley PM, et al. (eds) Fields virology. Lippincott Williams and Wilkins, Philadelphia, pp 609–654
Saarloos MN, Sullivan BL, Czerniewski MA, Parameswar KD, Spear GT (1997) Detection of HLA-DR associated with monocytotropic, primary, and plasma isolates of human immunodeficiency virus type 1. J Virol 71:1640–1643
Sakurai H, Williamson RA, Crowe JE, Beeler JA, Poignard P, Bastidas RB, Chanock RM, Burton DR (1999) Human antibody responses to mature and immature forms of viral envelope in respiratory syncytial virus infection: significance for subunit vaccines. J Virol 73:2956–2962

Sanner MF, Olson AJ, Spehner JC (1996) Reduced surface: an efficient way to compute molecular surfaces. Biopolymers 38:305–320
Sattentau QJ, Dalgleish AG, Weiss RA, Beverley PC (1986) Epitopes of the CD4 antigen and HIV infection. Science 234:1120–1123
Sattentau QJ, Moore JP (1995) Human immunodeficiency virus type 1 neutralization is determined by epitope exposure on the gp120 oligomer. J Exp Med 182:185–196
Sattentau QJ, Zolla-Pazner S, Poignard P (1995) Epitope exposure on functional, oligomeric HIV-1 gp41 molecules. Virology 206:713–717
Schønning K, Lund O, Lund OS, Hansen JES (1999) Stoichiometry of monoclonal anitbody neutralization of T-cell line-adapted human immunodefiency virus type 1. J. Virol 73:8364–8370
Schutten M, Andeweg AC, Bosch ML, Osterhaus AD (1995) Enhancement of infectivity of a non-syncytium inducing HIV-1 by sCD4 and by human antibodies that neutralize syncytium inducing HIV-1. Scand J Immunol 41:18–22
Scicluna LA, McCullough KC (1999) Rapidity of specific antibody-antigen interactions. Scand J Immunol 50:167–176
Smith TJ (2001) Antibody interactions with rhinovirus; lessons for mechanisms of neutralization and the role of immunity in viral evolution. Current Topics Microbiol Immunol 260:1–28
Smith TJ, Chase ES, Schmidt TJ, Olson NH, Baker TS (1996) Neutralising antibody to human rhinovirus 14 penetrates the receptor-binding canyon. Nature 383:350–354
Smith TJ, Olson NH, Cheng RH, Liu H, Chase ES, Lee WM, Leippe DM, Mosser AG, Rueckert RR, Baker TS (1993) Structure of human rhinovirus complexed with Fab fragments from a neutralizing antibody. J Virol 67:1148–1158
Stewart PL, Chiu C, Huang S, Muir T, Zhao Y, Chait B, Mathias P, Nemerow GR (1997) Cryo-EM visualization of an exposed RGD epitope on adenovirus that escapes antibody neutralization. EMBO J 16:1189–1198
Stewart PL, Nemerow GR (1997) Recent structural solutions for antibody neutralization of viruses. Trends Microbiol 5:229–233
Stott EJ (1994) Towards a vaccine against AIDS: lessons from simian immunodeficiency virus vaccines. Curr Top Microbiol Immunol 188:221–237
Sullivan BL, Takefman DM, Spear GT (1998a) Complement can neutralize HIV-1 plasma virus by a C5-independent mechanism. Virology 248:173–181
Sullivan N, Sun Y, Binley J, Lee J, Barbas III CF, Parren PWHI, Burton DR, Sodroski J (1998b) Determinants of human immunodeficiency virus type 1 envelope glycoprotein activation by soluble CD4 and monoclonal antibodies. J Virol 72:6332–6338
Sullivan N, Sun Y, Li J, Hofmann W, Sodroski J (1995) Replicative function and neutralization sensitivity of envelope glycoproteins from primary and T-cell line-passaged human immunodeficiency virus type 1 isolates. J Virol 69:4413–4422
Takeda A, Tuazon CU, Ennis FA (1988) Antibody-enhanced infection by HIV-1 via Fc receptor-mediated entry. Science 242:580–583
Taylor HP, Armstrong SJ, Dimmock NJ (1987) Quantitative relationships between an influenza virus and neutralizing antibody. Virology 159:288–298
Truneh A, Buck D, Cassatt DR, Juszczak R, Kassis S, Ryu SE, Healey D, Sweet R, Sattentau Q (1991) A region in domain 1 of CD4 distinct from the primary gp120 binding site is involved in HIV infection and virus-mediated fusion. J Biol Chem 266:5942–5948
Ugolini S, Mondor I, Parren PWHI, Burton DR, Tilley SA, Klasse PJ, Sattentau QJ (1997) Inhibition of virus attachment to $CD4^+$ target cells is a major mechanism of T cell line-adapted HIV-1 neutralization. J Exp Med 186:1287–1298
Universal virus database. International Committee on Taxonomy of Viruses. http://life.anu.edu.au/viruses/Images/index.htm
van Kammen A (1999) Beijerinck's contribution to the virus concept–an introduction. Arch Virol Suppl 15:1–8
Vanderplasschen A, Hollinshead M, Smith GL (1997) Antibodies against vaccinia virus do not neutralize extracellular enveloped virus but prevent virus release from infected cells and comet formation. J Gen Virol 78:2041–2048
Vernon SK, Neurath AR, Rubin BA (1972) Electron microscopic studies on the structure of rabies virus. J Ultrastruct Res 41:29–42
Vzorov AN, Compans RW (2000) Effect of the cytoplasmic domain of the simian immunodeficiency virus envelope protein on incorporation of heterologous envelope proteins and sensitivity to neutralization. J Virol 74:8219–8225

Wagner RR, Rose JK (1996) Rhabdoviridae: the viruses and their replication. In: Fields BN, Knipe DM, Howley PM, et al. (eds) Fields virology. Lippincott Williams and Wilkins, Philadelphia, pp 1121–1159

Wallis C, Melnick JL (1967) Virus aggregation as the cause of the non-neutralizable persistent fraction. J Virol 1:478–488

Wallis C, Melnick JL (1970) Herpesvirus neutralization: induction of the persistent fraction by insufficient antibody. Virology 42:128–137

Webster RG, Laver WG (1967) Preparation and properties of antibody directed specifically against the neuraminidase of influenza virus. J Immunol 99:49–55

Weiland E, Wieczorek-Krohmer M, Kohl D, Conzelmann KK, Weiland F (1999) Monoclonal antibodies to GP5 of porcine reproductive and respiratory syndrome virus are more effective in virus neutralization than monoclonal antibodies to GP4. Vet Microbiol 66:171–186

Westaway EG (1965) The neutralization of arboviruses. I. Neutralization in homologous virus-serum mixtures with two group B arboviruses. Virology 26:517–527

Wild F, Cathala F, Huppert J (1975–1976) Vesicular stomatitis virus (measles) pseudotypes: tool for demonstrating defective measles infections. Intervirology 6:185–189

Wolf K, Fisher E, Hackstadt T (2000) Ultrastructural analysis of developmental events in Chlamydia pneumoniae-infected cells. Infect Immun 68:2379–2385

Wu H, Kwong PD, Hendrickson WA (1997) Dimeric association and segmental variability in the structure of human CD4. Nature 387:527–530

Zavada J, Rosenbergova M (1972) Phenotypic mixing of vesicular stomatitis virus with fowl plague virus. Acta Virol 16:103–114

Zebedee SL, Lamb RA (1988) Influenza A virus M2 protein: monoclonal antibody restriction of virus growth and detection of M2 in virions. J Virol 62:2762–2772

# Antibody-Mediated Enhancement of Viral Disease

N.J. Sullivan



## 1 Introduction

Viral disease is a major cause of morbidity and mortality throughout the world. Unlike bacterial infections for which a wide range of specific and broadly acting antibiotics are available, viral infections continue to pose a daunting challenge in the quest for effective strategies for disease prevention. Although some success has been realized with antiviral therapies, effective treatment or prevention of viral

Vaccine Research Center, National Institutes of Health, Building 40, Room 4614B, 40 Convent Drive, Bethesda, MD 20892, USA infection usually requires the development of host immunity, acquired either naturally or through the administration of vaccines. Since the experiments of Edward Jenner in the late eighteenth century in which crude virus extracts were used to immunize against smallpox infection, it has been shown that protective host immune responses can be induced by antigens administered in a variety of forms. Live attenuated viruses produce both intracellular and cell-free antigens generated by host cells and, therefore, elicit both humoral and cellular responses. Likewise, recombinant DNA and viral vector vaccines present viral antigens in their native form because viral protein synthesis occurs intracellularly. In contrast, killed virus preparations and recombinant protein vaccines, which present antigen in extracellular form, induce mostly humoral immunity.

## 2 Immune Enhancement of Viral Infection

Despite major vaccine successes of the past century in the prevention of viral diseases, such as those caused by smallpox and poliovirus, many viruses remain refractory to vaccine-induced immune control. Indeed, there are cases in which vaccination has been reported to predispose individuals to exaggerated or enhanced disease upon exposure to virus. This was observed in the 1960s in epidemiological studies of certain viral infections, most notably in childhood infections caused by dengue or respiratory syncytial (RSV) viruses. Because of the severity of respiratory infections in infants and children caused by RSV, an immunization study with a formalin-inactivated RSV vaccine was initiated. Subsequently, an outbreak of RSV infections occurred in Washington, DC, and severe cases clustered in a group of children who were enrolled in the immunization program (Kapikian et al. 1969; Kim et al. 1969). The vaccine induced an antibody response in 97% of the children, yet 60% of the vaccinees still became infected. Furthermore, a more severe form of the illness resulting in pneumonia was observed in nearly 70% of the immunized children compared with 9% in nonvaccinated children, all residing in the same group home. The investigators noted that this pattern of exacerbated disease severity was similar to the increased incidence of infection observed in infants with passively acquired maternal antibodies (Chanock et al. 1970), and suggested that serum antibody may play an active role in the pathogenesis of RSV disease. Similar observations of disease enhancement were reported at about the same time in infants born to dengue virus immune mothers in Thailand (Halstead et al. 1969, 1970), and the term "antibody-dependent enhancement" (ADE) of viral infection was formally introduced to describe the phenomenon of immune-enhanced disease.

There are several examples of viruses subject to ADE, and they span multiple virus families. Prior to the epidemiological studies linking dengue and RSV to ADE, other viruses in the family Flaviviridae were implicated in ADE. In the course of evaluating serum antibody effects on virus replication in chick embryo fibroblasts, Hawkes serendipitously discovered an enhancing activity in avian sera reactive with West Nile virus (WNV), Murray Valley encephalitis (MVE), and Japanese encephalitis virus (JEV) (HAWKES 1964; HAWKES and LAFFERTY 1967). It was shown that the replication of some viruses was enhanced rather than inhibited in the presence of low-titer, homotypic hyperimmune serum. These results were confirmed in subsequent studies of flavivirus ADE for WNV (CARDOSA et al. 1983; PEIRIS and PORTERFIELD 1979), yellow fever virus (SCHLESINGER and BRANDRISS 1981), JEV (CECILIA and GHOSH 1988), and MVE (KLIKS and HALSTEAD 1983). Subsequently, ADE has been demonstrated for viruses from other genetically diverse families including Togaviridae (Ross River virus) (LIDBURY and MAHALINGAM 2000; LINN et al. 1996), Bunyaviridae (Rift Valley fever virus) (PEIRIS and PORTERFIELD 1981), Retroviridae (human immunodeficiency virus, simian immunodeficiency virus, equine infectious anemia virus, caprine arthritis-encephalitis virus) (FUST 1997; JOLLY et al. 1989; MONTEFIORI et al. 1990; ROBINSON et al. 1988; TAKEDA et al. 1988; WANG et al. 1994), Coronaviridae (feline infectious peritonitis virus) (HOHDATSU et al. 1991; OLSEN et al. 1992), Rhabdoviridae (rabies) (BLANCOU et al. 1980; PRABHAKAR and NATHANSON 1981; SIKES et al. 1971), Orthomyxoviridae (influenza A) (OCHIAI et al. 1992) and Paramyxoviridae (RSV) (GIMENEZ 1990; KAPIKIAN et al. 1969; KIM et al. 1969).

Although many of these viruses demonstrating ADE in vitro have also been associated with higher morbidity or mortalitiy in cases of prior immunity, the link between in vitro observations of ADE and the in vivo clinical phenomenon of disease enhancement is difficult to establish. This review will focus on two viruses for which ADE has been studied in great detail, dengue and human immunodeficiency virus (HIV).

## 3 Dengue Virus

Dengue virus is a single-stranded plus sense RNA virus belonging to the family Flaviridae. Genes encoding three structural proteins are located at the 5′ end of the viral genome, and open reading frames for the eight nonstructural proteins follow in the 3′ RNA. Dengue virions are spherical, surrounded by a lipid membrane that contains the envelope glycoprotein (E) and membrane protein (M) exposed on the outer surface. Virus attachment and internalization are mediated by the envelope glycoprotein, which is the primary neutralization target and is capable of eliciting a long-lasting antibody response (SE-THOE et al. 1999). Antibody reactivity has been used to classify the dengue viruses (DEN) into four distinct serotypes, DEN-1, DEN-2, DEN-3 and DEN-4, that are 60–80% homologous at the protein level.

## 4 Clinical Observations and Dengue Pathogenesis

Dengue virus is the major cause of arthropod-borne disease in tropical areas, and its importance as a pathogen was recognized nearly a half century ago in Southeast Asia (MORENS 1994). Dengue infection is associated with a range of symptoms from a mild flu-like syndrome to the most severe form of the disease, dengue hemorrhagic fever and dengue shock syndrome (DHF/DSS). Infection with any one of the four dengue viruses produces high-titer antibodies and homotypic immunity in the host.

Despite a strong immune response in exposed persons, there is an alarming incidence of severe DHF/DSS worldwide. While there are several virus and host factors that can influence pathogenesis (CHUNGUE et al. 1995; GREEN et al. 1999; ROSEN 1977; ROTHMAN and ENNIS 1999), early population-based studies suggested immune-mediated effects on disease severity. RUSSELL et al. (1967) introduced the secondary infection hypothesis as an explanation for severe dengue disease resulting from DEN-2 infection. It was noted that DHF/DSS occurred mostly during secondary infections with an antigenic subtype distinct from the original infecting virus, and that cross-reacting antibodies were usually present in these severe cases (HALSTEAD 1970; HALSTEAD et al. 1969; WINTER et al. 1969). Immune enhancement was also suggested in studies of the age-incidence of severe disease in infants, in which DHF/DSS cases peaked in the population between 5 and 10 months of age (HALSTEAD et al. 1969). Since these cases represented primary infections, it was hypothesized that immune enhancement was related to the presence of maternally acquired dengue antibodies. In a prospective population-based study from the 1980 epidemic in Thailand, it was shown that the major risk factor for DHF/DSS was secondary infection with DEN-2 (SANGKAWIBHA et al. 1984). More recent prospective studies have demonstrated that increased disease severity by heterotypic antibodies is directly related to higher peak dengue virus titers in humans and established a role for increased dengue virus load in the pathogenesis of DHF (VAUGHN et al. 2000).

## 5 In Vitro Studies of Dengue Antibody-Dependent Enhancement

The first attempts to use in vitro methods to explain a possible immunopathologic mechanism for DHF/DSS in secondary infection were reported in a series of studies using human and primate peripheral blood leukocytes (PBL), in which it was demonstrated that virus replication enhancement in isolated PBL was associated with the immune enhancement phenomenon that was observed in vivo. Specifically, DEN-2 virus replicated to high titers in PBL from DEN-1, -3, or -4 immune monkeys, while cells from non-immune monkeys were only marginally permissive for infection (HALSTEAD et al. 1973; MARCHETTE et al. 1976). Since in human epidemiological studies it had been shown that the immune status of the host prior to infection correlated significantly with the severity of disease (Halstead et al. 1970; Nimmannitya et al. 1969), these investigators next examined the effect of dengue immune sera on virus replication in cells from nonimmune humans or monkeys. The results demonstrated that non-immune PBL could be rendered permissive for infection by the addition of heterotypic, or low-titer-homotypic, dengue antibodies (Halstead and O'Rourke 1977a,b). These experiments defined specific cellular and serum factors involved in dengue virus replication enhancement, and they will be considered in some detail here.

Halstead and coworkers undertook a comprehensive examination of the in vitro features of dengue ADE, evaluating variables such as target cell type, immune serum specificity, and antibody donor species (Halstead and O'Rourke 1977a,b). Dengue infectivity enhancement in monkey or human PBL was tested using immune serum of mouse, rabbit, monkey or human origin, with specificity for DEN-1, DEN-2, DEN-3, DEN-4 or combined DEN1–4. In experiments on DEN-2 replication, homotypic antisera neutralized at high concentrations of immune serum but produced significant (>10-fold) enhancement at serum dilutions beyond the neutralizing endpoint. Antisera from all species, regardless of the anti-dengue specificity, enhanced the replication of DEN-2 in human PBL cultures in a dose-dependent manner, but DEN-2 serum required 10–100 times higher dilution than heterotypic antisera to promote enhancement. Generally, immune serum dilutions for infection enhancement exceeded neutralization titers by 20- to 100-fold for homologous antisera and 50- to 500-fold for heterologous sera.

Conditions for enhancement were not only dependent on serum dose, but also required that antiserum be present at the time virus was added to cells. If antiserum was preincubated with cells and washed away before virus addition, enhancement did not occur unless antiserum was readded at the time of virus-cell incubation. These results implied a dependence on the formation of immune complexes and suggested that immunoglobulin was the enhancing factor present in immune serum. This hypothesis was confirmed by fractionating serum components and showing that purified IgG alone could enhance DEN-2 replication.

Marchette et al. (1979) undertook similar in vitro PBL infection analyses to examine the role of maternal antibodies in DSS, which occurs in infants. Cord blood leukocytes (CBL) from neonates and PBL from 2- to 10-month-old infants were obtained from babies delivered to dengue immune and nonimmune mothers whose antibody status was determined at the time of delivery, and these cells were infected with DEN-2. Virus replication was greatest in blood cell cultures from infants less than 2 months of age born to dengue immune mothers and was equivalent to that observed in PBL from older children who had actively-acquired anti-dengue antibodies. PBL from infants in the 2–4-month-old group also replicated DEN-2 but at a lower level, as did the 5–10-month-old infant PBL, which yielded virus titers 100 times lower than in CBL from newborns. There was no DEN-2 replication in CBL cultures from infants born to dengue nonimmune mothers; however, the addition of autologous monkey anti-DEN-4 serum significantly increased the permissiveness of these cells for infection. It was proposed from these results that maternal anti-dengue antibodies were carried over into the cultures from newborns and infants, and that the age-dependence of enhancement results from progressive decrease of maternal antibodies in infants over time. However, the possibility of other enhancing factors in cord blood could not be excluded. To explain the observation that DEN-2 replication in CBL from infants of dengue immune mothers could be enhanced further by the addition of heterotypic antibody, it was postulated that two populations of dengue-permissive cells existed in these cultures: One contained cells with specific virus receptors and residual bound maternal antibodies, and a second population contained neither virus receptors nor bound antibody, but was capable of being infected by virus-antibody complexes.

Other in vitro studies have focused on elucidating the viral genetic and antibody determinants of dengue ADE. Infection enhancement was examined using a panel of DEN-2 monoclonal antibodies (mAb) and virus strains (HALSTEAD et al. 1984; MORENS et al. 1987). As described earlier, there were quantitative differences in the ability of each antibody to enhance the various virus strains. However, in this study, a subset of antibodies showed no enhancing activity on certain strains, indicating that in some instances enhancement was epitope-specific and that multiple enhancing determinants might be distributed across various virus strains. Despite the genetic variability in enhancing epitopes across virus isolates, no specific determinant was associated with increased disease severity. However, it was noted in low-passage isolates of DEN-2 that virus strains associated with severe DHF contained a greater number of epitopes involved in ADE (MORENS and HALSTEAD 1987). Subsequent experiments demonstrated that enhancement was determined not by dengue serotype but rather by antibody:virus ratios, with neutralization occurring at high ratios and enhancement at low ratios (MORENS et al. 1987). Interestingly, in clinical studies in which maternal dengue antibody titers were measured, there was a strong correlation between anti-dengue antibody titers and infant age at the time of onset of severe illness, which corresponded to the time predicted for maternal antibodies to diminish to low concentration (KLIKS et al. 1988). Therefore, in vivo, changing antibody:virus ratios results in a dual role for maternal antibodies, which protect early in infancy when the antibody titer is high (low ratio) but later increase the risk of developing DHF/DSS when antibody titers drop (high ratio). This dose-response pattern for antibody mediated neutralization and enhancement (Fig. 1) seems to be uniformly observed across virus families. High concentrations of antibodies override antibody-enhancing effects and neutralize viruses, as predicted by a protective humoral immune response down to some minimum threshold titer. At dilutions exceeding the endpoint neutralization titer, antibodies are capable of enhancing virus replication and exacerbating the symptoms of disease.

## 6 Mechanisms of Dengue Antibody-Dependent Enhancement

Conceptually, it is relatively simple to reconcile the notion that distinct classes of antibodies may effect virus replication differently, but the conclusion that a single

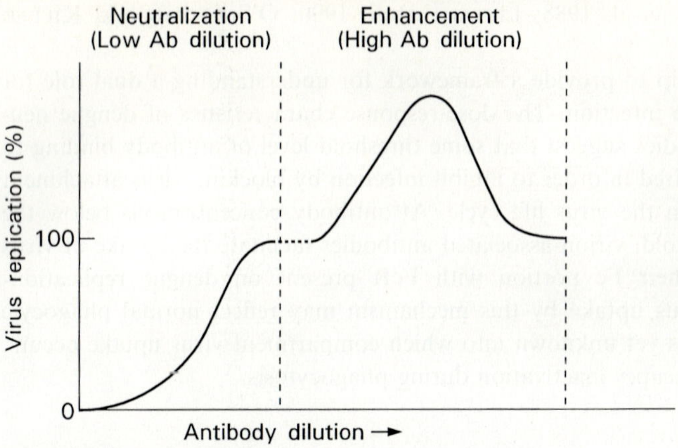


**Fig. 1.** Dose-response for antibody-mediated inhibition and enhancement of virus replication. At high concentrations of virus-specific antibody (*Low Ab Dilution*), the percent occupancy of antibody binding sites is sufficiently high to inhibit critical steps in the virus life cycle. As the antibody becomes more dilute, its inhibitory activity becomes attenuated; at some point below the neutralization endpoint, antibody binding at subneutralizing concentrations (*High Ab Dilution*) enhances virus infectivity

antibody can cause either neutralization or enhancement of infection seems paradoxical. However, reports describing the mechanism of enhancement by dengue antibodies have helped to resolve the apparent inconsistency of antibody effects.

Early in vitro studies of ADE showed that enhancement was mediated by the IgG fraction of immune serum and not by IgM fractions (Halstead and O'Rourke 1977; Halstead and O'Rourke 1977). Furthermore, these studies demonstrated that ADE required the Fc portion of immunoglobulin; $F(ab)_2$ preparations did not enhance DEN-2 infection. These findings suggested that Fc receptors (FcR) present on the surface of subpopulations of PBL, mostly phagocytic cells, might facilitate enhanced uptake of virus-antibody complexes into cells permissive for virus replication. Consistent with these observations, experiments showed that the kinetics of virus-antibody internalization were similar to the rate of phagocytosis by macrophages. In a companion paper, cell fractionation techniques and immunofluorescence were employed to demonstrate that mononuclear phagocytes were indeed the cells that were permissive for antibody-mediated dengue infection (Halstead et al. 1977).

Proof of FcR-mediated ADE for dengue virus was verified in studies using purified monocytes that expressed high levels of FcR (Daughaday et al. 1981). Similar to Halstead's results with PBL, monocyte infection was not enhanced by dengue virus-anti-virus $F(ab)_2$ complexes. However, in these studies, addition of Fc-containing anti-Fab antibodies to preformed virus-Fab complexes restored infection enhancement, and pretreatment of monocytes with aggregated gammaglobulin significantly reduced ADE. Subsequently, others demonstrated that augmentation of dengue virus infection occurred under activating conditions that caused the up-regulation of FcR expression, and in cells expressing either Fc$\gamma$RI or FcγRII (Kontny et al. 1988; Littaua et al. 1990; O'Sullivan and Killen 1994).

These results help to provide a framework for understanding a dual role for antibodies in dengue infection. The dose-response characteristics of dengue neutralization by antibodies suggest that some threshold level of antibody binding to virions is likely required in order to inhibit infection by blocking virus attachment or some step later in the virus life cycle. At antibody concentrations below the neutralization threshold, virion-associated antibodies facilitate the uptake of virus by interaction of their Fc portion with FcR present on dengue replication-permissive cells. Virus uptake by this mechanism may reflect normal phagocyic processes, but it is as yet unknown into which compartment virus uptake occurs, and how the virus escapes inactivation during phagocytosis.

## 7 In Vivo Studies of Dengue Infection Enhancement

While in vitro studies are useful for identifying possible cellular mechanisms of ADE, it is often difficult to demonstrate the in vivo relevance of such findings. Therefore, numerous strategies have been pursued to define the in vivo ADE phenomenon in a more controlled manner than is possible in population-based studies. Animal studies provide the means to control variables such as immune serum composition, and infecting virus type and inoculum. Halstead used monkeys to test the secondary infection hypothesis of human dengue disease, and attempted to reproduce the phenomenon observed in humans by sequentially inoculating monkeys with various dengue virus types (Halstead et al. 1973). Since it had been reported that secondary infection with DEN-2 was most often associated with DHF/DSS in humans (Russell et al. 1968; Winter et al. 1968), monkeys were inoculated with DEN-2 after primary infection with either DEN-1, DEN-3, DEN-4, or no infection. The results showed that animals with secondary DEN-2 infection circulated more virus than did monkeys who were nonimmune at the time of DEN-2 infection. Thus it was demonstrated definitively in an animal model that heterotypic immunity predisposed subjects to exacerbated infection. Although this conclusion was consistent with the demonstration in vitro that dengue immune sera and purified IgG enhanced virus replication, it was not possible from these studies to attribute the in vivo observations in monkeys exclusively to the presence of anti-dengue antibodies. Therefore, passive transfer experiments were conducted (Halstead 1979) in order to observe the in vivo infection-enhancing capability of dengue antibodies in the absence of other host factors that might be induced during secondary infection. Since enhancing antibodies had recently been identified in human cord blood (Marchette et al. 1979), monkeys were injected intravenously with cord blood to obtain the predicted enhancing antiserum dilution of 1:300, and blood samples were removed to confirm the transfer efficiency of infection-enhancing activity. The fold-enhancement activity in plasma from a recipient monkey was equivalent to that in the source dengue-immune cord blood, with peak enhancement levels of up to 18-fold. These monkeys were then infected with DEN-2 virus immediately after the passive transfer of antibodies, and dengue virus titers were evaluated over the course of several days. Animals receiving dengue-immune cord blood exhibited significantly higher levels and a longer duration of viremia than animals that received nonimmune cord blood. Cumulative enhancement measurements for matched pairs of monkeys revealed about 3- to 50-fold enhancement of viremia in recipients of dengue immune serum. These experiments provided strong in vivo relevance for the in vitro ADE observations, and supported the notion that immune complexes found between anti-dengue antibodies and virus may cause enhanced disease in humans. However, the presence of other enhancing activities in immune serum cannot be ruled out in the passive transfer experiments, and similar investigations using purified immunoglobulin will be required to provide definitive evidence that dengue antibodies cause enhanced disease in vivo.

## 8 Human Immunodeficiency Virus (HIV)

Human (HIV) and simian (SIV) immunodeficiency viruses are double-stranded RNA viruses from the family Retroviridae. These lentiviruses are characterized by disease progression that is prolonged over the course of many months to years. HIV-1 causes severe immunodeficiency that predisposes the host to a wide array of opportunistic infections considered to be the hallmark of acquired immunodeficiency syndrome (AIDS). HIV-1 infects cells of the immune system that express the primary CD4 receptor, found primarily on T lymphocytes and monocytes, and the secondary chemokine receptors CCR5 and CXCR4. Virus attachment and pH-independent entry into target cells are mediated by the SU (gp120) and TM (gp41) envelope glycoproteins. Virus association with the CD4 receptor facilitates conformational changes in SU that allow contact with the secondary receptor and subsequent fusion of the virus and cell membranes. Formation of the mature envelope glycoproteins requires proteolytic processing of the gp160 precursor protein. SU and TM then associate noncovalently in the form of trimers on the surface of the virion or infected cell. Sequence analysis of multiple HIV-1 isolates reveals a high degree of genetic variability in SU, with five hypervariable regions (V1–V5) interspersed between regions that are relatively conserved (C1–C5). Sequence diversity in SU results in the generation of phenotypic differences in coreceptor usage and virulence among virus isolates. HIV was originally isolated by passage in tissue culture, generating molecular clones that replicate to high titers in T cell lines and use CXCR4 as a coreceptor. These T-cell-line-adapted (TCLA) viruses do not replicate well in cells of the monocyte/macrophage lineage. Primary HIV-1 strains that have been isolated directly from patient PBL use either the CXCR4 or CCR5 coreceptors and display either T cell-, macrophage-, or dual-tropism depending on the coreceptor preference of each strain.

## 9 HIV Clinical Observations

HIV infection leads to an initial burst of viremia followed by a gradual decline in virus replication to low or undetectable levels for a period of years. The early spike in viremia triggers both cellular and humoral immune responses in infected patients, which in most cases are sufficient to reduce viremia and control acute disease. After an extended asymptomatic period, CD4-positive T cell counts decline, concurrent with an increase in viral load, and patients develop the pathologic symptoms associated with AIDS. A small percentage of infected individuals, termed long-term nonprogressors (LTNP), remain asymptomatic and viremia appears to be controlled indefinitely.

Despite the fact that high titer neutralizing antibodies to HIV are routinely detected in vitro from infected patient sera, seropositive individuals eventually progress to AIDS. One explanation for this apparent paradox lies in the inability of in vitro assays to accurately predict virus behavior in vivo. For example, neutralization assays were originally performed using TCLA HIV-1 isolates, but it is now known that these isolates do not adequately represent the diversity of isolates that result from genetic variation of HIV over time in patients (Hahn et al. 1985, 1986). TCLA isolates differ from primary isolates not only in their tropism, but also in their sensitivity to neutralization by antibodies. TCLA viruses tend to be sensitive to neutralization by antibodies, whereas primary viruses exhibit a spectrum of neutralization phenotypes (Cheng-Mayer et al. 1988), with some isolates being refractory to neutralization even at very high concentrations of HIV-1 patient sera. It is thought that the latter isolates likely predominate early in infection when the host is capable of mounting a robust humoral response. Indeed, it was shown that the replicative ability of these isolates is even enhanced by HIV patient sera obtained throughout the course of infection (Gras et al. 1988; Homsy et al. 1989; Robinson et al. 1988; Takeda et al. 1988).

## 10 In Vitro Studies on the Mechanism of HIV Antibody-Dependent Enhancement

In one of the earliest reports of antibody enhancement of HIV infection, Takeda et al. (1988) observed an enhancing activity that was present in heat-inactivated sera and appeared to be mediated by immunoglobulin. These investigators used a TCLA strain of HIV to assess infection enhancement in the U937 monocytic cell line, and replication was monitored directly by measuring yields of the viral protein p24. Reminiscent of the observations with dengue ADE, low dilutions of serum neutralized HIV-1, whereas virus replication was enhanced at higher, nonneutralizing serum dilutions. IgG fractions, but not $F(ab)_2$ fragments, from HIV immune serum enhanced HIV infection at the same serum dilution equivalents, and heat-aggregated gammaglobulin blocked infection. These data suggested that enhancement of HIV replication was mediated by FcR. This was confirmed in a subsequent study, undertaken to identify the specific cell surface receptors responsible for ADE of HIV-1, by selectively blocking each receptor with specific antibodies (Takeda et al. 1990). Flow cytometric analysis revealed that, while U937 cells express both FcγRI and FcγRII, enhancement was blocked by antibodies against only FcγRI. The authors also showed that FcR-mediated ADE of HIV infection required binding to the CD4 receptor; preincubation of cells with either a soluble form of the receptor, sCD4, or an anti-CD4 antibody effectively blocked HIV infection as well as ADE (Takeda et al. 1990). The CD4-dependence of FcR-mediated ADE was verified in cultured human macrophages, which may be more physiologically relevant than the U937 cell line.

Similar results were obtained in studies conducted independently by several other investigators (Homsy et al. 1989; Jouault et al. 1991; Perno et al. 1990). Matsuda et al. (1989) used subclones of U937 exhibiting differential expression of FcRs and varying degrees of phagocytic activity. Here it was shown that, although high levels of phagocytic activity are necessary for antibody-enhanced infection, significant phagocytosis alone was insufficient to promote HIV-1 ADE. Enhancement was not observed in a U937 subclone with low FcγRI surface expression and phagocytic activity, nor in the highly phagocytic, CD4-negative, macrophage cell line P388D1. In a U937 phagocytic clone expressing FcγRI and CD4, patient serum that enhanced virus replication also enhanced the binding of $^{125}$I-labeled gp120 to cells, and at a similar serum dilution. These data implied that enhancement of infection results not from phagocytosis, but from increased virus-cell attachment that facilitates the interaction of gp120 with its specific receptor(s).

Concurrent with the reports of CD4-dependent, FcR-mediated ADE, other investigations detected a form of FcR-mediated ADE that did not require CD4 (CD4-independent FcR-ADE) (Homsy et al. 1989). These experiments were conducted in bulk cultures of primary human PBMC and in isolated subpopulations enriched for macrophages or $CD4^+$ T lymphocytes, which may reproduce more closely the infection conditions that exist in vivo. Immune sera from guinea pigs, chimpanzees or humans were capable of enhancing the replication in macrophages of HIV-1 and HIV-2 by up to ten-fold. Blocking of FcγRIII, but not FcγRI or FcγRII, eliminated enhancement by human or chimpanzee immune sera, confirming other studies that suggested the importance of Fc receptors for HIV ADE, but also illustrating flexibility in the class of Fc receptor used: FcγRIII in primary human macrophages and FcγRI in monocytic cell lines which lack FcγRIII. Beyond demonstrating cell-type-specific differences in FcR usage, these experiments showed that neither sCD4 nor antibodies which block the CD4 receptor had any effect on the observed enhancement. This result suggested that under some conditions, ADE can circumvent the use of HIV receptors. Interestingly, in the $CD4^+$ lymphocytes used for these studies, enhancement was observed to be independent of both CD4 and Fc receptors, leaving open the possibility that other serum factors and/or cellular receptors may contribute to immune-enhanced HIV infection.

Indeed, ROBINSON et al. (1988) reported in vitro enhancement of HIV infection by human serum and identified two components that contribute to enhancement, one being present only in HIV immune sera and the other being a component of both immune and nonimmune sera. These investigators studied antibody-enhanced infectivity of TCLA HIV-1 isolates by measuring the induction of cytopathic effect in the T cell line MT-2. Immune serum at dilutions from 1:4 to 1:64 enhanced HIV-induced cytopathic effect when preincubated with virus before addition to cells. One immune-serum enhancing component was heat stable and removable by protein-A sepharose chromatography, so the authors concluded that this was likely an anti-HIV immunoglobulin, consistent with other reports. However, if sera were incubated at 60°C or treated with complement-pathway-inactivating cobra venom factor prior to addition to cells, the enhancing activity was lost and some sera became neutralizing. The heat-inactivated enhancing activity could be restored by the addition of normal human serum but not by some complement-deficient sera. The authors proposed that enhancement proceeds by the alternative pathway of complement, whereby HIV-specific immunoglobulin and complement facilitate virus attachment and uptake, possibly by binding to complement receptors, or by influencing a later step in the virus life cycle. Interestingly, complement-mediated antibody-dependent enhancement (C′-ADE) was apparent only in the initial infection; chronically infected cells or cells that were infected prior to the addition of serum did not support serum-enhanced infection.

ROBINSON et al. (1989) also conducted studies to define the cell surface receptor(s) required for C′-ADE. A variety of T cell lines were examined by flow cytometry for the expression of complement receptors (CR) and CD4. Those cell lines exhibiting significant levels of CD4 and CR on the cell surface demonstrated C′-ADE, whereas CR-negative cell lines did not. In a subsequent report, the same investigators used receptor-blocking antibodies to show that, in MT-2 cells, C′ADE requires both CD4 and CR2 (ROBINSON et al. 1990). The U937 monocytic cell line exhibited a similar requirement for both CR and CD4, but these cells used CR3 (REISINGER et al. 1990), demonstrating, as in the case with FcR-mediated ADE, the flexibility in receptor requirements between various cell types. Normal B lymphocytes demonstrated the broadest range of CR usage, with either CR1 or CR2 acting with CD4 to mediate C-ADE, whereas in a transformed B cell line enhancement was mediated by CD4 and CR2 only (GRAS and DORMONT 1991). More recently, Prohaszka and colleagues provided evidence for an even broader repertoire of complement receptors involved in C′-ADE than was previously suspected (PROHASZKA et al. 1997). They showed that gp41 mAbs increased binding of C1q to solid-phase gp41 in a dose-dependent manner. Moreover, C1q deposition on MT-4 cells was also dose dependent. Depletion of C1q from serum reduced the enhancing activity of normal human serum containing gp41 mAbs. There was, however, some residual enhancement that was shown to be mediated by CR2.

There is some evidence that complement-enhanced HIV infection can proceed in the absence of HIV-specific antibodies. While the studies described above demonstrated enhancement using HIV immune serum or purified antibodies, others have reported complement-mediated enhancement in the absence of HIV-specific antibodies (Boyer et al. 1991; Gras and Dormont 1991; Legendre et al. 1996; Robinson et al. 1989). Furthermore, in some cells, complement-mediated, antibody-independent enhancement (C′AIE) occurred independently of the CD4 receptor (Boyer et al. 1991, 1992), implying a mechanism of direct uptake of complement-coated virus. More recently, investigators have developed a reconstituted complement system in an attempt to precisely define the serum components necessary to enhance HIV-infection via complement receptors (Tacnet-Delorme et al. 1999). In these experiments the classical pathway was reconstituted with purified complement proteins C1q, C1r, C1 s, C2, C3, C4, factor H and factor I. Complement proteins, without added serum or antibodies, enhanced HIV-RF infection of MT-2 cells to the same degree as did normal human serum. These results provide direct evidence for an antibody-independent mechanism of complement-enhanced infection in some cell lines.

## 11 Fc Receptor-Independent, Complement Receptor-Independent, Antibody-Dependent Enhancement of HIV

Each form of immune enhancement described thus far proceeds through an initial interaction of antibody- or complement-coated virus with Fc or complement receptors, but the most recent reports of HIV immune enhancement have described antibody-mediated increases in virus replication that do not involve Fc or complement (Schutten et al. 1995, 1997; Sullivan et al. 1995, 1998). These studies utilized a reporter gene-containing recombinant retrovirus that can be pseudotyped with envelope glycoproteins dervived from TCLA or primary patient isolates. The replication assay measures early steps in the virus life cycle in a single round of replication, and allows precise quantitation of neutralization and enhancement in a single target cell type, irrespective of the parental isolate tropism, thus controlling for variables such as target cell receptor expression (Helseth et al. 1990; Sullivan et al. 1995). These studies examined the effects of a panel of antibodies, directed against a variety of SU epitopes, on the replication of TCLA viruses and primary HIV-1 patient isolates. As reported previously, HIV-1 isolates exhibited a range of neutralization phenotypes (Cheng-Mayer et al. 1988) when replication was measured in either SupT1 or Molt 4 T-cell lines. While most of the primary isolates were relatively resistant to neutralization, infection by some primary isolates was enhanced six- to eight-fold by all of the antibodies irrespective of the binding epitope. It is unlikely that the observed enhancement was mediated via Fc receptors because the target cells used in these studies do not express Fc receptors. Moreover, Fab antibody fragments representing diverse epitope-binding specificities on SU mediated enhancement as efficiently as complete immunoglobulins. The use of purified antibody preparations lacking complement ruled out the possibility that complement receptors were responsible for enhanced HIV-1 replication in this system.

The various mechanisms of HIV-1 ADE described above are summarized in Fig. 2. In one category of ADE (Fig. 2, top), virus-bound complement and/or antibodies associate with either complement or Fc receptors and facilitate the association of HIV-1 envelope glycoproteins with virus receptors, resulting in enhanced infection by the normal route of virus entry. This mechanism has been shown to be CD4-dependent and presumably also involves the HIV-1 coreceptor, although most of the studies described were conducted prior to discovery of chemokine coreceptors (Alkhatib et al. 1996; Feng et al. 1996). FcR-mediated ADE and C′-ADE can also proceed by a CD4-independent route which resembles phagocytosis of opsonized virus (Fig. 2, bottom left). It has not been determined how virus escapes inactivation in the phagocytic pathway (Fig. 2, bottom right). The most recent mechanism to be described for HIV-1 ADE uses neither Fc nor complement receptors. Here, antibody binding to SU induces conformational changes in the envelope glycoproteins that are thought to resemble virus receptor-activated modifications of SU structure that enable association with chemokine coreceptors to promote downstream membrane-fusion events.

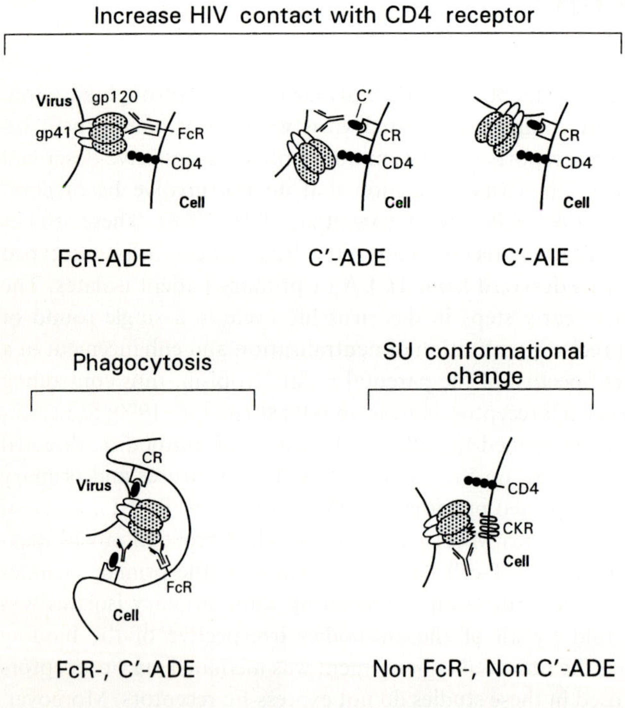


**Fig. 2.** Proposed mechanisms for antibody-mediated enhancement of HIV infection. Antibody or complement binding to the HIV-1 SU (gp120) or TM (gp41) envelope glycoproteins facilitates attachment of HIV-1 to the cell surface. Phagocytosis or contact with CD4 and/or chemokine (*CKR*) receptors enhances virus entry into the cell

## 12 Target Cell Specificity in Antibody-Dependent Enhancement of HIV

The range of receptors shown to promote ADE under different conditions suggests that the system employed to evaluate mechanisms of enhancement can, in and of itself, influence the resulting observations. This is due in part to the choice of target cells used to assess virus replication, and the cell-type specificity of Fc and complement receptor distribution. These receptors are expressed on the surface of most phagocytic cells, but have also been detected in numerous other cell types (Table 1). It may be expected that highly phagocytic cells, such as the P388 D1 macrophage cell line used in studies of flavivirus enhancement (PEIRIS and PORTERFIELD 1981), are especially sensitive indicators of ADE. Consideration of the in vivo relevance of studies using transformed cell lines must, therefore, include an evaluation of the level of receptor expression in vitro compared to that in the physiologically relevant target cell(s). It has been shown for the CD4 receptor and chemokine coreceptors that quantitative differences in cell surface receptor expression can yield qualitative differences in HIV replication phenotypes (KOZAK et al. 1997; PLATT et al. 1998). Indeed, altering virus adhesion by a variety of means has been demonstrated to influence HIV replication enhancement and sensitivity to neutralization (LUND et al. 1995; RIZZUTO and SODROSKI 1997). Furthermore, it has been shown that the CD4 dependence for HIV-1 attachment to target cells is highly cell-line-specific (MONDOR et al. 1998). It is tempting to generalize that increased virus-cell adhesion will always result in higher virus replication and relative resistance to neutralization, but it has been found that mAbs that enhance virus attachment retain their neutralizing potency (MONDOR et al. 1998). Because the expression of receptors and adhesion molecules varies with cell type and activation level, studies of ADE should include quantitative evaluations of cell surface receptor expression. Ideally, ADE should be assessed in the relevant in vivo target cells.

**Table 1.** Expression of Fc and complement receptors. The distribution of Fc and complement receptors is shown for several cell types

| Receptor | Cell type | | | | | | |
|---|---|---|---|---|---|---|---|
| | mϕ | DC | Monocyte | T | B | NK | PMN |
| FcγRI | + | + | + | | | | + |
| FcγRII | + | + | + | | + | | + |
| FcγRIII | + | | | | | + | + |
| FcαR | + | | | + | + | + | + |
| FcεRI | | | | | | | + |
| FcεRII | + | + | + | | + | | + |
| FcμRI | | | | + | | | |
| CR1 | + | + | + | | + | | + |
| CR2 | | + | | | + | | |
| CR3 | + | | + | | | | + |
| CR4 | + | | + | | | | + |
| ClqR | + | | + | | + | | |

+, Expression has been detected on either activated or resting cells; mϕ, macrophage; DC, dendritic cell; T, T lymphocyte; B, B lymphocyte; NK, natural killer cell; PMN, polymorphonuclear phagocyte.

## 13 Epitope Specificity of HIV Antibody-Dependent Enhancement

Although complement interacts with both gp120 and gp41 (Hidvegi et al. 1993; Prohaszka et al. 1995; Stoiber et al. 1995), the epitopes most often reported to mediate C′-ADE and C′AIE reside in the gp41 TM envelope protein (Robinson 1991; Prohaszka et al. 1997) but rarely in gp120 (Jiang et al. 1991). Four of eight human mAbs to gp41 were identified that could enhance HIV type-1 (HIV-1) infection in vitro by C′-ADE. These enhancing human mAbs were mapped to two distinct domains on the HIV-1 gp41 transmembrane glycoprotein by using synthetic peptides (Ebenbichler et al. 1991; Robinson et al. 1991). One region represents the immunodominant region of gp41 (residues 590–620) and contains the putative interaction site for C1q (Thielens et al. 1993). This region also partially overlaps the major SU/TM interactive region (Helseth et al. 1991), and it has been proposed that recognition of gp41 by complement, like SU, is driven by hydrophobic interactions (Quinkal et al. 1999).

FcR-mediated ADE does not appear to exhibit the level of epitope specificity observed for C′-ADE because enhancement has been demonstrated using antibodies directed against a variety of epitopes, though most studies of FcR-mediated ADE have used purified antibodies directed against the SU gp120 protein. However, some degree of epitope specificity was suggested in an early study (Takeda et al. 1992) using two mAbs recognizing distinct conformation-dependent epitopes and a third antibody that binds a linear determinant in the V3 region of SU. One of the two antibodies directed against a discontinuous epitope enhanced HIV replication by approximately 1.5-fold, whereas the other antibodies were neutralizing in the same concentration range. The differential effects of these antibodies was not due to differences in Fc receptor binding because all antibodies were of the IgG1 subclass, but differences in antibody binding affinities for SU cannot be ruled out. The enhancing antibody caused only modestly enhanced (less than two-fold) virus replication, so it is possible that the lack of neutralizing activity by this antibody simply reflected poor recognition of virion-associated envelope glycoproteins. It has been demonstrated in several studies that the degree of virus neutralization is determined by the strength of antibody association with the virion envelope glycoprotein trimer (Moore et al. 1995; Sattentau and Moore 1995; Sullivan et al. 1995).

## 14 HIV Envelope Glycoprotein Determinants of Antibody-Dependent Enhancement

Several pieces of evidence suggest it is most likely that the antigenic structure of the envelope glycoprotein, rather than epitope specificity of the antibody, is the major determinant of infectivity enhancement. Data demonstrating that enhancement is dictated by features specific to particular HIV-1 isolates were reported in a comprehensive study that examined the neutralization susceptibility of over 1,000 combinations of immune sera and primary HIV-1 isolates (KOSTRIKIS et al. 1996; MOORE et al. 1996). Infectivity enhancement was observed in 28% of the serum-HIV isolate combinations, and there was no segregation of enhancing activity with particular sera. Instead the degree of enhancement, measured as negative neutralization values, correlated best with certain HIV-1 isolates.

Other studies examined the precise envelope glycoprotein determinants that predispose particular HIV-1 isolates to antibody-enhanced infection. A neutralizing antibody directed against the V3 domain of HIV-1 exhibited differential effects on the replication of three virus isolates that differed by only one amino acid in the defined antibody-binding epitope (KLIKS et al. 1993). While there was no antibody effect on infection of one isolate, another was enhanced and the third was efficiently neutralized. The enhancement phenotype could be reversed by changing a single residue in the V3 domain to match that of the neutralization-sensitive isolate. Other investigators evaluated panels of antibodies directed against multiple recognition sites on SU and used envelope glycoproteins with modified variable loops to show that the enhancement phenotype was influenced primarily by the V3 domain, although the configuration of the V1/V2 domain also contributed (SULLIVAN et al. 1998). Interestingly, receptor blocking experiments indicated a dependence on binding the CCR5 coreceptor for antibody-enhanced infection of a primary patient isolate molecular clone. Moreover, determinants of antibody-enhanced replication were shown to overlap with the same regions involved in the receptor-activation of HIV-1 envelope glycoproteins that lead to virus-cell fusion. These results suggest that FcR/CR-independent HIV-1 ADE proceeds by a mechanism fundamentally similar to the CD4-receptor-activated virus entry pathway, whereby ligand (CD4 or antibody) binding to SU induces the conformational changes necessary to permit coreceptor binding (RIZZUTO and SODROSKI 2000; SULLIVAN et al. 1998; WU et al. 1996) and subsequent virus-cell fusion.

## 15 In Vivo Relevance of Antibody-Dependent Enhancement

The relationship of in vitro neutralization and enhancement to the clinical course of HIV-1 infection remains to be determined. Numerous studies have been undertaken to define correlates of protection in HIV-1-infected patients and most have detected associations between HIV-1 disease progression and in vitro immunological markers. WEBER et al. (1987) reported that asymptomatic patients have higher neutralizing antibody responses than do patients with AIDS. Longitudinal studies in cohorts of AIDS patients have also suggested a relationship between the presence of neutralizing antibodies and retarded disease progression (CAROTENUTO et al. 1998; CECILIA et al. 1999; CHEINGSONG-POPOV et al. 1991). Similarly, others have reported an inverse correlation between HIV-1-specific CTL and disease progression (HARRER et al. 1996; KLEIN et al. 1995; RIVIERE et al. 1995). Enhancing antibodies have been detected in serum from patients at both early and late stages in the clinical course of HIV-1 infection (JOLLY and WEISS 2000), and HOMSY et al. (1990) reported that enhancing antibodies increase over time with clinical progression to AIDS. However, JOLLY and WEISS (2000) showed that both enhancing and neutralizing antibodies are present at late time points in stable patients and suggested that ADE is not associated with progression. Likewise, MONTEFIORI et al. (1991, 1996) demonstrated C′-ADE in sera from both progressing and non-progressing HIV-1 patients and concluded that the presence of enhancing antibodies does not predict disease progression.

One limitation with studies of immune correlates of protection is that it can be difficult to attribute cause and effect. Other approaches have attempted to circumvent this problem by manipulating immune parameters at defined points in the infection process. SCHMITZ et al. (1999) administered cobra venom factor to deplete complement to <5% of control values at different points after infection of monkeys with SIV. Treated monkeys experienced a higher viral load and accelerated clinical course of disease, suggesting that complement proteins may constitute part of a protective immune response. Therefore, complement-mediated immune clearance mechanisms may override potential enhancing effects.

A protective role for neutralizing antibodies has also now been established in an in vivo setting. Two recent studies utilized passive transfer techniques to evaluate the neutralization capacity of antibodies in a primate model of HIV-1 infection (BABA et al. 2000; MASCOLA et al. 1999, 2000). Both studies used chimeric SIV/HIV viruses (SHIV), which contain HIV-1 envelope glycoproteins on a SIV background virus and are pathogenic in macaque monkeys. A combination of antibodies that neutralize HIV-1 in vitro were administered to monkeys immediately prior to either intravenous or mucosal challenge with SHIV and were shown to offer protection against infection. These encouraging results suggest that any enhancing effect of antibodies is offset by their neutralizing capacity. It should be noted, however, that the virus challenge in these experiments was timed to coincide with maximum antibody titers. It remains unclear whether in this model, as in vitro, virus challenge under conditions of subneutralizing titers of HIV-1 antibodies would lead to enhanced infection.

## 16 Antibody-Dependent Enhancement and HIV Vaccine Development

There is as yet no effective vaccine against HIV-1 infection, and the detection of enhancing antibodies in the sera from HIV-1 vaccinees has raised concern about the potential for disease enhancement by vaccine-induced antibodies (MASCOLA et al. 1993). The results of passive transfer studies described herein suggest that HIV-1 neutralizing antibodies present in sufficient quantity at the time of virus challenge are unlikely to exacerbate disease progression. Several caveats must, however, temper optimism with respect to vaccine-induced antibodies. First, a vaccine must induce a humoral response capable of generating sustained high-titer antibodies since infection can occur any length of time post-vaccination, and the effects of neutralizing antibodies on established infection in vivo remains to be determined. Second, although HIV-1 ADE is antibody-epitope-independent, the type of antibodies elicited by a vaccine remains important, to the extent that antibody binding affinity may be an important determinant of virus neutralization vs enhancement. Indeed, as suggested by the dose-response curves for virus neutralization and enhancement by antibodies, the percent occupancy of antibody binding sites on envelope glycoprotein spikes predicts neutralization efficiency (Parren et al. 1998). Unfortunately, much of the humoral immune response produced in natural HIV-1 infection appears to be misdirected toward viral debris, generating antibodies that show poor recognition for the functional envelope glycoproteins associated with infectious virions (Burton and Parren 2000; Parren et al. 2000). It will be important in the design of immunogens and vaccination strategies to ensure that immunization elicits high-titer antibodies with strong affinity for virion-associated envelope glycoproteins.

*Acknowledgement.* The author wishes to thank Dr. John Mascola for critical reading of the manuscript.

## References

Alkhatib G, Combadiere C, Broder CC, Feng Y, Kennedy PE, Murphy PM, Berger EA (1996) CC CKR5: a RANTES, MIP-1alpha, MIP-1beta receptor as a fusion cofactor for macrophage-tropic HIV-1. Science 272:1955–1958

Baba TW, Liska V, Hofmann-Lehmann R, Vlasak J, Xu W, Ayehunie S, Cavacini LA, Posner MR, Katinger H, Stiegler G, Bernacky BJ, Rizvi TA, Schmidt R, Hill LR, Keeling ME, Lu Y, Wright JE, Chou TC, Ruprecht RM (2000) Human neutralizing monoclonal antibodies of the IgG1 subtype protect against mucosal simian-human immunodeficiency virus infection. Nat Med 6:200–206

Blancou J, Andral B, Andral L (1980) A model in mice for the study of the early death phenomenon after vaccination and challenge with rabies virus. J Gen Virol 50:433–435

Boyer V, Delibrias C, Noraz N, Fischer E, Kazatchkine MD, Desgranges C (1992) Complement receptor type 2 mediates infection of the human CD4-negative Raji B-cell line with opsonized HIV. Scand J Immunol 36:879–883

Boyer V, Desgranges C, Trabaud MA, Fischer E, Kazatchkine MD (1991) Complement mediates human immunodeficiency virus type 1 infection of a human T cell line in a CD4- and antibody-independent fashion. J Exp Med 173:1151–1158

Burton DR, Parren PW (2000) Vaccines and the induction of functional antibodies: time to look beyond the molecules of natural infection? Nat Med 6:123–125

Cardosa MJ, Porterfield JS, Gordon S (1983) Complement receptor mediates enhanced flavivirus replication in macrophages. J Exp Med 158:258–263

Carotenuto P, Looij D, Keldermans L, de Wolf F, Goudsmit J (1998) Neutralizing antibodies are positively associated with $CD4^+$ T-cell counts and T-cell function in long-term AIDS-free infection. AIDS 12:1591–1600

Cecilia D, Ghosh SN (1988) Antibody dependent plaque enhancement by monoclonal antibodies against Japanese encephalitis virus. Indian J Med Res 87:521–525

Cecilia D, Kleeberger C, Munoz A, Giorgi JV, Zolla-Pazner S (1999) A longitudinal study of neutralizing antibodies and disease progression in HIV-1-infected subjects. J Infect Dis 179:1365–1374

Chanock RM, Kapikian AZ, Mills J, Kim HW, Parrott RH (1970) Influence of immunological factors in respiratory syncytial virus disease. Arch Environ Health 21:347–355

Cheingsong-Popov R, Panagiotidi C, Bowcock S, Aronstam A, Wadsworth J, Weber J (1991) Relation between humoral responses to HIV gag and env proteins at seroconversion and clinical outcome of HIV infection. BMJ 302:23–26

Cheng-Mayer C, Homsy J, Evans LA, Levy JA (1988) Identification of human immunodeficiency virus subtypes with distinct patterns of sensitivity to serum neutralization. Proc Natl Acad Sci USA 85:2815–2819

Chungue E, Cassar O, Drouet MT, Guzman MG, Laille M, Rosen L, Deubel V (1995) Molecular epidemiology of dengue-1 and dengue-4 viruses. J Gen Virol 76:1877–1884

Daughaday CC, Brandt WE, McCown JM, Russell PK (1981) Evidence for two mechanisms of dengue virus infection of adherent human monocytes: trypsin-sensitive virus receptors and trypsin-resistant immune complex receptors. Infect Immun 32:469–473

Ebenbichler CF, Thielens NM, Vornhagen R, Marschang P, Arlaud GJ, Dierich MP (1991) Human immunodeficiency virus type 1 activates the classical pathway of complement by direct C1 binding through specific sites in the transmembrane glycoprotein gp41. J Exp Med 174:1417–1424

Feng Y, Broder CC, Kennedy PE, Berger EA (1996) HIV-1 entry cofactor: functional cDNA cloning of a seven-transmembrane, G protein-coupled receptor. Science 272:872–877

Fust G (1997) Enhancing antibodies in HIV infection. Parasitology 115:S127–140

Gimenez B (1990) Antibody-mediated enhancement of respiratory syncytial virus infection. J Infect Dis 162:1211

Gras G, Strub T, Dormont D (1988) Antibody-dependent enhancement of HIV infection. Lancet:1285

Gras GS, Dormont D (1991) Antibody-dependent and antibody-independent complement-mediated enhancement of human immunodeficiency virus type 1 infection in a human, Epstein-Barr virus-transformed B-lymphocytic cell line. J Virol 65:541–545

Green S, Vaughn DW, Kalayanarooj S, Nimmannitya S, Suntayakorn S, Nisalak A, Rothman AL, Ennis FA (1999) Elevated plasma interleukin-10 levels in acute dengue correlate with disease severity. J Med Virol 59:329–334

Hahn BH, Gonda MA, Shaw GM, Popovic M, Hoxie JA, Gallo RC, Wong-Staal F (1985) Genomic diversity of the acquired immune deficiency syndrome virus HTLV-III: different viruses exhibit greatest divergence in their envelope genes. Proc Natl Acad Sci USA 82:4813–4817

Hahn BH, Shaw GM, Taylor ME, Redfield RR, Markham PD, Salahuddin SZ, Wong-Staal F, Gallo RC, Parks ES, Parks WP (1986) Genetic variation in HTLV-III/LAV over time in patients with AIDS or at risk for AIDS. Science 232:1548–1553

Halstead SB (1979) In vivo enhancement of dengue virus infection in rhesus monkeys by passively transferred antibody. J Infect Dis 140:527–533

Halstead SB (1970) Observations related to pathogensis of dengue hemorrhagic fever VI. Hypotheses and discussion. Yale J Biol Med 42:350–362

Halstead SB, Chow JS, Marchette NJ (1973) Immunological enhancement of dengue virus replication. Nature 243:24–26

Halstead SB, Nimmannitya S, Cohen SN (1970) Observations related to pathogenesis of dengue hemorrhagic fever IV. Relation of disease severity to antibody response and virus recovered. Yale J Biol Med 42:311–328

Halstead SB, Nimmannitya S, Margiotta MR (1969) Dengue and chikungunya virus infection in man in Thailand, 1962–1964 II. Observations on disease in outpatients. Am J Trop Med Hyg 18: 972–983

Halstead SB, O'Rourke EJ (1977) Antibody-enhanced dengue virus infection in primate leukocytes. Nature 265:739–741

Halstead SB, O'Rourke EJ (1977) Dengue viruses and mononuclear phagocytes I. Infection enhancement by non-neutralizing antibody. J Exp Med 146:201–217

Halstead SB, O'Rourke EJ, Allison AC (1977) Dengue viruses and mononuclear phagocytes II. Identity of blood and tissue leukocytes supporting in vitro infection. J Exp Med 146:218–229

Halstead SB, Scanlon JE, Umpaivit P, Udomsakdi S (1969) Dengue and chikungunya virus infection in man in Thailand, 1962–1964 IV. Epidemiologic studies in the Bangkok metropolitan area. Am J Trop Med Hyg 18:997–1021

Halstead SB, Shotwell H, Casals J (1973) Studies on the pathogenesis of dengue infection in monkeys II. Clinical laboratory responses to heterologous infection. J Infect Dis 128:15–22

Halstead SB, Venkateshan CN, Gentry MK, Larsen LK (1984) Heterogeneity of infection enhancement of dengue 2 strains by monoclonal antibodies. J Immunol 132:1529–1532

Harrer T, Harrer E, Kalams SA, Elbeik T, Staprans SI, Feinberg MB, Cao Y, Ho DD, Yilma T, Caliendo AM, Johnson RP, Buchbinder SP, Walker BD (1996) Strong cytotoxic T cell and weak neutralizing antibody responses in a subset of persons with stable nonprogressing HIV type 1 infection. AIDS Res Hum Retroviruses 12:585–592

Hawkes RA (1964) Enhancement of the infectivity of arboviruses by specific antisera produced in domestic fowls. Austr J Exp Biol Med Sci 42:465–482

Hawkes RA, Lafferty KJ (1967) The enhancement of virus infectivity by antibody. Virology 33:250–261

Helseth E, Kowalski M, Gabuzda D, Olshevsky U, Haseltine W, Sodroski J (1990) Rapid complementation assays measuring replicative potential of human immunodeficiency virus type 1 envelope glycoprotein mutants. J Virol 64:2416–2420

Helseth E, Olshevsky U, Furman C, Sodroski J (1991) Human immunodeficiency virus type 1 gp120 envelope glycoprotein regions important for association with the gp41 transmembrane glycoprotein. J Virol 65:2119–2123

Hidvegi T, Prohaszka Z, Ujhelyi E, Thielens NM, Dierich MP, Hampl H, Arlaud G, Nagy K, Fust G (1993) Studies on the mechanism of complement-mediated inhibition of antibody binding to HIV gp41. Clin Exp Immunol 94:490–493

Hohdatsu T, Nakamura M, Ishizuka Y, Yamada H, Koyama H (1991) A study on the mechanism of antibody-dependent enhancement of feline infectious peritonitis virus infection in feline macrophages by monoclonal antibodies. Arch Virol 120:207–217

Homsy J, Meyer M, Levy JA (1990) Serum enhancement of human immunodeficiency virus (HIV) infection correlates with disease in HIV-infected individuals. J Virol 64:1437–1440

Homsy J, Meyer M, Tateno M, Clarkson S, Levy JA (1989) The Fc and not CD4 receptor mediates antibody enhancement of HIV infection in human cells. Science 244:1357–1360

Jiang SB, Lin K, Neurath AR (1991) Enhancement of human immunodeficiency virus type 1 infection by antisera to peptides from the envelope glycoproteins gp120/gp41 [published erratum appears in J Exp Med 1992 Feb 1;175(2):621]. J Exp Med 174:1557–1563

Jolly PE, Huso D, Hart G, Narayan O (1989) Modulation of lentivirus replication by antibodies. Non-neutralizing antibodies to caprine arthritis-encephalitis virus enhance early stages of infection in macrophages, but do not cause increased production of virions. J Gen Virol 70:2221–2226

Jolly PE, Weiss HL (2000) Neutralization and enhancement of HIV-1 infection by sera from HIV-1 infected individuals who progress to disease at different rates. Virology 273:52–59

Jouault T, Chapuis F, Bahraoui E, Gluckman JC (1991) Infection of monocytic cells by HIV1: combined role of FcR and CD4. Res Virol 142:183–188

Kapikian AZ, Mitchell RH, Chanock RM, Shvedoff RA, Stewart CE (1969) An epidemiologic study of altered clinical reactivity to respiratory syncytial (RS) virus infection in children previously vaccinated with an inactivated RS virus vaccine. Am J Epidemiol 89:405–421

Kim HW, Canchola JG, Brandt CD, Pyles G, Chanock RM, Jensen K, Parrott RH (1969) Respiratory syncytial virus disease in infants despite prior administration of antigenic inactivated vaccine. Am J Epidemiol 89:422–434

Klein MR, van Baalen CA, Holwerda AM, Kerkhof Garde SR, Bende RJ, Keet IP, Eeftinck-Schattenkerk JK, Osterhaus AD, Schuitemaker H, Miedema F (1995) Kinetics of Gag-specific cytotoxic T lymphocyte responses during the clinical course of HIV-1 infection: a longitudinal analysis of rapid progressors and long-term asymptomatics. J Exp Med 181:1365–1372

Kliks SC, Halstead SB (1983) Role of antibodies and host cells in plaque enhancement of Murray Valley encephalitis virus. J Virol 46:394–404

Kliks SC, Nimmanitya S, Nisalak A, Burke DS (1988) Evidence that maternal dengue antibodies are important in the development of dengue hemorrhagic fever in infants. Am J Trop Med Hyg 38: 411–419

Kliks SC, Shioda T, Haigwood NL, Levy JA (1993) V3 variability can influence the ability of an antibody to neutralize or enhance infection by diverse strains of human immunodeficiency virus type 1. Proc Natl Acad Sci USA 90:11518–11522

Kontny U, Kurane I, Ennis FA (1988) Gamma interferon augments Fc gamma receptor-mediated dengue virus infection of human monocytic cells. J Virol 62:3928–3933

Kostrikis LG, Cao Y, Ngai H, Moore JP, Ho DD (1996) Quantitative analysis of serum neutralization of human immunodeficiency virus type 1 from subtypes A, B, C, D, E, F, and I: lack of direct correlation between neutralization serotypes and genetic subtypes and evidence for prevalent serum-dependent infectivity enhancement. J Virol 70:445–458

Kozak SL, Platt EJ, Madani N, Ferro FE, Jr., Peden K, Kabat D (1997) CD4, CXCR-4, and CCR-5 dependencies for infections by primary patient and laboratory-adapted isolates of human immunodeficiency virus type 1. J Virol 71:873–882

Legendre C, Gras G, Krzysiek R, Galanaud P, Richard Y, Dormont D (1996) Mechanisms of opsonized HIV entry in normal B lymphocytes. FEBS Lett 381:227–232

Lidbury BA, Mahalingam S (2000) Specific ablation of antiviral gene expression in macrophages by antibody-dependent enhancement of ross river virus infection. J Virol 74:8376–8381

Linn ML, Aaskov JG, Suhrbier A (1996) Antibody-dependent enhancement and persistence in macrophages of an arbovirus associated with arthritis. J Gen Virol 77:407–411

Littaua R, Kurane I, Ennis FA (1990) Human IgG Fc receptor II mediates antibody-dependent enhancement of dengue virus infection. J Immunol 144:3183–3186

Lund O, Hansen J, Soorensen AM, Mosekilde E, Nielsen JO, Hansen JE (1995) Increased adhesion as a mechanism of antibody-dependent and antibody-independent complement-mediated enhancement of human immunodeficiency virus infection. J Virol 69:2393–2400

Marchette NJ, Halstead SB, Chow JS (1976) Replication of dengue viruses in cultures of peripheral blood leukocytes from dengue-immune rhesus monkeys. J Infect Dis 133:274–282

Marchette NJ, Halstead SB, O'Rourke T, Scott RM, Bancroft WH, Vanopruks V (1979) Effect of immune status on dengue 2 virus replication in cultured leukocytes from infants and children. Infect Immun 24:47–50

Mascola JR, Lewis MG, Stiegler G, Harris D, VanCott TC, Hayes D, Louder MK, Brown CR, Sapan CV, Frankel SS, Lu Y, Robb ML, Katinger H, Birx DL (1999) Protection of Macaques against pathogenic simian/human immunodeficiency virus 89.6PD by passive transfer of neutralizing antibodies. J Virol 73:4009–4018

Mascola JR, Mathieson BJ, Zack PM, Walker MC, Halstead SB, Burke DS (1993) Summary report: workshop on the potential risks of antibody-dependent enhancement in human HIV vaccine trials. AIDS Res Hum Retroviruses 9:1175–1184

Mascola JR, Stiegler G, VanCott TC, Katinger H, Carpenter CB, Hanson CE, Beary H, Hayes D, Frankel SS, Birx DL, Lewis MG (2000) Protection of macaques against vaginal transmission of a pathogenic HIV-1/SIV chimeric virus by passive infusion of neutralizing antibodies. Nat Med 6:207–210

Matsuda S, Gidlund M, Chiodi F, Cafaro A, Nygren A, Morein B, Nilsson K, Fenyo EM, Wigzell H (1989) Enhancement of human immunodeficiency virus (HIV) replication in human monocytes by low titres of anti-HIV antibodies in vitro. Scand J Immunol 30:425–434

Mondor I, Ugolini S, Sattentau QJ (1998) Human immunodeficiency virus type 1 attachment to HeLa CD4 cells is CD4 independent and gp120 dependent and requires cell surface heparans. J Virol 72:3623–3634

Montefiori DC, Lefkowitz LB, Jr., Keller RE, Holmberg V, Sandstrom E, Phair JP (1991) Absence of a clinical correlation for complement-mediated, infection-enhancing antibodies in plasma or sera from HIV-1-infected individuals. Multicenter AIDS Cohort Study Group. AIDS 5:513–517

Montefiori DC, Pantaleo G, Fink LM, Zhou JT, Zhou JY, Bilska M, Miralles GD, Fauci AS (1996) Neutralizing and infection-enhancing antibody responses to human immunodeficiency virus type 1 in long-term nonprogressors. J Infect Dis 173:60–67

Montefiori DC, Robinson WE, Jr., Hirsch VM, Modliszewski A, Mitchell WM, Johnson PR (1990) Antibody-dependent enhancement of simian immunodeficiency virus (SIV) infection in vitro by plasma from SIV-infected rhesus macaques. J Virol 64:113–119

Moore JP, Cao Y, Leu J, Qin L, Korber B, Ho DD (1996) Inter- and intraclade neutralization of human immunodeficiency virus type 1: genetic clades do not correspond to neutralization serotypes but partially correspond to gp120 antigenic serotypes. J Virol 70:427–444

Moore JP, Cao Y, Qing L, Sattentau QJ, Pyati J, Koduri R, Robinson J, Barbas CF, 3rd, Burton DR, Ho DD (1995) Primary isolates of human immunodeficiency virus type 1 are relatively resistant to neutralization by monoclonal antibodies to gp120, and their neutralization is not predicted by studies with monomeric gp120. J Virol 69:101–109

Morens DM (1994) Antibody-dependent enhancement of infection and the pathogenesis of viral disease. Clin Infect Dis 19:500–512

Morens DM, Halstead SB (1987) Disease severity-related antigenic differences in dengue 2 strains detected by dengue 4 monoclonal antibodies. J Med Virol 22:169–174

Morens DM, Halstead SB, Marchette NJ (1987) Profiles of antibody-dependent enhancement of dengue virus type 2 infection. Microb Pathog 3:231–237

Morens DM, Larsen LK, Halstead SB (1987) Study of the distribution of antibody-dependent enhancement determinants on dengue 2 isolates using dengue 2-derived monoclonal antibodies. J Med Virol 22:163–167

Nimmannitya S, Halstead SB, Cohen SN, Margiotta MR (1969) Dengue and chikungunya virus infection in man in Thailand, 1962–1964. I. Observations on hospitalized patients with hemorrhagic fever. Am J Trop Med Hyg 18:954–971

Ochiai H, Kurokawa M, Matsui S, Yamamoto T, Kuroki Y, Kishimoto C, Shiraki K (1992) Infection enhancement of influenza A NWS virus in primary murine macrophages by anti-hemagglutinin monoclonal antibody. J Med Virol 36:217–221

Olsen CW, Corapi WV, Ngichabe CK, Baines JD, Scott FW (1992) Monoclonal antibodies to the spike protein of feline infectious peritonitis virus mediate antibody-dependent enhancement of infection of feline macrophages. J Virol 66:956–965

O'Sullivan MA, Killen HM (1994) The differentiation state of monocytic cells affects their susceptibility to infection and the effects of infection by dengue virus. J Gen Virol 75:2387–2392

Parren PW, Mondor I, Naniche D, Ditzel HJ, Klasse PJ, Burton DR, Sattentau QJ (1998) Neutralization of human immunodeficiency virus type 1 by antibody to gp120 is determined primarily by occupancy of sites on the virion irrespective of epitope specificity. J Virol 72:3512–3519

Parren PW, Poignard P, Ditzel HJ, Williamson RA, Burton DR (2000) Antibodies in human infectious disease. Immunol Res 21:265–278

Peiris JS, Porterfield JS (1981) Antibody-dependent enhancement of plaque formation on cell lines of macrophage origin – a sensitive assay for antiviral antibody. J Gen Virol 57:119–125

Peiris JS, Porterfield JS (1979) Antibody-mediated enhancement of Flavivirus replication in macrophage-like cell lines. Nature 282:509–511

Perno CF, Baseler MW, Broder S, Yarchoan R (1990) Infection of monocytes by human immunodeficiency virus type 1 blocked by inhibitors of CD4-gp120 binding, even in the presence of enhancing antibodies. J Exp Med 171:1043–1056

Platt EJ, Wehrly K, Kuhmann SE, Chesebro B, Kabat D (1998) Effects of CCR5 and CD4 cell surface concentrations on infections by macrophagetropic isolates of human immunodeficiency virus type 1. J Virol 72:2855–2864

Prabhakar BS, Nathanson N (1981) Acute rabies death mediated by antibody. Nature 290:590–591

Prohaszka Z, Hidvegi T, Ujhelyi E, Stoiber H, Dierich MP, Susal C, Fust G (1995) Interaction of complement and specific antibodies with the external glycoprotein 120 of HIV-1. Immunology 85:184–189

Prohaszka Z, Nemes J, Hidvegi T, Toth FD, Kerekes K, Erdei A, Szabo J, Ujhelyi E, Thielens N, Dierich MP, Spath P, Ghebrehiwet B, Hampl H, Kiss J, Arlaud G, Fust G (1997) Two parallel routes of the complement-mediated antibody-dependent enhancement of HIV-1 infection. AIDS 11:949–958

Quinkal I, Hernandez JF, Chevallier S, Arlaud GJ, Vernet T (1999) Mapping of the interaction between the immunodominant loop of the ectodomain of HIV-1 gp41 and human complement protein C1q. Eur J Biochem 265:656–663

Reisinger EC, Vogetseder W, Berzow D, Kofler D, Bitterlich G, Lehr HA, Wachter H, Dierich MP (1990) Complement-mediated enhancement of HIV-1 infection of the monoblastoid cell line U937. AIDS 4:961–965

Riviere Y, McChesney MB, Porrot F, Tanneau-Salvadori F, Sansonetti P, Lopez O, Pialoux G, Feuillie V, Mollereau M, Chamaret S, et al. (1995) Gag-specific cytotoxic responses to HIV type 1 are associated with a decreased risk of progression to AIDS-related complex or AIDS. AIDS Res Hum Retroviruses 11:903–907

Rizzuto C, Sodroski J (2000) Fine definition of a conserved CCR5-binding region on the human immunodeficiency virus type 1 glycoprotein 120. AIDS Res Hum Retroviruses 16:741–749

Rizzuto CD, Sodroski JG (1997) Contribution of virion ICAM-1 to human immunodeficiency virus infectivity and sensitivity to neutralization. J Virol 71:4847–4851

Robinson WE, Jr., Gorny MK, Xu JY, Mitchell WM, Zolla-Pazner S (1991) Two immunodominant domains of gp41 bind antibodies which enhance human immunodeficiency virus type 1 infection in vitro. J Virol 65:4169–4176

Robinson WE, Jr., Montefiori DC, Gillespie DH, Mitchell WM (1989) Complement-mediated, antibody-dependent enhancement of HIV-1 infection in vitro is characterized by increased protein and RNA syntheses and infectious virus release. J Acquir Immune Defic Syndr 2:33–42

Robinson WE, Jr., Montefiori DC, Mitchell WM (1988) Antibody-dependent enhancement of human immunodeficiency virus type 1 infection. Lancet 1:790–794

Robinson WE, Jr., Montefiori DC, Mitchell WM (1990) Complement-mediated antibody-dependent enhancement of HIV-1 infection requires CD4 and complement receptors. Virology 175:600–604

Robinson WE, Jr., Montefiori DC, Mitchell WM, Prince AM, Alter HJ, Dreesman GR, Eichberg JW (1989) Antibody-dependent enhancement of human immunodeficiency virus type 1 (HIV-1) infection in vitro by serum from HIV-1-infected and passively immunized chimpanzees. Proc Natl Acad Sci USA 86:4710–4714

Rosen L (1977) The Emperor's New Clothes revisited, or reflections on the pathogenesis of dengue hemorrhagic fever. Am J Trop Med Hyg 26:337–343

Rothman AL, Ennis FA (1999) Immunopathogenesis of Dengue hemorrhagic fever. Virology 257:1–6

Russell PK, Udomsakdi S, Halstead SB (1967) Antibody response in dengue and dengue hemorrhagic fever. Jpn J Med Sci Biol 20:103–108

Russell PK, Yuill TM, Nisalak A, Udomsakdi S, Gould DJ, Winter PE (1968) An insular outbreak of dengue hemorrhagic fever. II. Virologic and serologic studies. Am J Trop Med Hyg 17:600–608

Sangkawibha N, Rojanasuphot S, Ahandrik S, Viriyapongse S, Jatanasen S, Salitul V, Phanthumachinda B, Halstead SB (1984) Risk factors in dengue shock syndrome: a prospective epidemiologic study in Rayong, Thailand. I. The 1980 outbreak. Am J Epidemiol 120:653–669

Sattentau QJ, Moore JP (1995) Human immunodeficiency virus type 1 neutralization is determined by epitope exposure on the gp120 oligomer. J Exp Med 182:185–196

Schlesinger JJ, Brandriss MW (1981) Antibody-mediated infection of macrophages and macrophage-like cell lines with 17D-yellow fever virus. J Med Virol 8:103–117

Schmitz JE, Lifton MA, Reimann KA, Montefiori DC, Shen L, Racz P, Tenner-Racz K, Ollert MW, Forman MA, Gelman RS, Vogel CW, Letvin NL (1999) Effect of complement consumption by cobra venom factor on the the course of primary infection with simian immunodeficiency virus in rhesus monkeys. AIDS Res Hum Retroviruses 15:195–202

Schutten M, Andeweg AC, Bosch ML, Osterhaus AD (1995) Enhancement of infectivity of a non-syncytium inducing HIV-1 by sCD4 and by human antibodies that neutralize syncytium inducing HIV-1. Scand J Immunol 41:18–22

Schutten M, Andeweg AC, Rimmelzwaan GF, Osterhaus AD (1997) Modulation of primary human immunodeficiency virus type 1 envelope glycoprotein-mediated entry by human antibodies. J Gen Virol 78:999–1006

Se-Thoe SY, Ng MM, Ling AE (1999) Retrospective study of Western blot profiles in immune sera of natural dengue virus infections. J Med Virol 57:322–330

Sikes RK, Peacock GV, Acha P, Arko RJ, Dierks R (1971) Rabies vaccines: duration-of-immunity study in dogs. J Am Vet Med Assoc 159:1491–1499

Stoiber H, Schneider R, Janatova J, Dierich MP (1995) Human complement proteins C3b, C4b, factor H and properdin react with specific sites in gp120 and gp41, the envelope proteins of HIV-1. Immunobiology 193:98–113

Sullivan N, Sun Y, Binley J, Lee J, Barbas CF, 3rd, Parren PW, Burton DR, Sodroski J (1998) Determinants of human immunodeficiency virus type 1 envelope glycoprotein activation by soluble CD4 and monoclonal antibodies. J Virol 72:6332–6338

Sullivan N, Sun Y, Li J, Hofmann W, Sodroski J (1995) Replicative function and neutralization sensitivity of envelope glycoproteins from primary and T-cell line-passaged human immunodeficiency virus type 1 isolates. J Virol 69:4413–4422

Sullivan N, Sun Y, Sattentau Q, Thali M, Wu D, Denisova G, Gershoni J, Robinson J, Moore J, Sodroski J (1998) CD4-Induced conformational changes in the human immunodeficiency virus type 1 gp120 glycoprotein: consequences for virus entry and neutralization. J Virol 72:4694–4703

Tacnet-Delorme P, Boyer V, Thielens NM, Hernandez JF, Bally I, Sim RB, Desgranges C, Arlaud GJ (1999) In vitro analysis of complement-dependent HIV-1 cell infection using a model system. J Immunol 162:4088–4093

Takeda A, Robinson JE, Ho DD, Debouck C, Haigwood NL, Ennis FA (1992) Distinction of human immunodeficiency virus type 1 neutralization and infection enhancement by human monoclonal antibodies to glycoprotein 120. J Clin Invest 89:1952–1957

Takeda A, Sweet RW, Ennis FA (1990) Two receptors are required for antibody-dependent enhancement of human immunodeficiency virus type 1 infection: CD4 and Fc$\gamma$ R. J Virol 64:5605–5610

Takeda A, Tuazon CU, Ennis FA (1988) Antibody-enhanced infection by HIV-1 via Fc receptor-mediated entry. Science 242:580–583

Thielens NM, Bally IM, Ebenbichler CF, Dierich MP, Arlaud GJ (1993) Further characterization of the interaction between the C1q subcomponent of human C1 and the transmembrane envelope glycoprotein gp41 of HIV-1. J Immunol 151:6583–6592

Vaughn DW, Green S, Kalayanarooj S, Innis BL, Nimmannitya S, Suntayakorn S, Endy TP, Raengsakulrach B, Rothman AL, Ennis FA, Nisalak A (2000) Dengue viremia titer, antibody response pattern, and virus serotype correlate with disease severity. J Infect Dis 181:2–9

Wang SZ, Rushlow KE, Issel CJ, Cook RF, Cook SJ, Raabe ML, Chong YH, Costa L, Montelaro RC (1994) Enhancement of EIAV replication and disease by immunization with a baculovirus-expressed recombinant envelope surface glycoprotein. Virology 199:247–251

Weber JN, Clapham PR, Weiss RA, Parker D, Roberts C, Duncan J, Weller I, Carne C, Tedder RS, Pinching AJ, et al. (1987) Human immunodeficiency virus infection in two cohorts of homosexual men: neutralising sera and association of anti-gag antibody with prognosis. Lancet 1:119–122

Winter PE, Nantapanich S, Nisalak A, Udomsakdi S, Dewey RW, Russell PK (1969) Recurrence of epidemic dengue hemorrhagic fever in an insular setting. Am J Trop Med Hyg 18:573–579

Winter PE, Yuill TM, Udomsakdi S, Gould D, Nantapanich S, Russell PK (1968) An insular outbreak of dengue hemorrhagic fever. I. Epidemiologic observations. Am J Trop Med Hyg 17:590–599

Wu L, Gerard NP, Wyatt R, Choe H, Parolin C, Ruffing N, Borsetti A, Cardoso AA, Desjardin E, Newman W, Gerard C, Sodroski J (1996) CD4-induced interaction of primary HIV-1 gp120 glycoproteins with the chemokine receptor CCR-5. Nature 384:179–183

# The Role of the Antibody Response in Influenza Virus Infection

W. GERHARD



## 1 Introduction

Many components of the host's innate (STAEHELI 1990; GARCIA-SASTRE et al. 1998; RIMMELZWAAN et al. 1999; READING et al. 1997; SARANEWA et al. 1998) and adaptive (reviewed in CROWE 1999; DOHERTY et al. 1997; GERHARD et al. 1997) defense have been shown to participate in the control of influenza virus infection. However, the MHC class-I-restricted $CD8^+$ cytotoxic T cell (Tc) response is still considered by many investigators to be of paramount importance for clearing an established influenza virus infection in the immunologically intact host. Support for this has come from studies showing that: (1) a vigorous virus-specific Tc response can be detected in the infected lung, (2) the rise of this response coincides with virus clearance, and (3) adoptive transfer of activated Tc decreased virus titers in the infected recipient. To some extent, it may also have been inferred from observations

The Wistar Institute, 3601 Spruce Street, Philadelphia, PA 19104-4268, USA made in infections by non-cytotoxic viruses like lymphocytic choriomeningitis virus (Oldstone et al. 1986; Moskophidis et al. 1987), in which Tc were found to be essential, both for clearance and immune-mediated pathology. However, in the case of influenza virus, infection itself results in the eventual death of the host cell (Mori et al. 1995; Balachandran et al. 2000). In addition, it spreads solely by virus released into extracellular fluid that may contain virus-neutralizing antibody. Therefore, an efficient mechanism for killing infected cells may be dispensable for the control of influenza virus infection. This has been confirmed by showing that Tc-deficient β2m(−/−) mice (Eichelberger et al. 1991a; Bender et al. 1992) and athymic (nu/nu) mice after adoptive transfer of $CD4^+$ virus-specific T cell (Th) clones (Scherle et al. 1992) could indeed recover from moderately severe infections. Subsequent studies indicated that recovery in the absence of Tc was dependent on the concomitant presence of Th and B cells (Mozdzanowska et al. 1997b; Topham and Doherty 1998), suggesting that the Th-dependent antibody response played a significant role in recovery. In the following, evidence regarding the importance and mode of action of B cells and antiviral antibodies in the control of this infection in the naive and immune host is reviewed.

## 2 Course of Infection and Immune Response

Epithelial cells along the entire airway appear to support a productive infection cycle. In the mouse model, an initial nasal infection can be produced by administration of a small inoculum (2–3μl per nare) to anesthetized mice or a larger inoculum to non-anesthetized mice. Over the next several days, the infection may then descend into the lower respiratory tract (RT), although this descent is sufficiently slow with some virus strains that it can be prevented by the host's immune defense. An initial total RT infection is produced if a large bolus inoculum (20–50μl) is administered to the nares of anesthetized mice or if non-anesthetized mice are exposed to a small-particle aerosol. Total RT infection with 10–100 $MID_{50}$ (50% mouse infectious dose) of mouse-adapted viruses results in peak virus titers of $\sim 10^6$–$10^7$ $TCID_{50}$ per lung within 2–3 days. By day 5, virus titers begin to decline and usually become undetectable within the next 2–4 days. In mammals, productive infections with human viruses of H1, H2 and H3 subtypes remain confined to the airway epithelium, in part because cleavage of these hemagglutinin (HA) subtypes is dependent on locally secreted enzymes (Kido et al. 1992), although viral genetic material can be detected by PCR at early time points of infection in extrapulmonary sites (Eichelberger et al. 1991b, Mori et al. 1995). There are exceptions to this, as shown by some virus strains of H5 subtype, which were capable of producing productive infections also in extrapulmonary tissues (Gubareva et al. 1998; Lu et al. 1999).

Infection induces a very rapid response in draining lymph nodes (LN). After total RT infection, the cellularity of mediastinal LN may increase, within 3 days, by 10- to 20-fold, from ~0.5 million to 6–12 million, and may peak by day 6 to day 7 at 15–20 million (Allen et al. 1990; our unpublished observations). Antibody-secreting cells (ASC) become detectable in draining LN between days 4 and 5 after infection and reach peak numbers of a few thousand ASC per $10^6$ cells between days 7 and 10 (Harling-McNabb et al. 1999; Justewicz et al. 1995; Marshall et al. 1999; Tamura et al. 1998) or a few days later (Fazekas et al. 1994). Antibodies of various heavy chain isotypes are generated, IgM usually dominating early and IgG and IgA at later time points of the response (Harling-McNabb et al. 1999; Marshall et al. 1999). Apart from a fraction of the IgM secreting ASC, the response in draining LN appears to be largely Th-dependent (Justewicz et al. 1995). ASC in the spleen arise with a few days delay relative to the LN response and are dominated by the IgM isotype (Fazekas et al. 1994; Sangster et al. 1995). ASC in draining LN and the spleen appear to produce the majority of the antiviral antibodies in serum, where a significant rise in titer is usually seen between 5 and 7 days after initiation of infection (Palladino et al. 1995).

Serum antibody may contribute to the control of the infection after transudation into the interstitial extravascular space of the RT, from where it may further transudate through basal membrane and intercellular spaces into RT fluid or, in the case of polymeric IgA and IgM, may be actively transported through epithelial cells by polymeric Ig-receptors (pIgR) (Brandtzaeg 1989; Lamm 1997; Corthesy and Kraehenbuhl 1999). The efficacy of these processes differs along the airway, pIgR-mediated transcytosis being more effective in upper than lower RT, and transudation more effective in lower than upper RT (Kaltreider 1976; Tamura et al. 1991). In the upper RT, active transport of dimeric IgA has been estimated to be 10–100 times more effective than passive transudation of IgG (Renegar and Small 1991b). In the lower RT, transudation has been estimated to result in a serum:RT concentration gradient of ~200:1 during an ongoing respiratory syncytial virus infection (Prince et al. 1987), but it could be several-fold larger in the non-infected lung. In this context, it is important to note that the rate of antibody transport into RT secretions may change greatly in the course of an infection, as transcytosis can be enhanced by cytokines (Kvale et al. 1988; Sollid et al. 1987) and transudation by inflammation-induced increases in vascular permeability (Owens et al. 1981). The difference in the rates of transcytosis and transudation along the RT underlies the finding that serum antibody provides stronger protection against pulmonary than tracheal or nasal infection (Ramphal et al. 1979; Gerhard et al. 1997), while secretory IgA is the dominant protective antibody in the nasal cavity (Renegar and Small 1991a).

In distinction to the systemic response, ASC present in the submucosa of the airways give rise to the local antibody response (Brandtzaeg et al. 1999; Corthesy and Kraehenbuhl 1999). ASC become detectable in RT-associated tissue 5–7 days after infection and the response reaches peak frequencies of $10^2$–$10^3$ ASC per million cells by day 10 to day 20 (Jones and Ada 1986; Fazekas et al. 1994; Tamura et al. 1998; Harling-McNabb et al. 1999). IgA- and IgM-secreting ASC usually dominate the early response. Some of these ASC may have been recruited from the systemic response while some may have been induced locally, perhaps in a Th-independent fashion. Significant local induction appears to occur in the more organized naso-pharyngeal lymphoid tissue (TAMURA et al. 1998). A large part of the locally produced antibodies may find their way into RT secretions by transudation and transcytosis.

Antibody responses have also been studied by measuring the antibody content in nasal and bronchoalveolar lavage (BAL) fluid. In spite of the fact that RT secretions become greatly diluted by wash fluid upon sampling, a rise in antiviral antibody content has been detected 5–8 days after infection (OWENS et al. 1981; TAMURA et al. 1998) and has been shown to correlate with the presence of opsonized virus in wash fluid (TAMURA et al. 1998).

Taken together, the kinetics of the primary antibody response is consistent with its participation in the control of infection in the naive immunocompetent host. Of course, infection also induces a vigorous virus-specific Tc response that coincides with virus clearance (DOHERTY et al. 1997).

# 3 Role of B Cells in Recovery from Primary Infection

Although several studies have shown that infections in immuno-incompetent mice could be cured by treatment with HA-specific antibodies (VIRELIZIER et al. 1975; SCHERLE et al. 1992; PALLADINO et al. 1995), this was not consistently observed (KRIS et al. 1988). In addition, it was difficult to determine whether the passive antibody treatment mimicked the active response and therefore whether the active response would display the same efficacy. Less ambiguous information was obtained by comparing the course of infection in mice that differed in distinct components of the host defense (Table 1).

## 3.1 Infection in B-Cell-Deficient Mice

Initial experiments were performed with mice whose B cell responses had been suppressed by chronic treatment with IgM-specific antisera. One of these studies showed that such mice could not recover from a moderately severe total RT challenge with A/RR/8/34/H1N1 (PR8) (IWASAKI and NOZIMA 1977), while the second study reported complete but delayed recovery from a less severe nasal infection with the less pathogenic A/Port Chalmers/1/73/H3N2 (PC) virus (KRIS et al. 1985). More recent studies, performed with genetically B-cell-deficient μMT mice, showed that they succumbed to a ~100-fold smaller challenge with A/Japan/305/57/H2N2 (JAP) (GRAHAM and BRACIALE 1997) and a 10- to 30-fold smaller challenge with PR8 than intact mice (GERHARD et al. 1997). In surviving mice, clearance of PR8 virus was delayed by several days (MOZDZANOWSKA et al. 1997b). On the other hand, B-cell-deficient mice were found to be capable of clearing

Table 1. Role of distinct components of host defense in recovery from primary influenza virus infection

| Host defense | | | | Recovery from total RT infection with PR8 | | |
|---|---|---|---|---|---|---|
| Innate | $CD8^+$ | $CD4^+$ | B | 50–200 $TCID_{50}$[a] | | $10^7$ $TCID_{50}$[b] |
| | | | | Clearance (days)[c] | Survival (%)[d] | Survival (%)[d] |
| + | + | + | + | 7–10 | 100 | 80 |
| + | − | + | + | 10–14 | 100 | 10* |
| + | − | + | − | > 20 | 0* | |
| + | − | − | + | > 20 | 0* | |
| + | + | + | − | 10–14 | 35–85 | |
| + | + | − | − | > 14 | 20* | |
| + | + | − | + | 10–14 | 90 | |
| + | − | − | − | > 20 | 0* | |

RT, respiratory tract; $TCID_{50}$, 50% mouse infectious dose.
[a] Data from Scherle et al. 1992; Gerhard et al. 1997; Mozdzanowska et al. 1997, 2000.
[b] Data from Bender et al. 1992.
[c] Days after infection at which virus titers in the lung became undetectable in most mice by tissue culture assay ($< 10^{2.1}$ $TCID_{50}$) or egg inoculation ($< 10^{1.3}$ egg infectious doses).
[d] The asterisk (*) indicates significant difference in survival compared to intact mice.

infections with the less pathogenic viruses B/Ann Arbor (Epstein et al. 1998) and A/X31/H3N2 (X31) (Topham et al. 1996, 1998) with no or minimal delay. Thus, it appeared that the requirement for B cells depended on the severity of the virus challenge, in that B cell deficiency significantly impaired the control of relatively severe infections with pathogenic viruses like PR8 and JAP but had little impact on the control of infections with virus strains of low pathogenicity. Furthermore, as neither virus-specific Th (Gerhard et al. 1997; Topham et al. 1998) nor Tc responses (Topham et al. 1996; Mozdzanowska et al. 2000) were measurably diminished in μMT compared to intact mice, these studies indicated that B cells contributed directly by secretion of antibody or other factors, rather than indirectly through stimulation of Tc and Th responses.

## 3.2 Evidence that Recovery Can be Achieved in the Absence of Tc

Indirect evidence for a role of the B cell response in the resolution of an active infection was provided by studies showing that β2m(−/−) mice, which cannot generate significant antiviral Tc responses, were capable of resolving infection with viruses of low pathogenicity without significant delay (Eichelberger et al. 1991; Bender et al. 1992). However, as shown in Table 1, such mice showed increased mortality when subjected to a severe challenge with PR8 (Bender et al. 1992). Similarly, athymic nude (nu/nu) mice that had been reconstituted with virus-specific Th clones were capable of clearing infections with a low dose of PR8 (Scherle et al. 1992). Thus, Tc-mediated activities were not an absolute requirement for the control and resolution of an influenza virus infection and could be substituted quite effectively by Th- and B-cell-dependent activities.

Additional experiments showed that mice which possessed only B cells (Wells et al. 1981; Scherle et al. 1992; Doherty et al. 1997) or only Th cells (Mozdzanowska et al. 1997b; Scherle et al. 1992; Topham and Doherty 1998; Epstein et al. 1998) could not resolve infections. Thus, virus clearance in the absence of Tc appeared to be dependent on a collaborative activity between Th and B cells, presumably the promotion of a Th-dependent antiviral antibody response, which was consistent with the finding that established infections in athymic nude (Virelizier 1975) and SCID (Palladino et al. 1995) mice could be cured by treatment with HA-specific antibodies. In addition, Th and B cells appeared to also make non-collaborative contributions. In the case of Th cells, this was supported by the finding that virus clearance in β2m (−/−) mice was delayed if the mice were treated with anti-IFN-γ antibody (Sarawar et al. 1994), and the observation that Tc- and B-cell-deficient mice occasionally managed to clear infection with X31 virus (Topham and Doherty 1998). Similarly, a Th-independent contribution of still unknown nature by B cells was indicated by the finding (see below) that transfer of naïve B cells into Th-cell-depleted μMT mice greatly improved their recovery from infection (Mozdzanowska et al. 2000).

### 3.3 Poor Control of the Infection by Tc in the Absence of Th and B Cells

Recent studies performed with mice that have only innate defense and Tc cells available for controlling the infection have provided ambiguous data. On the one hand, clearance of B/Ann Arbor (B/AA) from the lung of Th-depleted B-cell-deficient [$J_H$(−/−) and Cκ (−/−)] mice of C57BL/6 background was only minimally delayed compared to intact mice (Epstein et al. 1998). On the other hand, clearance of the apathogenic X31 virus from lungs of Th-depleted μMT mice of BALB/c background was delayed by ~5 days and clearance of the same virus from nasal epithelium by ~15 days (Mozdzanowska et al. 2000). Upon challenge with the pathogenic PR8 virus, the mice exhibited progressive and severe morbidity. The outcome of the PR8 infection was usually (80%) fatal in spite of the fact that the infection dose was very small (~10 $MID_{50}$) and resulted only in minimal morbidity and no mortality in immunologically intact mice (Mozdzanowska et al. 2000). Importantly, the Th-depleted μMT mice made a vigorous virus-specific Tc response that showed no measurable difference from the Tc response of intact mice (Mozdzanowska et al. 2000). Thus, a virus-specific Tc response of normal size and composition cleared non-pathogenic virus infections with variable delay, but was hardly capable of controlling the more pathogenic PR8 virus infection. Importantly, adoptive transfer of ~$10^7$ T-cell-depleted spleen cells from naive mice into Th-depleted μMT mice 1 day before virus challenge greatly improved the ability of the mice to control and recover from PR8 infection. Since the Th-cell-depleted B-cell-reconstituted mice did not develop measurable Th-dependent antiviral antibody responses, it appears that B cells contributed to virus control in this situation by a Th-independent activity. "Natural antibodies" and Th-independent antibody responses have been shown in other virus systems to participate in virus control (Szomolanyi-Tsuda and Welsh 1996; Bachmann and Zinkernagel 1997; Ochsenbein et al. 1999), but it is presently not certain that the naive B cells contributed to influenza virus control through one of these mechanisms.

In summary, it is clear that many components of the host's defense contribute to the control of the influenza virus infection. Amongst these are Th-dependent and Th-independent contributions by B cells that appear to be highly relevant for recovery of the intact host from primary influenza virus infection.

## 4 Role of the Memory B Cell Response in Protection of the Immune Host Against Infection

The crucial role played by the memory B cell response in protection of the immune host is well accepted and forms the basis of current, inactivated whole-virus and subunit vaccines (for review see Couch and Kasel 1983; Ada and Jones 1986; Murphy and Clements 1989; Murphy and Webster 1996). Protection can be attributed primarily to antibodies specific for the HA molecule (see also below) which, when present in high enough concentration, may provide "sterilizing immunity", i.e. prevent initiation of infection. Neuraminidase (NA)-specific immunity is also antibody-mediated (Epstein et al. 1993), but is generally weaker than HA-specific immunity and is not consistently seen. This immunity cannot prevent infection, but may suppress subsequent virus replication and reduce morbidity and mortality (Johansson et al. 1989; Murphy and Clements 1989). In agreement with the long duration of the memory B cell response (Manz et al. 1998; Slifka et al. 1998), this antibody-mediated protection is long-lasting. For example, BALB/c mice still possessed sterilizing immunity against PR8 challenge 15 months after convalescence from PR8 infection. Similarly, the finding that A/USSR/77/H1N1 virus showed decreased attack rates in humans who may have experienced infection with a similar or identical virus strain 20–25 years previously attests to the potency and long duration of this immunity in humans (Couch and Kasel 1983). The main shortcoming of HA- and NA-specific immunity is a progressive decrease of its protective efficacy against virus strains that exhibit increasing antigenic drift in HA and NA relative to the immunizing strain (Fig. 1). In humans, immunity has been shown to be significant against strains that evolved over the subsequent 5–7 years and included two to three consecutive, epidemiologically important strains (Couch and Kasel 1983). In animal studies, differences in sequence of up to 15% between immunizing and challenge HA were still compatible with significant protection (Chambers et al. 1988; Hunt et al. 1988; Kodihalli et al. 1999). Although cross-protection is remarkably strong, constant adjustment of the vaccine strains to circulating epidemic strains is nevertheless important for optimal protection.

Immunity to the third viral transmembrane protein, matrix protein 2 (M2), (see also below), has been less extensively studied. Initial experiments performed with

M2-expressing recombinant vaccinia viruses provided no evidence for protection (Epstein et al. 1993; Jakeman et al. 1989), although passive antibody treatment was protective (Treanor et al. 1990; Mozdzanowska et al. 1999). More recent studies showed that vaccination of mice with recombinant M2 proteins provided some measure of protection, which was evidenced by reduced morbidity and mortality upon virus challenge (Slepushkin et al. 1995; Frace et al. 1999; Neirynck et al. 1999). The protective immunity appeared to be antibody-mediated as it could be passively transferred with immune serum (Neirynck et al. 1999) and, consistent with the sequence homology of the M2-ectodomains of human virus strains (Zebedee and Lamb 1988), was effective against viruses of different subtypes (Frace et al. 1999; Neirynck et al. 1999) (Fig. 1). M2-specific antibodies were not consistently detected in human convalescent sera (Black et al. 1993), indicating that this immunity is not or inconsistently operative in humans (which may, in part, explain the high sequence homology of M2 ectodomains). Convalescent BALB/c mouse sera contained low M2-specific antibody titers ($3 \pm 0.8 \mu g/ml$; our unpublished observation), which may be responsible for the antibody-mediated component of heterosubtypic protection seen in the mouse model (Liang et al. 1994; Gerhard et al. 1996).

Convalescence from infection results also in a strong humoral immunity to nucleoprotein (NP), but there is no evidence that NP-specific antibodies, or anti-

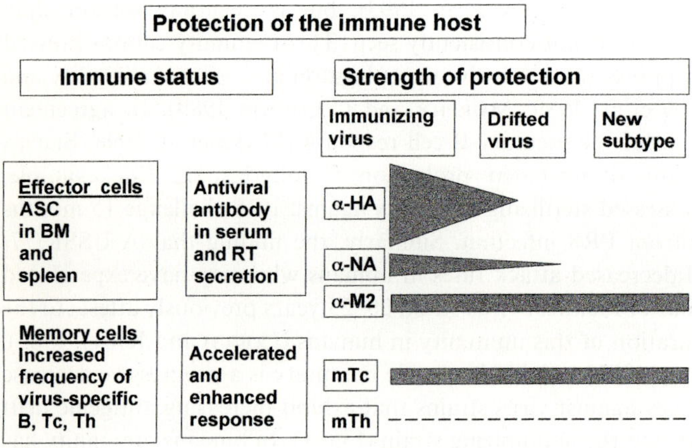


**Fig. 1.** The immune host exhibits increased resistance to infection which is due to extant transmembrane protein-specific antibodies and an increased frequency of memory T (*mT*) and B cells. Hemagglutinin (*HA*)-specific immunity provides the strongest protection. It is based mainly on virus neutralization (VN) activity, and to lesser degree yield reduction (YR) activity, and can provide sterilizing immunity, but is sensitive to antigenic drift and shift occurring in influenza virus in nature. Neuraminidase (*NA*)- and matrix protein 2 (*M2*)-specific immunity are based on antibody-mediated YR activity and are infection permissive. NA-specific immunity is sensitive to antigenic evolution occurring in influenza virus in nature, while M2-specific immunity appears to be more drift/shift-resistant, similar to mTc-mediated protection. However, M2-specific antibody-mediated protection would likely persist longer than mTc-mediated protection. *ASC*, antibody-secreting cells; *Tc*, cytotoxic T cells; *Th*, $CD4^+$ virus-specific T cells; *BM*, bone marrow; *RT*, respiratory tract

bodies to other nucleocapsid proteins, have any protective activity in vivo (VIRELIZIER et al. 1976; ASKONAS et al. 1982; STITZ et al. 1990; MOZDZANOWSKA et al. 1999). However, these proteins provide the dominant MHC class-I-restricted determinants (BENNINK and YEWDELL 1988), and NP-specific memory Tc have been shown to accelerate virus clearance and reduce morbidity and mortality upon virus challenge due to their increased frequency and improved responsiveness (DOHERTY et al. 1997). In general, memory T-cell-mediated immunity is always infection-permissive (as the memory T cells must be reactivated by infection before they can become active) and weaker and more easily overwhelmed by a severe virus challenge than antibody-mediated protection. In addition, protection provided by memory T cells appears to be of shorter duration than antibody-mediated protection (TANNOCK and PAUL 1987, WEBSTER and ASKONAS 1980; LIANG et al. 1994).

## 5 Virus Replication and Susceptibility to Antibody-Mediated Control In Vitro

The infectious virus particle (for review see LAMB and KRUG 1996; MURPHY and WEBSTER 1996) consists of a lipid envelope, which encloses the viral genome, in association with several viral proteins involved in its intracellular transport and transcription. On its outside, it displays multiple copies of three transmembrane proteins, HA, NA, and M2. HA is a homotrimer with a large ectodomain (3 × ~510 amino acids) which forms rod-like structures that are ~135Å long and 15–40Å wide and project radially from the viral surface (WILEY and SKEHEL 1987; WILSON and COX 1990). The distal globular domain of each trimer contains three to six binding sites that are specific for terminal sialic acid residues on glycosylated cell components, which serve as cellular ligands for virus attachment (SAUTER et al. 1992). A post-translational cleavage of HA in its stem region by host-derived enzymes generates a membrane-proximal C-terminal HA2 polypeptide with a highly hydrophobic N-terminal region that is required for fusion between viral and endosomal membrane. NA is a homotetramer whose ectodomain (4 × ~435 amino acids) consists of a 100-Å-long fibrous stem that supports a large globular domain (80Å × 80Å × 60Å) (COLMAN and WARD 1985). The globular domain contains four enzymatically active sites that cleave terminal sialic acid residues, thereby removing viral ligand. NA activity is required for the release of newly formed virus from the infected host cell and may be needed for inactivation of certain types of inhibitors present in biological fluids. An average spherical virus particle of ~1000Å diameter contains 500–600HA trimers and 50–100NA tetramers. The close proximity between spikes (50–100Å) (TIFFANY and BLOUGH 1970) may increasingly impair antibody access to determinants with increasing proximity to the membrane. The third transmembrane protein, M2, is a homotetramer (HOLSINGER and LAMB 1991; SUGRUE and HAY 1991) with a small (4 × 24 amino acids) ectodomain. The average virus particle has been estimated to contain approximately ten M2 tetramers. M2 has ion channel activity, which is required at two stages of the viral replication cycle. Within the endosome, it promotes acidification of the viral nucleocapsid complex, thereby enabling proper release of the nucleoprotein complex into the cytoplasm subsequent to fusion. In addition, M2 reduces acidification of transport vesicles during transfer of viral transmembrane proteins from Golgi to the plasma membrane and thereby prevents a premature acid-induced conformational change in acid-sensitive HA molecules.

Virus attachment to the host cell (and subsequent endocytosis and fusion) can be inhibited by antibodies bound to determinants surrounding the ligand-binding site of HA. These regions on the tip of HA are well accessible to antibody on the intact virus. They are usually referred to as sites B, A, and D (Wilson and Cox 1990) or Sb, Sa, and Ca in the H1 subtype (Caton et al. 1982). This antibody activity is usually measured by hemagglutination inhibition (HI) assay and has been found to be in the range of 0.2- to ~10nM (concentration for unit activity) for some antibodies binding to the above sites (Mozdzanowska et al. 1997 and unpublished observations; Schofield et al. 1997). Antibodies specific for regions that are membrane proximal of the sialic acid binding site often exhibit low or no measurable HI activity. This is exemplified by antibodies to site Cb on HA of PR8 (Mozdzanowska et al. 1997a) (corresponding to site E on the H3 subtype), to site C on H3 (Brown et al. 1990) and to determinants on HA2 (Becht et al. 1984; Kostolansky et al. 1989; Okuno et al. 1993; Sanchez-Fauquier et al. 1987; Vanlandschoot et al. 1998). In the case of Cb-specific antibodies, the low HI activity appears to be due to the low accessibility of this site to antibody on intact virus because the antibodies displayed 20- to 500-fold increased HI activity against glycoprotein rosettes in which a wider angle between spikes would be expected to increase the site's accessibility. Antibodies with similar properties have been described by other investigators (Kida et al. 1982; Yoden et al. 1982; Becht et al. 1984).

Attached virus (together with bound antibody that failed to inhibit attachment) is rapidly taken up into endosomes, where a decrease in pH triggers M2 ion-channel activity and the conformational change in HA required for fusion (for review see Lamb and Krug 1996). There is no evidence that M2 ion-channel activity can be inhibited at this stage by M2-specific antibody. By contrast, there are several reports of antibodies that are HI-negative but virus-neutralization (VN)-positive (Kida et al. 1982; Yoden et al. 1982; Okuno et al. 1993; Nagy et al. 1994) and believed to prevent infection by inhibiting intraendosomal fusion. The VN activity of such antibodies appears to be low, as the example described by Okuno et al. (1994) required ~700nM for neutralization of 100PFU of virus compared to 0.2–1.4nM in the case of an antibody binding to site Sb on HA of H1 subtype (Mozdzanowska et al. 1997) or 1–7nM for antibodies to sites A, B, and D on HA of H7 subtype (Schofield et al. 1997).

After fusion and prior to synthesis of viral proteins, antibody cannot modify any more the course of the infection. Although uptake of antibodies into cytoplasm and nuclei of living cells has been reported in some experimental systems (Zack et al. 1996), there is no evidence that this occurs in epithelial cells of the RT.

However, there is evidence that antibody can interfere with virus replication in the course of pIgR-mediated transcytosis (MAZANEC et al. 1995). Not much is known about the interactions occurring at this stage, but one would expect that antibodies to each viral transmembrane protein may encounter the specific antigen. This activity would obviously be restricted to polymeric IgA and IgM.

During the process of virus assembly at the apical plasma membrane, viral transmembrane proteins accumulate in patches from which constitutive host membrane proteins have been displaced and virus particles are being formed by a budding process. Reaction of antibodies with viral transmembrane protein expressed by the infected cell may result in a reduction of virus yield (YR). This activity is exhibited by antibodies to each viral transmembrane protein and apparently by the broadest range of antibodies within each specificity. In the case of HA-specific antibodies, YR activity has been seen with many HI-negative and VN-negative antibodies (WEBER et al. 1993; MOZDZANOWSKA et al. 1997a; VANLANDSCHOOT et al. 1998). HA(Cb)-specific antibodies displayed YR activity (75% reduction of yield in a 3-day assay) at 0.2–6nM concentration (MOZDZANOWSKA et al. 1997a). In the case of NA-specific antibodies, YR activity has been seen both with neuraminidase inhibition (NI)-positive and NI-negative antibodies (WEBSTER et al. 1988). Finally, YR activity is exhibited by some (ZEBEDEE and LAMB 1988) but not all (HUGHEY et al. 1995) M2-specific antibodies.

Several studies reported the presence of NP on intact infected cells (VIRELIZIER et al. 1977; YEWDELL et al. 1981; STITZ et al. 1990). However, NP-specific antibodies have no measurable effect on the course of infection in vivo (VIRELIZIER et al. 1976; ASKONAS et al. 1982; STITZ et al. 1990; GERHARD et al. 1997). Reports of expression of M1 on infected cells (BIDDISON et al. 1977; ADA and YAP 1979) were probably due to detection of M2, which shares a conserved 9-amino-acids-long N-terminal with M1 protein. M1-specific antibody also has no effect on the course of infection (VIRELIZIER et al. 1976; ASKONAS et al. 1982; GERHARD et al. 1997).

## 6 Antibody-Mediated Control Mechanisms In Vivo

Antibody can control an infection in vivo in two ways: (1) by impairing the ability of infectious virus to spread the infection to new host cells, and (2) by impairing maturation and release of progeny virus from infected host cells. In the following, activities that contribute to the former are collectively referred to as virus neutralization (VN) and those that contribute to the latter as yield reduction (YR).

### 6.1 Virus Neutralization

In vitro, VN activity is mediated by HA-specific antibodies that prevent either attachment of virus to the host cell or intra-endosomal fusion. VN activity in vivo is more complex, because virus-bound antibody of any specificity may result in VN through secondary phenomena like deposition of complement and/or interaction with rheumatoid factor and Fc-receptor (FcR) (Corbeil et al. 1995; Schlesinger and Chapman 1995). The influence of such secondary interactions is demonstrated by enhancement of VN activity in vitro in the presence of serum (Table 2). In the case of the antibodies shown in Table 2, the degree of enhancement in vitro varied, apparently depending on the antibody's specificity or avidity (likely Fc-dependent effects are not addressed with this group of antibodies which share the IgG2a isotype). VN activity in vivo can be assessed by determining the dose of antibody required for protecting 50% of mice from infection (50% prophylaxis) if given 1 day prior to challenge. As 50% prophylaxis requires less passive antibody than 50% cure (antibody given shortly after infection), it is clear that antibody operates in prophylaxis by neutralizing the initial virus inoculum rather than by resolving an established infection.

HA-specific antibodies that reacted equally well with native HA, as shown by binding to live infected cells, but exhibited vastly different VN activity in vitro gave 50% prophylaxis in the dose range of from 3µg to ~200µg per mouse, which corresponded to serum antibody concentrations of ~1.2µg/ml and 80µg/ml, respectively, at the time of virus challenge. As shown by the greatly different C/A ratios in Table 2 (see legend for definitions), the prophylactic activity in vivo was poorly predictable from the antibody's VN activity measured in vitro in the absence of serum. The relation improved if VN activity was measured in vitro in the presence of serum, but there was still a more than 20-fold difference between the VN activity ratios in vivo vs in vitro amongst the antibodies (Table 2, C/B). Assuming that the antibodies transudated at the same rate from serum into RT fluid, this ratio should be the same for each antibody, if the assay in vitro mimicked perfectly the interactions in vivo. The factors responsible for these differences in VN activity in vitro and in vivo are not known.

At the maximum dosage of 200µg per mouse, NA- and M2-specific antibodies showed no prophylactic activity, which is consistent with studies (see above) showing that NA- and M2-specific immunity induced by vaccination was infection-

**Table 2.** Relation between virus neutralization activity in vitro and passive serum concentration giving 50% prophylaxis in vivo[a]

| Antibody | | VN in vitro (nM) | | Serum conc. (nM) giving 50% prophylaxis | In vivo/in vitro | |
|---|---|---|---|---|---|---|
| | | 0% serum A | 1.6% serum B | C | C/A | C/B |
| H36-4-5.2 | G2a, Sb | 0.006 | 0.003 (2×) | 8 | 1300 | 2700 |
| L2-10C1 | G2a, Cb | 1.9 | 0.15 (13×) | 160 | 80 | 1100 |
| H35-C12 | G2a, Cb | 16.5 | 0.22 (75×) | 53 | 3 | 240 |
| H2-4C2 | G2a, Cb | ~70 | 4.6 (~15×) | ~500 | 7 | 110 |

[a] A, B, and C are arbitrary labels for the respective columns of data. C/A and C/B are the computed ratios of the data of the respective columns.

permissive. Thus, opsonization of virus at the density achieved by antibodies of these specificities is not sufficient for mediating VN in vivo.

## 6.2 Yield Reduction

Yield reduction activity can be assessed with antibodies that are capable of reacting with live infected cells but lack VN activity in vivo. M2- and NA-specific antibodies fulfilled this condition. Treatment of infected SCID mice with these antibodies resulted in significant reductions in lung virus titer compared to untreated mice, but the infection was not resolved, even though the antibodies were used at therapeutically saturating concentration (Mozdzanowska et al. 1999). The therapeutically saturating dose of the M2-specific antibody (IgG2a) was in the range of 50μg/mouse and resulted in an initial serum concentration of ~130nM and a final concentration of ~60nM 7 days later, when the experiments were terminated. Treatment of mice with the combination of both antibodies did not improve the therapeutic activity, indicating that each antibody had already saturated the therapeutic efficacy of the antibody-mediated YR mechanisms available to it. Thus, the data indicated that antibody-mediated YR activity, while capable of suppressing virus growth by 99%–99.9%, was not powerful enough to resolve the PR8 virus infection. Whether it would be more effective against a less pathogenic virus like X31, as seen with Tc-mediated activity (see above), remains to be tested.

Various mechanisms may contribute to YR activity in vivo. First, cell-bound antibodies may reduce virus yield by mechanisms similar to those operating in vitro. Second, they may target innate defense mechanisms like complement and FcγR-expressing effectors to the infected host cells. In the case of the M2-specific antibody, which did not exhibit YR in vitro against PR8 (Zebedee and Lamb 1988; Mozdzanowska et al. 1999), YR in vivo must have resulted from the targeting of innate defense to the infected host cells. This is consistent also with the finding that an IgG2a switch variant of the M2-specific antibody was therapeutically more effective than the parental IgG1 antibody, as both complement activation by the classic pathway (Neuberger and Rajewsky 1981) and FcγRI (Ravetch and Kinet 1991; Hulett and Hogarth 1994) display a strong preference for the IgG2a compared to the IgG1 isotype.

In the case of the NA-specific antibody, YR activity in vitro was effective at a concentration of ~50nM and delayed virus yield by 1–2 days relative to control cultures. However, it appears unlikely that a passive serum antibody concentration in the range of 250–500nM (the highest dose tested) would result by transudation in an antibody concentration of 50nM in RT secretion. Furthermore, YR in vitro merely delayed virus yield by 1–2 days, and it is difficult to envision how such a short delay in virus replication might account for a significant reduction in virus titer in vivo 7 days after infection. Therefore, the therapeutic activity of the NA-specific antibody may have resulted mainly from targeting innate defenses to infected cells. This is consistent also with the finding that treatment of mice with the combination of M2- and NA-specific antibodies did not improve therapeutic activity, indicating that both antibodies used the same effector mechanisms and had individually already saturated the efficacy of these mechanisms (Mozdzanowska et al. 1999).

YR activity may result also from pIgR-mediated transport of IgA and IgM through infected epithelial cells. While the efficacy of this mechanism in YR of influenza virus has been demonstrated in vitro (Mazanec et al. 1995) and has been proposed to underlie the IgA-mediated cure of rotavirus infection in SCID mice (Burns et al. 1996), its role in the control of influenza virus infection in vivo is not clear. For example, repetitive treatment of PR8-infected SCID mice with polymeric virus-specific IgA or IgM had little effect on pulmonary virus titers (Palladino et al. 1995), although both antibodies were prophylactically effective (Renegar and Small 1991b; Palladino et al. 1995), indicating that they were transported into RT secretions. However, given that therapy requires higher antibody concentration than prophylaxis, it is possible that the therapeutically effective concentration was not achieved and may indeed be difficult to achieve with these isotypes if they are administered systemically, as both have short half-lives in vivo and probably transudate poorly into the interstitial compartment of the RT because of their large size. Of note in this context is the recent finding that IgA deficiency did not measurably impair recovery of mice from influenza virus infection (Mbawuike et al. 1999).

### 6.3 Roles of Virus Neutralization and Yield Reduction in Anti-HA Antibody-Mediated Therapy

HA-specific antibodies, which display both YR and VN activity, were therapeutically very effective and could resolve massive pulmonary infections in SCID mice (Palladino et al. 1995). As the early infection could be resolved by intranasal administration of a Fab fragment that exhibited high VN but no YR activity (our unpublished observation), it is clear that VN activity is in principle sufficient for curing the infection. Similar findings have been made in respiratory syncytial virus infection (Crowe et al. 1994). Nevertheless, in the case of intact antibodies, YR activity is likely to contribute significantly to an antibody's therapeutic activity, particularly if its VN activity is low. This is supported by the finding that treatment of infected SCID mice with a combination of M2- and NA-specific antibodies and a therapeutically suboptimal dose of an HA-specific antibody resulted in significantly stronger suppression of virus growth ($10^{4.1}$-fold reduction relative to control) than treatment with the M2- and NA-specific ($10^{2.5}$-fold) or the HA-specific ($10^{0.6}$-fold) antibodies alone (Mozdzanowska et al. 1999).

## 7 Concluding Remarks

Current evidence indicates that B cells make both Th-independent and Th-dependent contributions in the control and resolution of influenza type A virus infection in the naive mouse. These contributions are significant not only in severe infections, in which survival usually requires that all components of the host's defense work in concert, but also in the control of non-severe infections by viruses of low pathogenicity. The importance of the HA- and NA-specific memory B cell response in protection of the immune host has long been recognized and shown to be both more protective and of longer duration than memory T-cell-mediated immunity. Its only shortcoming is the propensity of the virus to alter the relevant target structures by mutation (antigenic drift) and gene reassortment (antigenic shift), which necessitates continuous updating of vaccines. Recent evidence indicates that M2-specific immunity may provide broadly cross-reactive, antibody-mediated immunity. Whether influenza virus will be capable of escaping M2-specific immunity by antigenic drift in the M2-ectodomain (Zebedee and Lamb 1989) remains to be seen, and much remains to be learned about the precise mechanisms involved in antibody-mediated protection in vivo and their dependence on antibody isotype, avidity, and fine specificity.

## References


Ada GL, Yap KL (1979) The measurment of haemagglutinin and matrix protein present on the surface of influenza virus infected P815 mastocytoma cells. J Gen Virol 42:541–553

Ada GL, Jones PD (1986) The immune response to influenza infection. Curr Top Microbiol Immunol 128:1–54

Allan W, Tabi Z, Cleary A, Doherty PC (1990) Cellular events in the lymphnode and lung of mice with influenza. Consequences of depleting $CD4^+$ T cells. J Immunol 144:3980–3986

Askonas BA, McMicheal AJ, Webster RG (1982) In: Bear AS (ed) Basic and applied influenza research. CRC, Boca Raton, Fla., pp 159–188

Bachmann MF, Zinkernagel RM (1997) Neutralizing antiviral B cell responses. Annu Rev Immunol 15:235–270

Balachandran S, Roberts PC, Kipperman T, Bhalla KN, Compans R, Archer DR, Barber GN (2000) Alpha/beta interferons potentiate virus-induced apoptosis through activation of the FADD/Caspase-8 death signaling pathway. J Virol 74:1513–1523

Becht H, Huang RT, Fleischer B, Boschek CB, Rott R (1984) Immunogenic properties of the small chain HA2 of the hemagglutinin of influenza virus. J Gen Virol 65:173–183

Bender BS, Croghan T, Zhang L, Small PA (1992) Transgenic mice lacking class I major histocompatibility complex-restricted T cells have delayed viral clearance and increased mortality after influenza virus challenge. J Exp Med 175:1143–1145

Bennink JR, Yewdell JW (1988) Murine cytotoxic T lymphocyte recognition of individual influenza virus proteins. High frequency of nonresponder MHC class I alleles. J Exp Med 168:1935–1939

Biddison WE, Doherty PC, Webster RG (1977) Antibody to influenza virus matrix protein detects a common antigen on the surface of cells infected with type A influenza viruses. J Exp Med 146:690–679

Black RA, Rota PA, Gorodkova N, Klenk H-D, Kendal AP (1993) Antibody response to the M2 protein of influenza A virus expressed in insect cells. J Gen Virol 74:143–146

Brandtzaeg P (1989) Overview of the mucosal immune system. Curr Top Microbiol Immunol 146:13–24

Brandtzaeg P, Farstad IN, Johansen F-E, Morton HC, Norderhaug IN, Yamanaka T (1999) The B-cell system of human mucosae and exocrine glands. Immunol Rev 171:45–87

Brown LE, Murray JM, White DO, Jackson DC (1990) An analysis of the properties of monoclonal antibodies directed to epitopes on influenza virus hemagglutinin. Arch Virol 114:1–26

Burns JW, Siadat-Pajouh M, Krishnaney AA, Greenberg HB (1996) Protective effect of rotavirus VP6-specific IgA monoclonal antibodies that lack neutralizing activity. Science 272:104–107

Chambers TM, Kawaoka Y, Webster RG (1988) Protection of chickens from lethal influenza infection by vaccinia-expressed hemagglutinin. Virology 167:414–421

Colman PM, Ward CW (1985) Structure and diversity of influenza virus neuraminidase. Curr Top Microbiol Immunol 114:178–247

Corbeil S, Seguin C, Trudel M (1995) Involvement of the complement system in the protection of mice from challenge with respiratory syncytial virus long strain following passive immunization with monoclonal antibody 18A2B2. Vaccine 14:521–525

Corthesy B, Kraehenbuhl J-P (1999) Antibody-mediated protection of mucosal surfaces. Curr Top Micro Immunol 236:93–111

Couch RB, Kasel JA (1983) Immunity to influenza in man. Annu Rev Microbiol 37:529–549

Crowe JE, Murphy BR, Chanock RM, Williamson RA, Barbas CF, Burton DR (1994) Recombinant human respiratory syncytial virus (RSV) monoclonal antibody Fab is effective therapeutically when introduced directly into the lungs of RSV-infected mice. Proc Natl Acad Sci USA 91:1386–1390

Crowe JE (1999) Host responses to respiratory virus infection and immunization. Curr Top Micro Immunol 236:191–214

Doherty PC, Topham DJ, Tripp RA, Cardin RD, Brooks JW, Stevenson PG (1997) Effector $CD4^+$ and $CD8^+$ T-cell mechanisms in the control of respiratory virus infections. Immunol Rev 159:105–117

Eichelberger M, Allan W, Zijlstra M, Jaenisch R, Doherty PC (1991a) Clearance of influenza virus respiratory infection in mice lacking class I major histocompatibility complex-restricted CD8 T cells. J Exp Med 174:875–880

Eichelberger MC, Wang M, Allan W, Webster RG, Doherty PC (1991b) Influenza virus RNA in the lung and lymphoid tissue of immunologically intact and CD4-depleted mice. J Gen Virol 72: 1695–1698

Epstein SL, Misplon JA, Lawson CM, Subbarao EK, Connors M, Murphy BR (1993) β2-Microglobulin-deficient mice can be protected against influenza A infection by vaccination with vaccinia-influenza recombinants expressing hemagglutinin and neuraminidase. J Immunol 150:5484–5493

Epstein SL, Lo CY, Misplon JA, Bennink JR (1998) Mechanisms of protective immunity against influenza virus infection in mice without antibodies. J Immunol 160:322–327

Fazekas G, Rosenwirth B, Dukor P, Gergely J, Rajnavolgyi (1994) IgG isotype distribution of local and systemic immune responses induced by influenza virus infection. Eur J Immunol 24:3063–3067

Frace AM, Klimov AI, Rowe T, Black RA, Katz JM (1999) Modified M2 proteins produce heterotypic immunity against influenza A virus. Vaccine 17:2237–2244

Garcia-Sastre A, Durbin RK, Zheng H, Palese P, Gertner R, Levy DE, Durbin JE (1998) The role of interferon in influenza virus tissue tropism. J Virol 72:8550–8558

Gerhard W, Mozdzanowska K, Furchner M (1996) In: Brown LE, Hampson AW, Webster RG (eds) Options for the control of influenza III. Elsevier, pp 235–249

Gerhard W, Mozdzanowska K, Furchner M, Washko G, Maiese K (1997) Role of the B cell response in recovery of mice from primary influenza virus infection. Immunol Rev 159:95–104

Graham MB, Braciale TJ (1997) Resistance to and recovery from lethal influenza virus infection in B lymphocyte-deficient mice. J Exp Med 186:2063–2068

Gubareva LV, McCullers JA, Bethel RC, Webster RG (1998) Characterization of influenza A/Hong-Kong/156/97(H5N1) virus in a mouse model and protective effect of Zanamivir on H5N1 infection in mice. J Inf Dis 178:1592–1596

Harling-McNabb L, Delyannis G, Jackson DC, Gerondakis S, Grigoriadis G, Brown LE (1999) Mice lacking the transcription factor subunit Rel can clear an influenza infection and have functional antiviral cytotoxic T cells but do not develop an optimal antibody response. Int Immunol 11:1431–1439

Holsinger LJ, Lamb RA (1991) Influenza virus M2 integral membrane protein is a homotetramer stabilized by formation of disulfide bonds. Virology 183:32–43

Hughey PG, Roberts PC, Holsinger LJ, Zebedee SL, Lamb RA, Compans RW (1995) Effects of antibody to the influenza A virus M2 protein on M2 surface expression and virus assembly. Virology 212: 411–421

Hulett MD, Hogarth PM (1994) Molecular basis of Fc receptor function. Adv Immunol 57:1–98

Hunt LA, Brown DW, Robinson HL, Naeve CW, Webster RG (1988) Retrovirus-expressed hemagglutinin protects against lethal influenza virus infections. J Virol 62:3014–3019

Iwasaki T, Nozima T (1977) Defense mechanisms against primary influenza virus infection in mice. I. Roles of interferon and neutralizing antibodies and thymus dependence of interferon and antibody production. J Immunol 118:256–263

Jakeman KJ, Smith H, Sweet C (1989) Mechanism of immunity to influenza: maternal and passive neonatal protection following immunization of adult ferrets with a live vaccinia-influenza virus hemagglutinin recombinant but not with recombinants containing other influenza virus proteins. J Gen Virol 70:1523–1531

Johansson BE, Bucher DJ, Kilbourne ED (1989) Purified influenza virus hemagglutinin and neuraminidase are equivalent in stimulation of antibody response but induce contrasting types of immunity to infection. J Virol 63:1239–1246

Jones PD, Ada GL (1986) Influenza virus-specific antibody-secreting cells in the murine lung during primary influenza virus infection. J Virol 60:614–619

Justewicz DM, Doherty PC, Webster RG (1995) The B-cell response in lymphoid tissue of mice immunized with various antigenic forms of the influenza virus hemagglutinin. J Virol 69:5414–5421

Kaltreider HB (1976) Expression of immune mechanisms in the lung. Am Rev Resp Dis 113:347–379

Kida H, Brown LE, Webster RG (1982) Biological activity of monoclonal antibodies to operationally defined antigenic regions on the hemagglutinin molecule of A/Seal/Massachusetts/1/80 (H7N7) influenza virus. Virology 122.38–47

Kido H, Yokogoshi Y, Sakai K, Tashiro M, Kishino Y, Fukutomi A, Katunuma N (1992) Isolation and characterization of a novel trypsin-like protease found in rat bronchiolar epithelial Clara cells. A possible activator of the viral fusion glycoprotein. J Biol Chem 267:13573–13579

Kodihalli S, Goto H, Kobasa DL, Krauss S, Kawaoka Y, Webster RG (1999) DNA vaccine encoding hemagglutinin provides protective immunity against H5N1 influenza virus infection in mice. J Virol 73:2094–2098

Kostolansky F, Styk B, Russ G (1989) Inhibition of influenza virus haemolytic and haemagglutination activities by monoclonal antibodies to haemagglutinin glycopolypeptides HA1 and HA2. Acta Virol 33:504–512

Kris RM, Asofsky R, Evans CB, Small PA (1985) Protection and recovery in influenza virus-infected mice immunosuppressed with anti-IgM. J Immunol 134:1230–1235

Kris RM, Yetter RA, Cogliano R, Ramphal R, Small PA (1988) Passive serum antibody causes temporary recovery from influenza virus infection of the nose, trachea and lung of nude mice. Immunology 63:349–353

Kvale D, Lovhaug D, Sollid LM, Brandtzaeg P (1988) Tumor necrosis factor-α up-regulates expression of secretory component, the epithelial receptor for polymeric Ig. J Immunol 140:3086–3089

Lamb RA, Krug RM (1996) In: Fields BN, Knipe DM, Howley PM (eds) Fields virology, 3rd edn. Lippincott-Raven, Philadelphia

Lamm ME (1997) Interactions of antigens and antibodies at mucosal surfaces. Annu Rev Microbiol 51:311–340

Liang S, Mozdzanowska K, Palladino G, Gerhard W (1994) Heterosubtypic immunity to influenza type A virus in mice: effector mechanisms and their longevity. J Immunol 152:1653–1661

Lu X, Tumpey TM, Morken T, Zaki SR, Cox NJ, Katz JM (1999) A mouse model for the evaluation of pathogenesis and immunity to influenza A (H5N1) viruses isolated from humans. J Virol 72:5903–5911

Manz RA, Lohning M, Cassese G, Thiel A, Radbruch A (1998) Survival of long-lived plasma cells is independent of antigen. Int Immun 10:1703–1711

Marshall D, Sealy R, Sangster M, Coleclough C (1999) Th cells primed during influenza virus infection provide help for qualitatively distinct antibody responses to subsequent immunization. J Immunol 163:4673–4682

Mazanec MB, Courdet CL, Fletcher DR (1995) Intracellular neutralization of influenza virus by immunoglobulin A anti-hemagglutinin monoclonal antibodies. J Virol 69:139–1343

Mbawuike IN, Pacheco S, Acuna CL, Switzer KC, Zhang Y, Harriman GR (1999) Mucosal immunity to influenza virus without IgA: an IgA knockout mouse model. J Immunol 162:2530–2537

Mori I, Komatsu T, Takeuchi K, Nakakuki K, Sudo M, Kimura Y (1995) In vivo induction of apoptosis by influenza virus. J Gen Virol 76:2869–2873

Moskophidis D, Cobbold SP, Waldmann H, Lehmann-Grube F (1987) Mechanism of recovery from acute virus infection: treatment of lymphocytic choriomeningitis virus-infected mice with monoclonal antibodies reveals that Lyt-$2^+$ T lymphocytes mediate clearance of virus and regulate the antiviral antibody response. J Virol 61:1867–1874

Mozdzanowska K, Furchner M, Washko G, Mozdzanowski J, Gerhard W (1997a) A pulmonary influenza virus infection in SCID mice can be cured by treatment with hemagglutinin-specific antibodies that display very low virus-neutralizing activity in vitro. J Virol 71:4347–4355

Mozdzanowska K, Furchner M, Maiese K, Gerhard W (1997b) $CD4^+$ T cells are ineffective in clearing a pulmonary infection with influenza type A virus in the absence of B cells. Virology 239:217–225

Mozdzanowska K, Maiese K, Furchner M, Gerhard W (1999) Treatment of influenza virus-infected SCID mice with non-neutralizing antibodies specific for the transmembrane proteins matrix 2 and neuraminidase reduces the pulmonary virus titer but fails to clear the infection. Virology 254:138–146

Mozdzanowska K, Maiese K, Gerhard W (2000) Th cell-deficient mice control influenza virus infection more effectively than Th- and B cell-deficient mice: evidence for a Th-independent contribution by B cells to virus clearance. J Immunol 164:2635–2643

Murphy BR, Clements ML (1989) The systemic and mucosal immune response of humans to influenza A virus. Curr Top Microbiol Immunol 146:107–116

Murphy BR, Webster RG (1996) Orthomyxoviruses. In: Fields BN, Knipe DM, Howley PM, Chanock RM, Hirsch MS, Melnick JL, Monath TP, Roizman B, Strauss SE (eds) Fields virology. Lippincott-Raven, Philadelphia, pp 1397–1445

Nagy Z, Rajnavolgyi E, Hollosi M, Toth GK, Varadi G, Penke B, Toth I, Horvath A, Gergely J, Kurucz I (1994) The intersubunit region of the influenza virus hemagglutinin is recognized by antibodies during infection. Scand J Immunol 40:281–291

Neirynck S, Deroo T, Saelens X, Vanlandschoot P, Min Jou W, Fiers W (1999) A universal influenza A vaccine based on the extracellular domain of the M2 protein. Nat Med 5:1157–1163

Neuberger MS, Rajewsky K (1981) Activation of mouse complement by monoclonal mouse antibodies. Eur J Immunol 11:1012–1016

Ochsenbein AF, Fehr T, Lutz C, Suter M, Brompacher F, Hengartner H, Zinkernagel RM (1999) Control of early viral and bacterial distribution and disease by natural antibodies. Science 286:2156–2158

Okuno Y, Isegawa Y, Sasao F, Ueda S (1993) A common neutralizing epitope conserved between the hemagglutinins of influenza A virus H1 and H2 strains. J Virol 67:2552–2558

Okuno Y, Matsumoto K-I, Isegawa Y, Ueda S (1994) Protection against the mouse-adapted A/FM/1/47 strain of influenza A virus in mice by a monoclonal antibody with cross-neutralizing activity among H1 and H2 strains. J Virol 68:517–520

Oldstone MBA, Blount P, Southern PJ (1986) Cytoimmunotherapy for persistent virus infection reveals a unique clearance pattern from the central nervous system. Nature 321:239–243

Owens SL, Osebold JW, Zee YC (1981) Dynamics of B-lymphocytes in the lungs of mice exposed to aerosolized influenza virus. Inf Immun 33:231–238

Palladino G, Mozdzanowska K, Washko G, Gerhard W (1995) Virus-neutralizing antibodies of immunoglobulin G (IgG) but not IgM or IgA isotypes can cure influenza virus pneumonia in SCID mice. J Virol 69:2075–2081

Prince GA, Hemmig VG, Horswood RL, Baron PA, Chanock RM (1987) Effectiveness of topically administered neutralizing antibodies in experimental immunotherapy of respiratory syncytial virus infection in cotton rats. J Virol 61:1851–1854

Ramphal R, Cogliano RC, Shands JS, Small PA (1979) Serum antibody prevents lethal murine influenza pneumonitis but not tracheitis. Infect Immun 25:992–997

Ravetch JV, Kinet J-P (1991) Fc Receptors. Annu Rev Immunol 9:457–492

Reading PC, Morey LS, Crouch EC, Anders EM (1997) Collectin-mediated antiviral host defense of the lung: evidence from influenza virus infection of mice. J Virol 71:8204–8212

Renegar KB, Small PA (1991a) Immunoglobulin A mediation of murine nasal anti-influenza virus immunity. J Virol 65:2146–2148

Renegar KB, Small PA (1991b) Passive transfer of local immunity to influenza virus infection by IgA antibody. J Immunol 146:1972–1978

Rimmerlzwaan GR, Baars MMJW, Lijster P, Fouchier RAM, Osterhaus ADME (1999) Inhibition of influenza virus replication by nitric oxide. J Virol 73:8880–8883

Sanchez-Fauquier A, Villanueva N, Melero JA (1987) Isolation of cross-reactive, subtype-specific monoclonal antibodies against influenza virus HA1 and HA2 hemagglutinin subunits. Arch Virol 37:251–265

Sangster M, Hyland L, Sealy R, Coleclough C (1995) Distinct kinetics of the antibody-forming cell response to Sendai virus infection of mice in different anatomical compartments. Virology 207: 287–291

Sarawar SR, Sangster M, Coffman RL, Doherty PC (1994) Administration of anti-IFN-$\gamma$ antibody to $\beta$2-microglobulin-deficient mice delays influenza virus clearance but does not switch the response to a T helper cell 2 phenotype. J Immunol 153:1246–1253

Sareneva T, Matikainen S, Kurimoto M, Julkunen I (1998) Influenza A virus-induced IFN-$\alpha/\beta$ and IL-18 synergistically enhance IFN-$\gamma$ expression in human T cells. J Immunol 160:6032–6038

Sauter NK, Glick GD, Crowther RL, Park S-J, Eisen MB, Skehel JJ, Knowles JR, Wiley DC (1992) Crystallographic detection of a second ligand binding site in influenza virus hemagglutinin. Proc Natl Acad Sci USA 89:324–328
Scherle PA, Palladino G, Gerhard W (1992) Mice can recover from pulmonary influenza virus infection in the absence of class I-restricted cytotoxic T cells. J Immunol 148:212–217
Schlesinger JJ, Chapman S (1995) Neutralizing F(ab')$_2$ fragments of protective monoclonal antibodies to yellow fever virus (YF) envelope proteins fail to protect mice against lethal YF encephalitis. J Gen Virol 76:217–220
Schofield DJ, Stephenson JR, Dimmock NJ (1997) Variations in the neutralizing and haemagglutination-inhibiting activities of five influenza A virus-specific IgGs and their antibody fragments. J Gen Virol 78:2431–2439
Slepushkin VA, Katz JM, Black RA, Gamble WC, Rota PA, Cox NJ (1995) Protection of mice against influenza A virus challenge by vaccination with baculovirus-expressed M2 protein. Vaccine 23:1399–1402
Slifka MK, Antia R, Whitmire JK, Ahmed R (1998) Humoral immunity due to long-lived plasma cells. Immunity 8:363–372
Sollid LM, Kvale D, Brandtzaeg P, Markussen G, Thorsby E (1987) Interferon-γ enhances expression of secretory component, the epithelial receptor for polymeric immunoglobulins. J Immunol 138:4303–4306
Staeheli P (1990) Interferon-induced proteins and the antiviral state. Adv Virus Res 38:147–200
Stitz L, Schmitz C, Binder D, Zinkernagel R, Paoletti E, Becht H (1990) Characterization and immunological properties of influenza A virus nucleoprotein (NP): cell-associated NP isolated from infected cells or viral NP expressed by vaccinia recombinant virus do not confer protection. J Gen Virol 71:1169–1179
Sugrue RJ, Hay AJ (1991) Structural characterization of the M2 protein of influenza A viruses: evidence that it forms a tetrameric channel. Virology 180:617–624
Szomolanyi-Tsuda E, Welsh RM (1996) T cell-independent antibody-mediated clearance of polyoma virus in T cell-deficient mice. J Exp Med 183:403–411
Tamura S, Funato H, Hirabayashi Y, Suzuki Y, Nagamine T, Aizawa C, Kurata T (1991) Cross-protection against influenza A virus infection by passively transferred respiratory tract IgA antibodies to different hemagglutinin molecules. Eur J Immunol 21:1337–1344
Tamura S, Iwasaki T, Thompson AH, Asanuma H, Chen Z, Suzuki Y, Aizawa C, Kurata T (1998) Antibody-forming cells in the nasal-associated lymphoid tissue during primary influenza virus infection. J Gen Virol 79:291–299
Tannock GA, Paul JA (1987) Homotypic and heterotypic immunity of influenza A viruses induced by recombinants of the cold-adapted master strain A/AnnArbor/6/60-ca. Arch Virol 92:121–133
Tiffany JM, Blough HA (1970) Estimation of the number of surface projections on myxo- and paramyxoviruses. Virology 41:392–394
Topham DJ, Tripp RA, Hamilton-Easton AM, Sarawar SR, Doherty PC (1996) Quantitative analysis of the influenza virus-specific CD4+ T cell memory in the absence of B cells and Ig. J Immunol 157:2947–2952
Topham DJ, Doherty PC (1998) Clearance of an influenza A virus by CD4+ T cells is inefficient in the absence of B cells. J Virol 72:882–885
Treanor JJ, Thierney EL, Zebedee SL, Lamb RA, Murphy BR (1990) Passively transferred monoclonal antibody to the M2 protein inhibits influenza A virus replication in mice. J Virol 64:1375–1377
Vanlandschoot P, Beirnaert E, Barrere B, Calder L, Millar B, Wharton S, MinJou W, Fiers W (1998) An antibody which binds to the membrane-proximal end of influenza virus haemagglutinin (H3 subtype) inhibits the low-pH-induced conformational change and cell-cell fusion but does not neutralize virus. J Gen Virol 79:1781–1791
Virelizier JL (1975) Host defenses against influenza virus: the role of anti-hemagglutinin antibody. J Immunol 115:434–439
Virelizier JL, Oxford JS, Schild GC (1976) Mechanisms of immunity to influenza in experimental animals. Postgrad Med J 52:332–337
Virelizier JL, Allison AC, Oxford JS, Schild GC (1977) Early presence of ribonucleoprotein antigen on surface of influenza virus-infected cells. Nature 266:52–54
Weber C, Scheepers K, Linder D, Becht H (1993) Isolation and immunogenic properties of a monomeric form of the HA1 subunit of the influenza virus haemagglutinin from infected cells. J Gen Virol 74:2793–2797
Webster RG, Askonas BA (1980) Cross-protection and cross-reactive cytotoxic T cells induced by influenza virus vaccines in mice. Wur J Immunol 10:396–401

Webster RG, Reay PA, Laver WG (1988) Protection against lethal influenza with neuraminidase. Virology 164:230–237

Wells MA, Albrecht P, Ennis FA (1981) Recovery from a viral respiratory infection. I. Influenza pneumonia in normal and T-deficient mice. J Immunol 126:1036–1041

Wilson IA, Cox NJ (1990) Structural basis of immune recognition of influenza virus hemagglutinin. Annu Rev Immunol 8:737–771

Wiley DC, Skehel JJ (1987) The structure and function of the hemagglutinin membrane glycoprotein of influenza virus. Annu Rev Biochem 56:365–394

Yewdell JW, Evan F, Gerhard W (1981) Expression of influenza A virus internal antigens on the surface of infected P815 cells. J Immunol 126:1814–1819

Yoden S, Kida H, Yanagawa R (1982) An avian influenza virus of which infectivity is neutralized by antisera lacking hemagglutination-inhibition activity. Ach Virol 74:205–210

Zack DJ, Stempniak M, Won AL, Taylor C, Weisbart RH (1996) Mechanisms of cellular penetration and nuclear localization of an anti-double strand DNA autoantibody. J Immunol 157:2082–2088

Zebedee SL, Lamb RA (1988) Influenza A virus M2 protein: monoclonal antibody restriction of virus growth and detection of M2 in virions. J Virol 62:2762–2772

Zebedee SL, Lamb RA (1989) Growth restriction of influenza A virus by M2 protein antibody is genetically linked to the M1 protein. Proc Natl Acad Sci USA 86:1061–1065

# Role of Antibodies in Controlling Alphavirus Infection of Neurons

D.E. Griffin, S. Ubol, P. Desprès, T. Kimura, and A. Byrnes



## 1 Introduction

The successful control of virus infection of cells in the nervous system, particularly infection of neurons, is complicated by access of immune mediators generated in the periphery to the central nervous system (CNS), the limited expression of major histocompatibility complex (MHC) molecules, and the postmitotic state and irreplaceable nature of neurons. Although a major function of cytotoxic T lymphocytes (CTLs) is to clear virus-infected cells from a variety of tissues, the role of the cytotoxic function of CTLs in clearing virus infection from the CNS, especially neurons, is likely to be limited. Studies of the clearance of neuronotropic viruses that cause encephalomyelitis have indicated a primary role for antibody in the clearance process (Amor et al. 1996; Levine et al. 1991). For virus clearance, the effects of antibody are dependent on antibody specificity and on the biologic properties of the antiviral antibodies. Antibody-mediated clearance is likely to be augmented by other components of the innate and acquired immune system. Understanding this process is important for identifying mechanisms of control of virus infections of the CNS and recovery from viral encephalitis. The most extensively

W. Harry Feinstone Department of Molecular Microbiology and Immunology, Johns Hopkins University School of Hygiene and Public Health, Baltimore, MD 21205, USA studied neuronal infection for which antibody plays an important role in clearance is Sindbis-virus-induced encephalomyelitis in mice.

## 2 Overview of Sindbis Virus Replication and Pathogenesis

Sindbis virus (SV) is an alphavirus closely related to western equine encephalitis virus, an important cause of mosquito-borne human and equine encephalomyelitis in the Americas. Alphaviruses are enveloped plus-strand RNA viruses that enter cells by receptor-mediated endocytosis, replicate entirely in the cytoplasm and mature by budding from the plasma membrane. Replication of the minus-strand copy of genomic RNA is dependent on continued synthesis of the nonstructural proteins, and translation of these proteins ceases, along with cessation of synthesis of host proteins, a few hours after infection is initiated. Replication of plus-strand RNA and translation of the viral structural proteins continues for 24–48h with continuous virion production during this time (Fig. 1). The alphavirus virion is composed of three major proteins, two envelope glycoproteins, E1 and E2, and a capsid protein that surrounds the RNA. E1 and E2 heterodimerize in the ER and then trimerize to form the spikes on the surface of the virion. The E2 glycoprotein is responsible for virus binding to the cell surface. The tail of E2 interacts with the capsid protein during assembly and drives budding from the surface. The E1 glycoprotein is responsible for fusion of the virus envelope with the cell membrane after induction of a conformational change by exposure to the acid pH in the endosome. A fourth protein, 6K, is important for proper budding and is present in the virion in small amounts (Strauss and Strauss 1994).

During infection, $Na^+K^+$ATPase function is impaired so that concentrations of intracellular $Na^+$ rise, concentrations of intracellular $K^+$ fall and plasma membrane potential cannot be maintained (Fig. 1). These changes are thought to

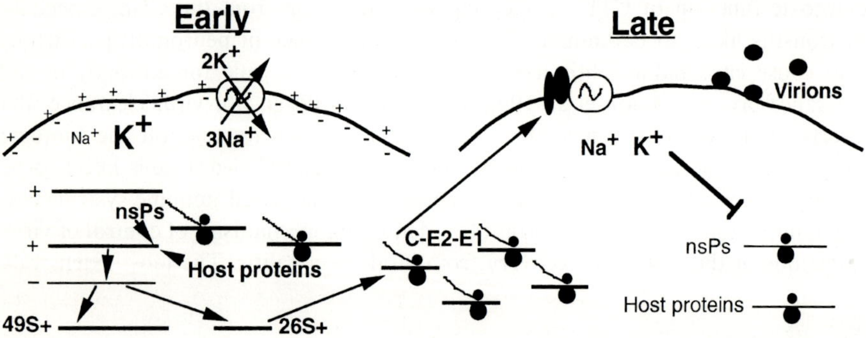


**Fig. 1.** The changes in intracellular cation concentration, membrane potential and protein synthesis during Sindbis virus infection. (Adapted from Griffin et al. 1997)

be important both for translation of the viral mRNA for structural proteins, in preference to translation of mRNAs for nonstructural proteins and mRNAs for host proteins, and for virus budding from the plasma membrane. The subgenomic RNA encoding the viral structural proteins, unlike most host mRNAs and viral genomic RNA, does not require high [$K^+$] for translation (GARRY et al. 1979). The need for a depolarized membrane for virus budding is not understood but probably relates to regulation of membrane deformation. The mechanism by which SV replication inhibits $Na^+K^+$ATPase function is not known, but correlates temporally with synthesis of the glycoproteins (ULUG and BOSE Jr. 1985; ULUG et al. 1989). It is possible that there is a physical interaction between one or more of the viral proteins in the plasma membrane and one or more subunits of the $Na^+K^+$ATPase. These alterations of cellular function probably contribute to cell death, but SV is also able to induce death by apoptosis in neurons and other cellular targets of infection (LEVINE et al. 1996a).

In cycling cells grown in tissue culture, SV induces apoptosis by activating sphingomyelinases to degrade sphingomyelin, release ceramide intracellularly, and initiate the apoptotic cascade (JAN et al. 2000). SV can also induce apoptosis in immature neurons, but as neurons mature they become increasingly resistant to virus-induced cell death (LEVINE et al. 1993). This is presumed to be the result of increased expression of cellular inhibitors of apoptosis as a part of the program of neuronal maturation. In this situation, since mature neurons survive infection, SV, as well as other neuronotropic viruses, can establish persistent nonfatal CNS infections if host mechanisms for control and clearance of virus are defective. Antibody plays a primary role in controlling SV replication in neurons and in clearing infectious virus from the CNS (LEVINE et al. 1991).

## 3 Role of Antibody in Clearing Sindbis Virus from the CNS

Initial studies indicating a major role for antiviral antibody in the control of SV replication in neurons came through the study of persistent SV infection established in the CNS of CB17 mice with severe combined immunodeficiency (SCID). These mice lack a recombinase activity that prevents the development of mature B or T cells. After infection with SV, neurons become persistently, productively, and asymptomatically infected. Passive transfer of immune serum leads to clearance of infectious virus from the CNS within 48h (Fig. 2) (LEVINE et al. 1991). This effect can be replicated in vitro in explant cultures of mature dorsal root ganglion (DRG) neurons persistently infected with SV, although the time course of clearance is slower than it is in vivo (Fig. 2). After antibody treatment, both in vivo and in vitro there is a decrease in the production of infectious virus followed over the next several days by a decrease in synthesis of viral protein and RNA. Subsequent experiments demonstrated that monoclonal antibodies reactive with at least two different epitopes on the E2 glycoprotein are able to control SV replication in

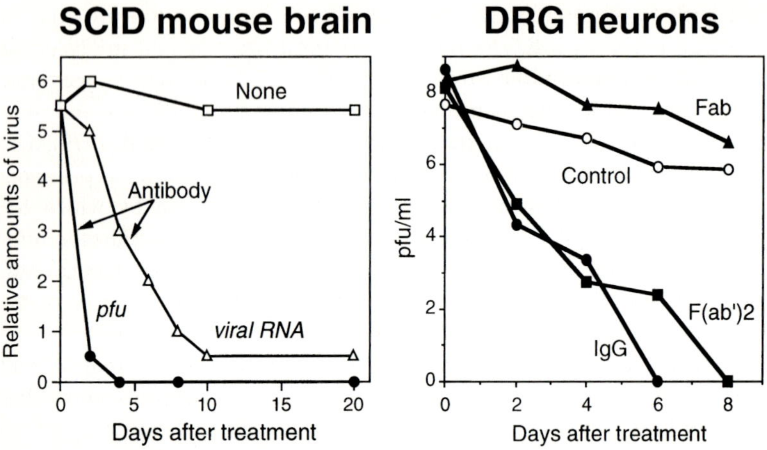


**Fig. 2.** In vivo and in vitro clearance of Sindbis virus (SV) from persistently infected neurons by antibody to the E2 glycoprotein. *Left*: Persistent replication of SV in the brains of immunodeficient CB17/SCID mice and the effect of passive transfer of monoclonal antibody to the SV E2 glycoprotein on clearance of infectious virus (plaque-forming units, *pfu*) and viral RNA. *Right*: Persistent replication of SV in mature dorsal root ganglia (*DRG*) neurons and the effect of treating the cultures with anti-E2 antibody, either intact IgG, Fab or $F(ab')_2$ IgG fragments, for 90min. Bivalent antibody is required to shut off virus production. (Adapted from Griffin et al. 1997 and data in Levine et al. 1991 and Ubol et al. 1995)

neurons in vivo and in persistently infected DRG neuron cultures (Levine et al. 1991). To be effective, the antibody needs to be bivalent but there is no requirement for a particular antibody isotype (Ubol et al. 1995). In addition, antibody does not need to be continuously present in culture since antibody remains bound to the cell surface and continues to exert its effects over a prolonged period (Levine et al. 1991; Ubol et al. 1995). These observations suggested that antibody cross-linking of E2 glycoproteins present on the cell surface leads to control of intracellular replication of SV.

The mechanism by which cross-linking of viral glycoproteins on the cell surface inhibits intracellular virus replication is not yet clear, but several potential mechanisms have been investigated. One possibility is that intracellular signaling occurs through phosphorylation of the cytoplasmic tails of E2 when E2 molecules are brought into close association by antibody. To determine a possible role for phosphorylation in this process, Levine and colleagues mutated the tyrosine, serine, and threonine residues in the E2 cytoplasmic tail to alanine. None of these amino acid changes affected the ability of antibody to E2 to down-regulate virus replication in persistently infected DRG neurons for any mutated virus that was viable (Levine et al. 1996b). These data suggested that the process signaling down-regulation of SV replication intracellularly is not dependent on phosphorylation events involving the cytoplasmic tail of E2.

Further examination of the intracellular consequences of treating persistently infected cells with antibody to E2 required a cell culture system that would become persistently infected, respond to antibody treatment by down-regulating virus replication, and provide sufficient numbers of cells for biochemical analysis. AT3 rat prostatic carcinoma cells over-expressing the antiapoptotic protein Bcl-2 (LEVINE et al. 1993) have these properties and have provided a system for beginning to analyze the molecular and biological consequences of antibody treatment of infected cells. SV-induced cell death is delayed in these cells due to the overexpression of Bcl-2. This allows analysis of antibody effects for several hours, but death eventually occurs so only the initial effects of antibody treatment can be assessed (Fig. 3) (UBOL et al. 1994).

The earliest effect of treatment of infected cells with antibody to E2 is the inhibition of virus budding from the cell surface, leading to accumulation of viral nucleocapsids in the cytoplasm (DESPRÈS et al. 1995a). This inhibition of virus budding is associated with a dose-dependent restoration of $K^+$ flux due to improved $N^+K^+$ATPase function. There is also a dose-dependent improvement of cellular protein synthesis (Fig. 4) (DESPRÈS et al. 1995a,b). Viral protein synthesis and RNA synthesis continue in these cells after antibody treatment, indicating no immediate effect on intracellular virus replication. Synthesis of the viral non-structural proteins is prolonged beyond the usual 3–4h after infection, as would be predicted by restoring more normal, intracellular cation concentrations to the infected cell. The E1 and E2 glycoproteins continue to be synthesized and transported to the plasma membrane, but, in the absence of virion assembly and release from the plasma membrane, these proteins are degraded and shed from the cell surface (DESPRÈS et al. 1995a). These data suggest that antibody treatment initially alters

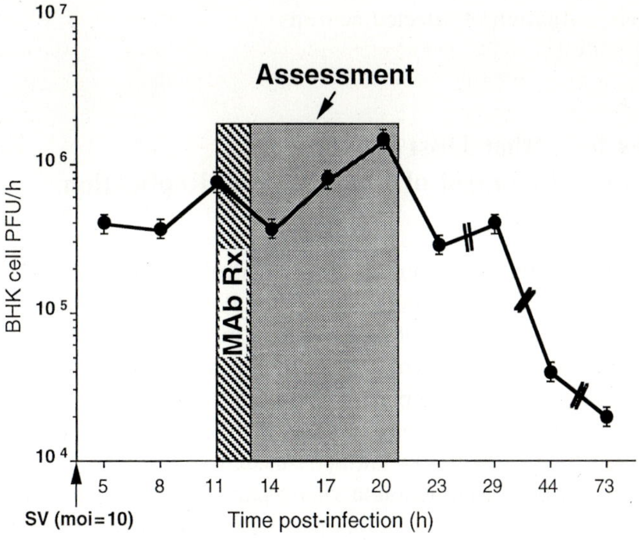


**Fig. 3.** Sindbis virus (SV) infection in AT3-Bcl-2 cells used for analysis of the effects of anti-SV E2 antibody treatment on SV replication. SV-induced cell death is delayed and the effects of treatment can be assessed over a period of persistent infection of 6–8h

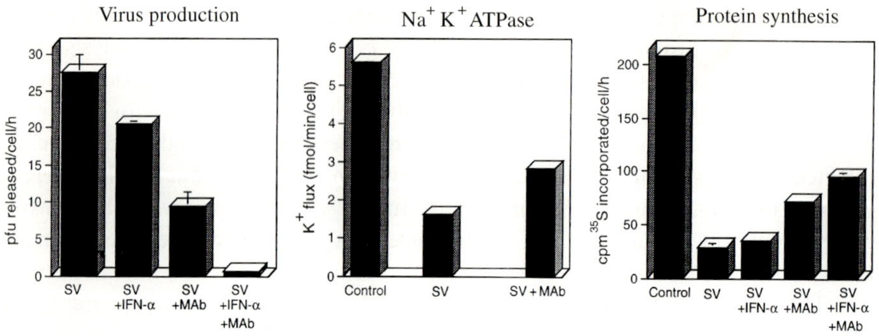


**Fig. 4.** The effects of treating infected AT3-Bcl-2 cells with monoclonal antibody (mAb) to E2 and with interferon-α (IFN-α). Four hours after mAb and/or IFN-α treatment, production of infectious virus, $Na^+K^+$ATPase-dependent $K^+$ influx and host-cell protein synthesis were measured. mAb inhibited virus production and enhanced $K^+$ influx and host protein synthesis. The effects of mAb on virus production were enhanced by simultaneous treatment with IFN-α. (Drawn from data in Desprès et al. 1995a,b)

intracellular cation concentrations leading to restoration of membrane potential and host protein synthesis and inhibition of virus release from the plasma membrane (Fig. 1). The mechanism by which cross-linking of E2 with antibody restores $Na^+K^+$ATPase function is not yet clear.

Studies of persistently infected neurons indicate that these changes, or additional intracellular effects of antibody treatment, eventually result in shut-off of viral structural protein synthesis and of viral plus-strand RNA synthesis. These longer term effects cannot be studied over the few hours available for analysis in AT3-Bcl-2 cells and will probably require a neuronal culture system that more closely mimics the in vivo situation of infected neurons.

## 4 Adjunctive Roles for Other Host Factors in Antibody-Mediated Control of Sindbis Virus Replication

Antibody to the E2 glycoprotein is clearly important for control of virus replication in the CNS, but may work in conjunction with other host antiviral defense mechanisms. To analyze the possible roles of phagocytic or mononuclear host cells (e.g. NK cells, monocytes) and of complement, persistently infected SCID mice were treated with cyclophosphamide, to deplete inflammatory cells, or with cobra venom factor, to deplete complement, at the same time that they were treated with antiviral antibody. Neither treatment affected antibody-mediated clearance (Levine et al. 1991). The ability of antibody to function in vitro using cultured neurons also indicates that neither complement nor monocytes plays an essential role in antibody-mediated control of SV replication in neurons.

Type 1 interferon (IFN-α/β), which is produced in normal amounts by infected SCID mice, is an important early antiviral defense mechanism and contributes in an essential way to initial control of virus replication in vivo. The replication of SV is highly sensitive to IFN-α/β in cultured cells, and SV rapidly induces the production of IFN in the CNS of infected animals (DESPRÈS, et al. 1995b; HACKBARTH et al. 1973; SHERMAN and GRIFFIN 1990; VILCEK 1964). Activation of IFN-α/β pathways that control the spread of virus is essential during the days after infection and prior to the appearance of antiviral antibody in the infected host. Mice that do not express IFNAR-I, an important component of the IFN-α/β receptor, and mice that do not express STAT-1, an essential molecule for IFN-α/β and IFN-γ signaling, are exquisitely susceptible to fatal infection with even the most benign strains of SV (BYRNES et al. 2000; RYMAN et al. 2000). These mice are unable to control virus replication early after infection, leading to very high titers of virus in the blood and CNS and rapid death. Although the mice are able to produce antibody normally in response to infection, they die before antibody is synthesized.

A number of studies have suggested that IFN-α/β also acts synergistically with antiviral antibody at later times to control infection in vivo (COPPENHAVER et al. 1995). In vitro studies that analyzed the control of virus replication in AT3-Bcl-2 cells show that IFN-α synergizes with anti-E2 antibody to control intracellular virus replication (DESPRÈS et al. 1995b). Antibody treatment restores the ability of infected cells to respond to IFN-α and results in further reduction of virus production and improvements in host protein synthesis (Fig. 4). Because there are a number of known IFN-induced mechanisms for regulating virus replication, we explored the possibility that the primary function of antibody was to restore the cellular response to IFN. The response to IFN-α is dependent in part on the presence of normal levels of intracellular [$K^+$] and on the ability of the cell to synthesis new proteins (NAKAGAWA et al. 1992). To determine if IFN is an essential component of the antibody-mediated control of virus replication, antibody was passively transferred to IFNAR-1 knockout mice 3–24h after infection. These mice were able to survive infection and clear infectious virus, indicating that response to IFN is not an absolute requirement for antibody to function in clearance (BYRNES et al. 2000) (Fig. 5).

## 5 Antibody Controls Virus Replication, but Does Not Eliminate Viral RNA

Prolonged follow-up of SCID mice persistently infected with SV that had received passive transfer of a single dose of antiviral antibody showed that virus was not permanently eliminated by the antibody treatment (LEVINE and GRIFFIN 1992). These SCID mice were unable to make their own immune response to the virus, and virus replication in the CNS resumed when antibody levels fell to below those needed for control. Viral RNA could be continuously detected in the CNS of antibody-treated SCID mice and also in the CNS of immunocompetent mice that had recovered from infection. These data indicate that antibody treatment does not

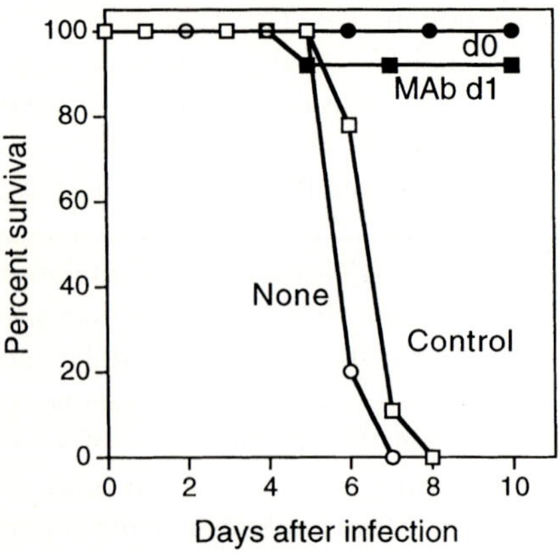


**Fig. 5.** Effect of deficiencies in the cellular response to interferon (IFN) on recovery from SV infection in the presence and absence of passively transferred antibody to the E2 glycoprotein of SV. A129 mice null for the IFNAR-1 receptor for IFN-α/β were infected subcutaneously with the Toto1101 strain of Sindbis virus (SV) and treated with an irrelevant monoclonal antibody (mAb) (control) or mAb to SV glycoprotein E2 at the time of infection or 24h later. (Data from BYRNES et al. 2000)

result in complete clearance of viral RNA from neurons and that virus-specific T cells, acting in conjunction with antibody in immunocompetent animals, also do not eliminate viral RNA (LEVINE and GRIFFIN 1992; TYOR et al. 1992). There is, therefore, a need for long-term antibody synthesis within the CNS for continued suppression of virus replication. B cells enter the CNS along with T cells and other mononuclear cells during the acute inflammatory response associated with SV-induced encephalomyelitis (MOENCH and GRIFFIN 1984; TYOR et al. 1989; TYOR and GRIFFIN 1993). B cells remain present and continue to synthesize antiviral antibody within the CNS for the life of the animal, providing control at the site where it is needed (TYOR et al. 1992).

Recent studies of SV infection and clearance in T-cell-deficient mice suggest a supplementary role for T cells in clearance of viral RNA (KIMURA and GRIFFIN 2000). As predicted by the outcome of antibody treatment of SCID mice (Fig. 2), mice deficient in CD8 T lymphocytes synthesize antibody normally and have normal clearance of infectious virus (Fig. 6). However, the decrease in brain viral

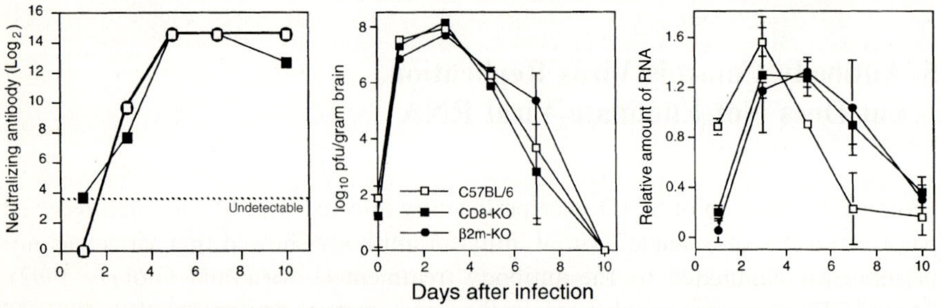


**Fig. 6.** Control of Sindbis virus (SV) infection in immunocompetent mice and in mice deficient in CD8 T cells. C57BL/6 mice lacking expression of CD8α or β-2 microglobulin (β2m) produce antibody and clear infectious virus as well as immunocompetent B6 mice. However, clearance of viral RNA is significantly slower, suggesting a noncytolytic role for CD8 T cells in control of SV replication in neurons. (Modified from KIMURA and GRIFFIN 2000)

RNA is slower in CD8 T-cell-deficient mice than in control mice with normal T cell function (Fig. 6). These observations suggest a previously unappreciated auxiliary role in virus RNA clearance for CD8 T cells in the CNS. Neurons are not harmed by this T-cell-mediated clearance mechanism, as they are not harmed by antibody, because there is no evidence of paralysis or neuronal cell death in recovering mice. This suggests that a soluble factor or factors produced by the CD8 T cells, rather than cytolytic activity, is most likely the mechanism for enhanced control of RNA synthesis.

## 6 Conclusions

Antibody specific for the E2 surface glycoprotein of SV is sufficient for the control of virus replication in persistently infected neurons. Neither phagocytic cells, complement, nor interferon is required for this effect. However, infected animals must survive long enough after infection for antibody to be produced, and this requires that neurons survive infection. IFN-$\alpha/\beta$ activity is essential for inhibition of virus replication and spread early after infection. Once the virus-specific immune response is generated, antiviral antibody is sufficient for viral clearance. However, clearance of viral RNA, but not infectious virus, is more rapid if virus-specific T lymphocytes are generated as well. Thus, virus clearance from the CNS, as from other tissues, involves overlapping mechanisms of antiviral host defenses. Further study is needed to determine how antibody signals intracellular shut-off of virus replication.

## References

Amor S, Scallan MF, Morris MM, Dyson H, Fazakerley JK (1996) Role of immune responses in protection and pathogenesis during Semliki Forest virus encephalitis. J Gen Virol 77:281–291

Byrnes AP, Durbin JE, Griffin DE (2000) Control of Sindbis virus infection by antibody in interferon-deficient mice. J Virol 74:3905–3908

Coppenhaver DH, Singh IP, Sarzotti M, Levy HB, Baron S (1995) Treatment of intracranial alphavirus infections in mice by a combination of specific antibodies and an interferon inducer. Trop Med Hyg 52:34–40

Desprès P, Griffin JW, Griffin DE (1995b) Antiviral activity of alpha interferon in Sindbis virus-infected cells is restored by anti-E2 monoclonal antibody treatment. J Virol 69:7345–7348

Desprès P, Griffin JW, Griffin DE (1995a) Effects of anti-E2 monoclonal antibody on Sindbis virus replication in AT3 cells expressing bcl-2. J Virol 69:7006–7014

Garry RF, Bishop JM, Park S, Westbrook K, Lewis G, Waite MRF (1979) Na+ and K+ concentrations and the regulation of protein synthesis in Sindbis virus-infected chick cells. Virology 96:108–120

Griffin D, Levine B, Tyor W, Ubol S, Desprès P (1997) The role of antibody in recovery from alphavirus encephalitis. Immunol Rev 159:155–161

Hackbarth SA, Reinarz ABG, Sagik BP (1973) Age-dependent resistance of mice to Sindbis virus infection: Reticuloendothelial role. J Reticulo Soc 14:405–424

Jan J-T, Chatterjee SB, Griffin DE (2000) Sindbis virus entry into cells triggers apoptosis by activating sphingomyelinase leading to the release of ceramide. J Virol 74:6425–6432

Kimura T, Griffin DE (2000) The role of CD8$^+$ T, cells and major histocompatibility complex class I, expression in the central nervous system of mice infected with neurovirulent Sindbis virus. J Virol 74:6117–6125

Levine B, Goldman JE, Jiang HH, Griffin DE, Hardwick JM (1996a) Bcl-2 protects mice against fatal alphavirus encephalitis. Proc Natl Acad Sci USA 93:4810–4815

Levine B, Griffin DE (1992) Persistence of viral RNA in mouse brains after recovery from acute alphavirus encephalitis. J Virol 66:6429–6435

Levine B, Hardwick JM, Trapp BD, Crawford TO, Bollinger RC, Griffin DE (1991) Antibody-mediated clearance of alphavirus infection from neurons. Science 254:856–860

Levine B, Huang Q, Isaacs JT, Reed JC, Griffin DE, Hardwick JM (1993) Conversion of lytic to persistent alphavirus infection by the bcl-2 cellular oncogene. Nature 361:739–742

Levine B, Jiang HH, Kleeman L, Yang G (1996b) Effect of E2 envelope glycoprotein cytoplasmic domain mutations on Sindbis virus pathogenesis. J Virol 70:1255–1260

Moench TR, Griffin DE (1984) Immunocytochemical identification and quantitation of mononuclear cells in cerebrospinal fluid, meninges, and brain during acute viral encephalitis. J Exp Med 159:77–88

Nakagawa Y, Petricoin EF, Akai H, Grimley PM, Rupp B, Larner AC (1992) Interferon-alpha-induced gene expression: evidence for a selective effect of ouabain on activation of the ISGF3 transcription complex. Virology 190:210–220

Ryman K, Klimstra W, Nguyen K, Biron C, Johnston R (2000) Alpha/beta interferon protects adult mice from fatal Sindbis virus infection and is an important determinant of cell and tissue tropism. J Virol 74:3366–3378

Sherman LA, Griffin DE (1990) Pathogenesis of encephalitis induced in newborn mice by virulent and avirulent strains of Sindbis virus. J Virol 64:2041–2046

Strauss JH, Strauss EG (1994) The alphaviruses: gene expression, replication and evolution. Microbiol Rev 58:491–562

Tyor WR, Griffin DE (1993) Virus specificity and isotype expression of intraparenchymal antibody-secreting cells during Sindbis virus encephalitis in mice. J Neuroimmunol 177:475–482

Tyor WR, Moench TR, Griffin DE (1989) Characterization of the local and systemic B cell response of normal and athymic nude mice with Sindbis virus encephalitis. J Neuroimmunol 24:207–215

Tyor WR, Wesselingh S, Levine B, Griffin DE (1992) Long term intraparenchymal Ig secretion after acute viral encephalitis in mice. J Immunol 149:4016–4020

Ubol S, Levine B, Lee S-H, Greenspan NS, Griffin DE (1995) Roles of immunoglobulin valency and the heavy-chain constant domain in antibody-mediated downregulation of Sindbis virus replication in persistently infected neurons. J Virol 69:1990–1993

Ubol S, Tucker PC, Griffin DE, Hardwick JM (1994) Neurovirulent strains of alphavirus induce apoptosis in bcl-2-expressing cells; role of a single amino acid change in the E2 glycoprotein. Proc Natl Acad Sci USA 91:5202–5206

Ulug ET, Bose Jr. HR (1985) Effect of tunicamycin on the development of the cytopathic effect in Sindbis virus-infected avian fibroblasts. Virology 143:546–557

Ulug ET, Garry RF, Bose Jr HR (1989) The role of monovalent cation transport in Sindbis virus maturation and release. Virology 172:42–50

Vilcek J (1964) Production of interferon by newborn and adult mice infected with Sindbis virus. Virology 22:651–652

# Role of Mucosal Antibodies in Viral Infections

H. KATO, R. KATO, K. FUJIHASHI, and J.R. MCGHEE



## 1 Introduction

Human mucosal tissues form a boundary with the outside world, with a surface area exceeding 400m$^2$ and inevitably, this becomes the primary site of pathogen invasion (MCGHEE and KIYONO 1999). Mucosal tissues are the first line of immunological defense against pathogens and foreign antigens encountered from the environment. Despite this protective barrier, over 95% of viruses, bacteria, and parasites causing common human infectious diseases enter the host by directly invading the epithelium that covers the mucosa. Since the mucosa is the portal of entry for pathogens, several host immune defense mechanisms are required for protection from them. At the same time, mucosal surfaces also absorb nutrients indispensable for life, and food proteins are actively taken up in these mucosal

Departments of Microbiology and Oral Biology, The Immunobiology Vaccine Center, The University of Alabama at Birmingham, Birmingham, AL 35294-2170, USA areas. This may, at first glance, appear to be contradictory; however, host defense mechanisms at the mucosa are carefully regulated, forming cellular and molecular networks in order to induce and maintain an appropriate immunological barrier against inhaled and ingested antigens, including allergens, food products, and pathogenic microorganisms. Further, compelling evidence has been provided that the immune systems of the mucosal and systemic compartments are largely independent from each other, although this is still matter of debate.

## 2 Overview of Innate Mucosal Immunity

Host defense at mucosal sites is mediated by both innate and acquired mechanisms for immune protection. Innate mechanisms include both physical and physiochemical components. Among these, peristalsis of mucosal smooth muscle helps to expel pathogens from the gastrointestinal and respiratory tracts, and respiratory epithelial cells express cilia that beat in unison to move foreign material up the respiratory tree. The mucus secreted from specialized mucosal epithelial cells, termed goblet cells, also acts as a physiochemical barrier, and the mucus layer plays a major role in mucosal host defense. Mucosal secretions contain digestive enzymes, e.g., mucin, lysozyme, α1-antitrypsin, lactoferrin and lactoperoxidase, all of which possess antimicrobial effects. Defensins, a family of antimicrobial peptides, are also major innate molecules that provide host defense. Defensins consist of 30–42 amino acids with an overall positive charge and contain six cysteine residues that form three intermolecular disulfide bonds. They are classified into either α- or β-defensins based upon the distribution of the cysteine residues. Thus, α-defensins have two contiguous cysteine residues, while β-defensins have cysteines separated by six amino acids (Bevins et al. 1999; Hecht 1999). Defensins are expressed by neutrophils, macrophages, and other cells and are found in mucosal tissues such as the gastrointestinal (GI) tract, in tracheal epithelial cells, cells of the mammary gland, the oral mucosa, the salivary glands and the epithelium of the reproductive tract (Jones and Bevins 1992; Schonwetter et al. 1995; Yarus et al. 1996; Zhao et al. 1996, 1999; Stolzenberg et al. 1997; Svinarich et al. 1997a,b; Quayle et al. 1998; Singh et al. 1998; Mathews et al. 1999; Mizukawa et al. 1999). In the small intestine, Paneth cells were shown to produce granules rich in α-defensins and additional antimicrobial peptides and to secrete them into the lumen (Jones and Bevins 1992; Ouellette 1999). The antimicrobial mechanisms of defensins are not fully understood; however, defensins have cytotoxic activity against bacteria, fungi, parasites, and viruses and influence complement activation (Panyutich et al. 1994). These molecules may also act on host cells and stimulate chemotactic activity for T lymphocytes and monocytes by direct effects or by induction of interleukin (IL)-8 synthesis (Van Wetering et al. 1997). Matrilysin-deficient mice, which lack α-defensins, were shown to have decreased antimicrobial activity, suggesting that defensins and matrilysins function in synergy for antimicrobial host defense

(WILSON et al. 1999). Further, defensins were shown to effect adaptive immunity by enhancing systemic antibody (Ab) responses through help provided by $CD4^+$ Th1- and Th2-type cytokines (LILLARD et al. 1999).

## 3 Induction of Mucosal Immune Responses

The acquired mucosal immune response consists of both humoral and cell-mediated components. Of these two forms, humoral or Ab-mediated immunity, represented by the production of secretory IgA (S-IgA) Abs, is a distinctive feature of the mucosal host defense system. IgA is the major isotype produced in humans, and represents more than twice the amount (approximately 70mg/kg) of other isotypes, including the four human IgG subclasses. Furthermore, IgA is dominant in mucosal tissues and secretions and plays a central role in mucosal host defense by preventing pathogen invasion across the epithelium.

Several clinically important viruses are transmitted by the oral (e.g., rotavirus and Norwalk-type viruses), nasal (e.g., influenza virus and respiratory syncytial virus), and genitourinary tract [e.g., human immunodeficiency virus (HIV)] routes. Initially, these viruses invade mucosal surfaces of the GI, upper and lower respiratory tract, and epithelial cells in the vagina of the reproductive tract, where they replicate and subsequently invade the systemic compartment. Therefore, it is clear that mucosal S-IgA Ab responses to viral pathogens are regulated to block viral entry and to diminish the level of viruses invading the host. In fact, a number of clinical and experimental studies have shown that induction of antigen-specific S-IgA Abs correlate with protection against GI and respiratory tract pathogens. S-IgA Abs are also thought to interfere with viral invasion of the systemic compartments in at least three characteristic pathways that involve polymeric IgA transport. (MAZANEC et al. 1993; LAMM 1997). Thus, considerable effort is currently being made to develop effective mucosal vaccines to preferentially induce S-IgA antiviral Ab responses. In addition to S-IgA, recent studies also suggest that systemic IgG Abs may protect mucosal sites from infection by the simian immunodeficiency virus (SIV) (BABA et al. 2000; MASCOLA et al. 2000).

Although the mucosal tissues extend throughout the body, they form an integrated and connected network that is well regulated at the tissue, cell, and molecular levels. From the standpoint of both structure and function, the mucosal immune system can be divided into two distinct compartments. The first consists of inductive sites, where antigen is encountered and initial immune responses such as induction of effector and memory lymphocytes occur. The second type consists of the effector sites, where high numbers of IgA-producing plasma cells are present and where mucosal antigen-specific humoral and cell-mediated immune responses counter bacterial and viral infections. The primary mucosal inductive sites include the gut-associated lymphoreticular tissues (GALT), nasopharyngeal-associated lymphoreticular tissues (NALT), and in nonhuman vertebrate models such as rats and mice, the bronchus-associated lymphoreticular tissues (BALT). Induction of mucosal immune responses is initiated following antigen uptake and transport across the epithelial barrier into the inductive site itself. For example, Peyer's patches, which are major components of GALT, are the most extensively studied inductive tissue. They are covered by a unique epithelium, termed follicle-associated epithelium (FAE), which is composed of unilaminate epithelial cells and forms the boundary between the luminal environment and intestinal lymphoid tissues. The FAE contains microfold or M cells, which have short microvilli, small cytoplasmic vesicles and few lysosomes (NEUTRA et al. 1996b). The M cells efficiently take up and transport a wide variety of antigens, ranging in size from macromolecules, particles, and viruses, to bacteria and even small parasites, by endocytosis and transcytosis mechanisms (BOCKMAN and STEVENS 1977; OWEN 1977; JOEL et al. 1978; WOLF et al. 1981; GRUTZKAU et al. 1990; JONES et al. 1994; NEUTRA et al. 1996a,b). The M cells themselves do not significantly degrade antigens but instead deliver them to the basolateral membrane side of the FAE (WOLF and BYE 1984; GEBERT et al. 1996). The basolateral surface of M cells are deeply invaginated to form an "intraepithelial pocket" where T and B lymphocytes and macrophages are resident (MESTECKY and MCGHEE 1987; GEBERT et al. 1996). Several pathogenic bacteria and viruses adhere to M cells at the apical membrane via specific receptors and invade into the tissue by exploiting transportation mechanisms of M cells (JONES et al. 1994). The origin of the M cell remains controversial. However, the finding that the number of M cells is rapidly up-regulated during infection suggest that they are differentiated from enterocytes through immunological mechanisms (SAVIDGE et al. 1991; CUVELIER et al. 1993; SAVIDGE and SMITH 1995). Further, recent in vitro studies showed that a human intestinal cell line (Caco-2) could be converted into M cells that retained transcytosis activity when cultured in the presence of Peyer's patch lymphocytes or the Raji B cell line, but not by splenocytes or the Jurkat T cell line (KERNEIS et al. 1997). Thus, at present, M cells are thought to originate from enterocytes during proinflammatory responses through a Peyer's patch B cell-dependent mechanism.

The interaction between transported antigens and antigen-presenting cells (APCs) in mucosal inductive sites primes and elicits the initial immune response. The Peyer's patches anatomically and functionally comprise three distinct regions, termed the dome, B cell follicle and T cell parafollicular area. Interestingly, a recent study clearly shows that the localization of T and B cells in Peyer's patches is controlled by the requirement of secondary lymphoid tissue chemokine (SLC) for their arrest in these tissues; that is to say, T cells preferentially arrest in high endothelial venules (HEVs) of the T cell areas displaying high levels of SLC, and this arrest is selectively inhibited by blocking the function of CC chemokine receptor 7 (CCR7), which is the receptor for SLC and macrophage inflammatory protein (MIP)-3β. In contrast, B cell arrest does not require SLC or MIP-3β and occurs in the follicular regions where the level of SLC expression is low or negative (WARNOCK et al. 2000). The region between the FAE and the B cell follicles is the dome, where APCs including dendritic cells (DCs) and macrophages (Mϕ) capture antigens immediately after transport through the M cells (SPENCER et al. 1986;

KELSALL and STROBER 1996a; HOPKINS and KRAEHENBUHL 1997). Based on the expression of specific molecules, the DCs present in the dome regions ($CD11c^+$, $M342^-$ and $NLDC\text{-}145^-$) are considered to be immature and actively capture antigens and process them (KELSALL and STROBER 1996b). These DCs most likely migrate into the T cell zone and become mature DCs ($CD11c^+$, $M342^+$ and $NLDC\text{-}145^+$), which support primary immune responses by presenting antigens to $CD4^+$ and $CD8^+$ T cells.

B cell isotype switching from surface $IgM^+$ ($sIgM^+$) B cells to $sIgA^+$ B cells occur in germinal center regions of the B cell follicle (BUTCHER et al. 1982; LEBMAN et al. 1987; GEORGE and CEBRA 1991; WEINSTEIN and CEBRA 1991; WEINSTEIN et al. 1991); however, unlike systemic lymph nodes and spleen, plasma cell development does not occur in the germinal center itself. Several in vitro studies showed that transforming growth factor (TGF)-β, which is generally considered to be immunosuppressive, is a major cytokine for $\mu \rightarrow \alpha$ isotype switching (COFFMAN et al. 1989; SONODA et al. 1989, 1992; LEBMAN et al. 1994). TGF-β has been shown to stimulate isotype switching to IgA in cultures of lipopolysaccharide (LPS)-stimulated B cells (COFFMAN et al. 1989). Further, a recent study showed that IgA levels in serum and external secretions and numbers of IgA-Ab-forming cells (AFC) were markedly decreased in TGF-β knockout mice (VAN GINKEL et al. 1999).

Since the respiratory and GI tracts are the major sites for primary infections by viruses, characterization of lymphoid tissues of the respiratory tract is under intense investigation. In rodents, NALT is considered to be the major inductive tissue for the upper respiratory tract (BELAI et al. 1977; REUMAN et al. 1989; SPIT et al. 1989; KUPER et al. 1990; ASANUMA et al. 1995). In humans, the Waldeyer's ring, which includes the tonsils and adenoids, may correspond to the NALT of rodents (GOERINGER and VIDIC 1987). M cells are present in the epithelial layer overlying NALT and the crypts of tonsils, and lymphoid follicles containing germinal centers and both B and T cell zones are observed in this tissue (SPIT et al. 1989; VAN DER VEN and SMINIA 1993; WU and RUSSELL 1993; ASANUMA et al. 1997; HERITAGE et al. 1997). The presence of M cells is characteristic of mucosal inductive sites such as Peyer's patches. In addition, the cell populations of NALT are similar to cells of Peyer's patches and exhibit a high frequency of $CD4^+$ Th0 type cells in this tissue (HIROI et al. 1998). These findings clearly suggest that NALT is the major mucosal inductive tissue for the upper respiratory tract.

Following antigen stimulation in NALT and GALT inductive sites, antigen-sensitized T and B cells exit, migrate to the regional lymph nodes, and then reach the systemic circulation through the thoracic duct. Ultimately, these lymphocytes enter remote exocrine glands (e.g., salivary, mammary, and lacrimal glands) and mucosa-associated tissues (e.g., GI, respiratory, and genitourinary tracts), termed effector sites (MCGHEE et al. 1989). In early studies, lymphocytes were shown to migrate from Peyer's patches and mesenteric lymph nodes (MLN) into the lamina propria of the GI tract, the vagina, and salivary and mammary glands (CRAIG and CEBRA 1971; CRAIG and CEBRA 1975; MCWILLIAMS et al. 1975; ROUX et al. 1977; HUSBAND and GOWANS 1978; MCDERMOTT and BIENENSTOCK 1979). Later studies indicated that lymphocyte homing into lymphoid tissues is regulated by the association of homing receptors present on B and T lymphocytes with adhesion molecules specifically expressed on HEV. For example, lymphocyte migration into peripheral lymph nodes (PLN) is largely regulated by the interaction of L-selectin on B and T cells with peripheral node addressin (PNAd) on HEV in PLN (GALLATIN et al. 1983; ARBONES et al. 1994). In contrast, monoclonal Abs (mAbs) to $\alpha_4$ and/or $\beta_7$ were shown to prevent lymphocyte binding to HEV via mucosal addressin cell adhesion molecule-1 (MAdCAM-1) in Peyer's patches in vitro and in vivo (HOLZMANN et al. 1989; BELL and ISSEKUTZ 1993; BERLIN et al. 1993; HAMANN et al. 1994). In support of this, $\beta_7$-gene knockout mice exhibit dramatically smaller Peyer's patches, with decreased cellularity and rudimentary follicles, and significantly decreased cell numbers in lamina propria and intestinal intraepithelial lymphocytes (WAGNER et al. 1996). These observations clearly demonstrated that trafficking into Peyer's patches is mainly mediated by $\alpha_4\beta_7$ integrin binding to MAdCAM-1 expressed on HEV in the Peyer's patch (GALLATIN et al. 1983; ROSEN et al. 1985; HOLZMANN et al. 1989; HOLZMANN and WEISSMAN 1989; ROSEN et al. 1989; WATSON et al. 1990; HU et al. 1992; WILLIAMS and BUTCHER 1997). In addition, MAdCAM-1 is also expressed in the GI tract lamina propria and on follicular dendritic cells (FDCs) in the GALT (STREETER et al. 1988; DOGAN et al. 1997; ROTT et al. 1997; SZABO et al. 1997). However, a recent study showed that NALT has unique characteristics that differ from Peyer's patches. Unlike Peyer's patches, all NALT HEV expressed PNAd, either associated with MAdCAM-1 or alone, and NALT FDCs expressed both MAdCAM-1 and vascular cell adhesion molecule (VCAM)-1 (CSENCSITS et al. 1999). Further, a monoclonal anti-L-selectin Ab blocked almost all naive lymphocyte binding to NALT HEV, indicating that the interaction of L-selectin and PNAd, rather than that of $\alpha_4\beta_7$ integrin and MAdCAM-1, is essential for NALT lymphocyte homing. This observation suggests that NALT and Peyer's patches exhibit important differences in terms of cell migration behavior for subsequent induction of immune responses in effector sites.

In mucosal effector sites, B cells clonally expand and differentiate into mature IgA-secreting plasma cells under the influence of certain cytokines, including IL-5 and IL-6 (MURRAY et al. 1987; BEAGLEY et al. 1988, 1989; MCGHEE et al. 1989; SONODA et al. 1989; RAMSAY et al. 1994a). In addition, IL-2 also enhances IgA secretion (LEBMAN et al. 1990), and IL-10 enhances IgA secretion in humans (DEFRANCE et al. 1992; BRIERE et al. 1994). Also, the reduction of S-IgA Ab levels in mucosal secretions and IgA AFC numbers in mucosal tissues was reported in TCR-$\delta^{-/-}$ mice, indicating that $\gamma\delta$ T cells play an important role on the regulation of mucosal IgA responses (FUJIHASHI et al. 1996a,b). This cell distribution pathway from IgA inductive tissues to effector sites is often termed the common mucosal immune system (CMIS) (MESTECKY 1987) (Fig. 1).

One unique characteristic of mucosal effector sites is the frequent presence of CMIS-independent IgA-producing plasma cells derived from so-called B-1 cells, which are not observed in inductive sites. For example, up to 40% of IgA-producing plasma cells in the murine GI tract are derived from B-1 cells and up to 50% of intestinal B cells are $CD5^+$ in humans (PETERS et al. 1989; KANTOR and HERZENBERG 1993). Further, other effector tissues such as submandibular and

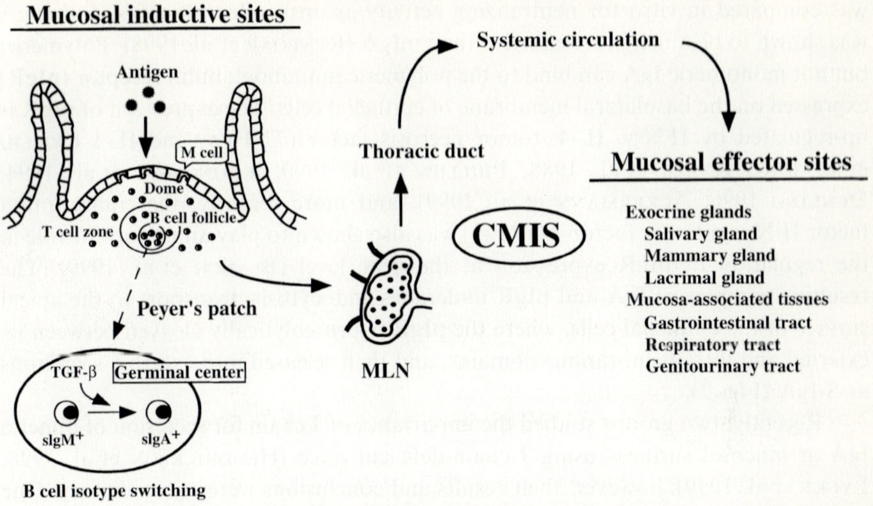


**Fig. 1.** Concept of inductive and effector sites for mucosal immune responses. Primary mucosal immune responses are initiated following M cell antigen transport across the epithelial barrier into the mucosal inductive sites. Isotype switching of antigen-sensitized B cells into surface-IgA-positive ($sIgA^+$) B cells occurs in germinal centers in the B cell follicle, presumably under the influence of TGF-β. Following antigen stimulation, antigen-sensitized T and B cells leave the inductive tissue and migrate through the mesenteric lymph node (*MLN*) and then into the thoracic duct to reach the systemic circulation. Finally, these cells enter remote exocrine glands and mucosa-associated tissues, termed mucosal effector sites, where terminal differentiation of B cells into IgA-secreting plasma cells occurs. This pathway from mucosal inductive tissues to effector sites is termed the "common mucosal immune system" (*CMIS*)

lacrimal glands contain a high frequency of IgA-committed B-1 cells (Hiroi et al. 1994; Saitoh-Inagawa et al. 2000). B-1 cells are distinguishable from conventional B cells (B-2 cells) by cell surface markers; B-1 cells express CD5, Mac-1 and $B220^{low}$, while B-2 cells are $B220^{high}$ and $CD5^-$ (Coffman and Weissman 1981; Murakami et al. 1995). In addition, B-1 cells are derived from the peritoneal cavity, and a portion of B-1 B cells develop through T cell-independent mechanisms. B-1 cells also react with commensal bacteria and orally inoculated virus (Bos et al. 1996; Franco and Greenberg 1997; Macpherson et al. 2000). Interestingly, a recent study using knockout mice clearly showed that the IL-5/IL-5 receptor signaling pathway is indispensable for differentiation of B-1 cells (Hiroi et al. 1999).

## 4 Structure and Function of Mucosal IgA Antibodies

IgA Abs secreted by plasma cells in effector sites mainly form dimers, trimers, or tetramers with a 15kDa polypeptide, termed J chain (Mestecky and McGhee 1987). The biological effect of monomeric IgA (mIgA) and polymeric IgA (pIgA) was compared in vitro for neutralizing activity against influenza virus, and pIgA was shown to be much more effective than mIgA (Renegar et al. 1998). Polymeric, but not monomeric IgA can bind to the polymeric immunoglobulin receptor (pIgR) expressed on the basolateral membrane of epithelial cells. The expression of pIgR is up-regulated by IFN-γ, IL-4, tumor necrosis factor (TNF)-α, and IL-1 (Sollid et al. 1987; Kvale et al. 1988; Phillips et al. 1990; Youngman et al. 1994; Denning 1996; Ackermann et al. 1999), and more recently, the transcription factor IFN regulatory factor-1 (IRF-1) was also shown to play an important role in the regulation of pIgR expression at the gene level (Blanch et al. 1999). The resulting complex of IgA and pIgR undergoes endocytosis, transport to the apical surface of the epithelial cells, where the pIgR is proteolytically cleaved between its external and intramembranous domains, and then released into external secretions as S-IgA (Fig. 2).

Recently, two groups studied the importance of J chain for secretion of dimeric IgA at mucosal surfaces using J chain-deficient mice (Hendrickson et al. 1996; Lycke et al. 1999); however, their results and conclusions were contradictory: One group concluded that J chain was not absolutely required for IgA transport into mucosal secretions. In particular, comparable levels of IgA Abs occured in mucosa/glandular secretions, but not in feces from J chain-deficient mice. In addition, those IgA Abs were not associated with secretory component (SC) and were transported by a pIgR-independent mechanism (Hendrickson et al. 1996). The second group found remarkable reductions in IgA Ab levels at mucosal surfaces of J chain knockout mice when sensitive methods were used for detection of mucosal IgA Abs (Lycke et al. 1999). At present, it is difficult to reconcile these different results. Nonspecific leakage of IgA from serum to mucosal surfaces may occur in J chain-deficient mice, since serum levels were about 30 times higher than those of normal

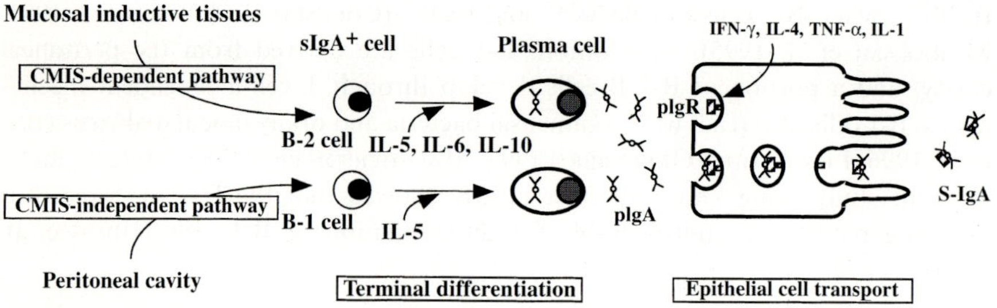


**Fig. 2.** Differentiation of surface IgA-positive (sIgA$^+$) B cells and IgA antibody (Ab) secretion in mucosal effector tissues. In these tissues, terminal differentiation of common mucosal immune system (*CMIS*)-dependent sIgA$^+$ B cells (*B-2* B cells) into mature IgA-secreting plasma cells occurs under the influence of IL-5, -6 and -10. IL-5 also induces the differentiation of CMIS-independent sIgA$^+$ B cells (*B-1* B cells). IgA Abs secreted by plasma cells mainly occur as dimers or other polymers with J chain. Polymeric IgA (*pIgA*) binds to the polymeric immunoglobulin receptor (*pIgR*) expressed on the basolateral membrane of epithelial cells. The pIgA-pIgR complex undergoes endocytosis and transport to the apical surface of the epithelial cells, where cleavage of pIgR occurs, with subsequent release into external secretions as S-IgA-containing secretory component (*SC*). The expression of pIgR is up-regulated by IFN-γ, IL-4, TNF-α and IL-1

mice. Furthermore, J chain is considered to be essential in host defense by IgA present in the mucosal secretions. In fact, the formation of S-IgA was shown to be particularly resistant to proteolytic degradation: and in the absence of J chain, IgA is more susceptible to degradation due to its lack of association with SC. A marked reduction in IgA Ab levels in the feces from J chain-deficient mice was shown in both reports. An indispensable role for pIgR in pIgA secretion was recently shown by using pIgR-deficient mice (Johansen et al. 1999; Shimada et al. 1999). These mice exhibit high levels of serum IgA; however, pIgA is not transported into external secretions. This pIgR transport process is highly efficient, and the normal human adult produces 2.8–3.0g of S-IgA per day in the GI tract alone.

## 5 Protective Mechanisms of Mucosal Antiviral Antibodies

The major human Ig isotypes in systemic immunity are IgM and IgG, where IgM and both IgG1 and IgG3 effectively activate complement. IgA is the predominant Ab in the mucosal compartment, and as shown in various clinical and experimental models, IgA plays an important role in antiviral host defense without activation of complement, that is, without inducing any serious inflammatory responses and tissue damage in the mucosal compartments. Up until now, S-IgA has been shown to interact with antigens in three anatomically distinct compartments and each step presumably interferes with viral invasion (Mazanec et al. 1993; Lamm 1997). Traditionally, S-IgA was thought to act as a mucosal barrier to prevent the attachment of viruses to mucosal surfaces. However, recent studies indicate that mucosal pIgA Abs possess at least two additional functions, "intraepithelial neutralization" and "excretory function", as antiviral host defense mechanisms. In addition to S-IgA Abs, recent studies suggest that systemic IgG Abs also play an important role in host defense against viral infections in mucosal compartments.

### 5.1 Mucosal Barrier Functions for S-IgA Antibodies

As described above, pIgA is efficiently transported across the epithelium and released as S-IgA into the external secretions. The importance of S-IgA as a mucosal barrier has been suggested by clinical and experimental observations that clearance of viral infections strongly correlates with the presence of antigen-specific IgA Abs in mucosal secretions (Smith et al. 1967; Mills et al. 1971; Scott and Sydiskis 1976; Liew et al. 1984; Kris et al. 1985; Offit and Clark 1985). In earlier studies, antigen-specific IgA Abs were shown to bind to the virus particle and prevent its attachment to cells in vitro (Williams and Gibbons 1972; Fubara and Freter 1973; Taylor and Dimmock 1985; Outlaw and Dimmock 1990). Further, passive transfer of mAbs to mucosal surfaces protected the host from viral infection in vivo (Mazanec et al. 1987; Weltzin et al. 1996; Ruggeri et al. 1998; Weltzin and

MONATH 1999). For example, it was shown that resistance to influenza virus infection could be abrogated by nasal instillation of anti-IgA but not anti-IgG or anti-IgM Abs (RENEGAR and SMALL 1991a,b). These studies clearly illustrate one major role for S-IgA Abs as a mucosal barrier to prevent initial viral infection [Fig. 3(1)]. There is ample evidence to suggest that passively transferred IgA Abs also act as mucosal barriers under more physiological conditions. Maternal S-IgA has been shown to mediate protective immunity of nursing infants (OGRA et al. 1983; SHERIDAN et al. 1984; KRAMER and CEBRA 1995; PICKERING et al. 1995; BISHOP et al. 1996; HODGINS et al. 1999). The mucosal immune system in the neonatal period is largely nonfunctional and no IgA is detectable in the mucosal secretions during the first few days after birth. By contrast, colostrum and milk produced by the mammary gland, which is a major mucosal effector site, contains specific antibodies against a wide range of viruses encountered through the enteric and respiratory mucous membranes (MATA and WYATT 1971; BOHL et al. 1972;

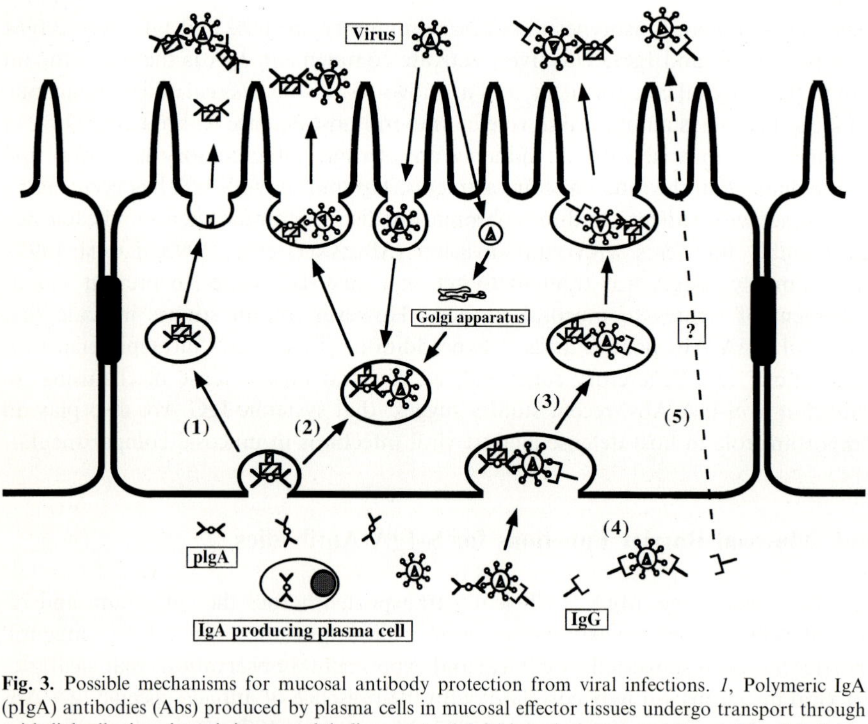


**Fig. 3.** Possible mechanisms for mucosal antibody protection from viral infections. *1*, Polymeric IgA (pIgA) antibodies (Abs) produced by plasma cells in mucosal effector tissues undergo transport through epithelial cells via polymeric immunoglobulin receptor (pIgR) into the lumen, and the resulting S-IgA Abs bind to viruses at mucosal surfaces. *2*, During transcytosis, IgA Abs interact with newly synthesized viral proteins and transcytosing virions within the epithelial cells and interfere with viral replication and/or assembly. This is followed by release of the immune complex into the lumen through an IgA transport mechanism. *3*, The pIgA Abs interact with viruses in mucosal effector tissues in the presence or absence of monomeric IgA or IgG, and the resulting immune complexes are excreted by an IgA transport mechanism. *4*, IgG Abs produced in systemic or mucosal effector tissues bind to viruses in mucosal effector sites and neutralize them. *5*, Further, IgG may leak or perhaps exhibit an additional transport mechanism for secretion toward the lumenal side when present in significant titers from the peripheral blood

Downham et al. 1976; McLean and Holmes 1980; Hanson and Soderstrom 1981; Ogra et al. 1983). In breast-fed infants, the acquisition of S-IgA Abs represents an ideal mechanism to compensate for a lack of mucosal immunity. A large amount of clinical and epidemiologic information, mostly from developing countries, has suggested that breast milk-fed infants are less prone than bottle-fed infants to the development of acute respiratory and enteric mucosal infections (Schoub et al. 1978; De Martino et al. 1987; Ali et al. 1989; Linhares et al. 1989; Minamishima et al. 1994; Naficy et al. 1999). Thus, both actively produced and passively transferred S-IgA Abs play important roles as a first line of defense in prevention of viral infections.

## 5.2 Intraepithelial Neutralization of Viruses

Generally, Ab responses function against both extracellular and cell surface antigens, whereas cell-mediated immunity forms a surveillance system for intracellular pathogens. However, in contrast to other isotypes, IgA Abs are actively transported, and these intracellular stages also represent opportunities for Abs to function. Based on these findings, it was hypothesized that IgA Abs may interfere with viral replication by binding viral proteins within infected epithelial cells. This is logical since viruses first enter the epithelium and replicate in these cells, and pIgA is also being transported through epithelial cells at the same time. This hypothesis was tested by use of a polarized epithelial monolayer system (Mazanec et al. 1992, 1995; Fujioka et al. 1998). The pIgR-expressing polarized epithelial cells were infected with Sendai virus via the apical surface, and monoclonal IgA Abs specific for the viral envelope glycoprotein hemagglutinin-neuraminidase (HN) were added to the basolateral compartment after virus infection. Antigen-specific pIgA, but not IgG, was co-localized with viral protein within the cells, as determined by immunofluorescence, and viral titers were reduced in the supernatants and cell lysates from monolayers treated with virus-specific IgA but not IgG (Mazanec et al. 1992). The addition of anti-IgA Abs to the supernatants did not interfere with the neutralizing activity of IgA (Mazanec et al. 1992). Additional immunoelectron microscopy studies showed that the interaction of viral proteins with IgA but not IgG occurred within the epithelial cells themselves (Fujioka et al. 1998).

In general, the viral life cycle consist of the stages of: (1) viral capsid uncoating, (2) transcription, and (3) translation of the viral genome, (4) posttranscriptional modification of synthesized viral protein, and (5) assembly and budding of virions from the epithelial cell. The HN glycoproteins of Sendai virus are synthesized in infected cells, glycosylated, processed through the Golgi apparatus, and transported to the apical surface, where assembly and budding of virions occur. Thus, it is most probable that IgA functions to interfere with viral replication by interacting with newly synthesized viral proteins within epithelial cells during the process of transcytosis. This function of IgA was also shown to occur for influenza-virus-infected polarized epithelial cells (Mazanec et al. 1995). Further, although the mucosal transmission system of HIV is known to differ from those of Sendai virus and influenza virus, intracellular neutralization of HIV was also reported in vitro (BOMSEL et al. 1998). HIV can be rapidly internalized by the epithelial cells, transcytosed toward the basolateral side of the cell, and released into the mucosal compartments when this virus is produced through the direct contact of HIV-infected peripheral blood mononuclear cells with the apical surface of the mucosal epithelial cells (BOMSEL 1997). HIV envelope-protein-specific dimeric IgA or IgM Abs were shown to inhibit HIV infection intracellularly during the transcytosis process by use of the polarized epithelial monolayer system. In addition, laser-scanning confocal microscopy analysis indicated that HIV-specific IgA or IgM Abs trapped the transcytosing virions within the apical recycling endosome. Further, this study showed that pIg-HIV immune complexes were released into the apical chamber, but not the basolateral medium, by using an IgA-transport mechanism.

Intraepithelial virus neutralization by IgA in vivo has been studied in a rotavirus infection model (BURNS et al. 1996). Rotaviruses are transmitted by the fecal-oral route and primarily affect the GI tract. They cause severe diarrhea in infants and young children and are the most significant factor in mortality due to infantile diarrhea in the third world (ESTES and COHEN 1989; GLASS et al. 1994). To determine whether an intraepithelial neutralization mechanism inhibits viral infection in vivo, a murine "backpack tumor" transplantation model was used. Hybridoma B cells producing mAbs against the outer capsid viral protein VP4 or against the major inner capsid viral protein VP6 of rotavirus were transplanted into the upper back of mice and analyzed for protection against subsequent viral infection. In this experiment, it was shown that several monoclonal IgA Abs to VP6, which did not have neutralizing activity in vitro or in vivo, when passively transferred to the luminal side of the intestinal tract, could prevent primary and could resolve chronic murine rotavirus infection. These results support the notion that the anti-rotavirus effect of the neutralizing IgA mAbs occurred during IgA transcytosis rather than as an extracellular event in the GI tract. These findings strongly suggested that antigen-specific IgA Abs inactivate viruses intracellularly to inhibit viral replication during transcytosis. Thus, in addition to a mucosal barrier function to block pathogens from invading the body, pIgA Abs also function inside epithelial cells of the mucosal compartment to prevent viral pathogens from spreading to systemic tissues by interactions with viral outer and/or inner capsid proteins [Fig. 3 (2)]. This intracellular neutralization mechanism by pIgA is thought to be an especially effective host defense mechanism against intracellular pathogens, especially viruses.

## 5.3 Excretory Antiviral Functions for pIgA

Antigen-sensitized B cells in mucosal inductive sites home to mucosal tissues and differentiate into IgA-producing plasma cells in the effector sites. Theoretically, it is possible that IgA interacts with antigens and forms immune complexes of pIgA and antigens in these tissues. However, since IgA does not activate complement and cannot lyse or inactivate viral antigens efficiently, the pIgA immune system appears to have evolved additional mechanisms to process these immune complexes. Thus, the hypothesis was raised that antigen-pIgA complexes may be excreted from the cell and onto the mucosal surface. Using a polarized epithelial monolayer system, it was shown that soluble immune complexes of dinitrophenyl (DNP)/biotin-bovine serum albumin and anti-DNP dimeric IgA could be transported from the basolateral to the apical surface and then released into the upper chamber (KAETZEL et al. 1994). This in vitro experiment clearly supports the possibility that transcytosis of antigen-IgA complexes also occurs across mucosal epithelial cells in vivo. In additional studies, it was shown that monomeric IgA or IgG antigen complexes were also transported in the pIgA-Ag complex (KAETZEL et al. 1994). It has not been shown whether this mechanism occurs for excretion of viruses that have invaded mucosal tissues. However, viruses have numerous chances to invade the host, since mucosal effector tissues are generally separated from the external environment by a monolayer of epithelial cells. It is probable that intact viruses or viral proteins can bind to pIgA in these effector sites and then be excreted intact from the host [Fig. 3(3)]. Thus, IgA Abs play important roles in the prevention of virus invasion of the systemic compartment without inducing serious complement-mediated inflammatory events.

## 5.4 Mucosal IgG Antiviral Antibodies

In humans, IgA deficiency is reported to occur in approximately 1 in 600 individuals of Northern European ancestry (BURROWS and COOPER 1997). Although a high incidence of infection does occur in both mucosal and systemic sites in these subjects, not all of them suffer from persistent GI or respiratory viral infections (BURROWS and COOPER 1997). A recent study using IgA knockout mice showed that aerosol challenge of IgA-deficient and normal mice with either sublethal or lethal doses of influenza virus resulted in similar levels of pulmonary virus infection and mortality (MBAWUIKE et al. 1999). In addition, adoptively transfered, influenza-virus-specific IgG1 and IgG2a Abs were shown to significantly neutralize mucosal-challenged virus and prevented their replication in vivo (MBAWUIKE et al. 1999). Further, intravenous infusion of human neutralizing monoclonal IgG1 Abs was shown to protect macaques from mucosal infection with an HIV/SIV chimera virus (SHIV) encoding the HIV *env* gene (BABA et al. 2000; MASCOLA et al. 2000). One group showed that neonatal macaques given systemic anti-HIV IgG mAbs were completely protected against oral SHIV challenge, although oral exposure of neonates to cell-free SHIV resulted in systemic infection (BABA et al. 2000). Others reported that treatment of female macaques with intravenous anti-HIV IgG mAbs protected from vaginal SHIV challenge (MASCOLA et al. 2000). In this study, infused IgG mAbs were shown to reach mucosal secretions. However, as discussed in this review, the thinning of the vaginal epithelium caused by progesterone treatment before vaginal challenge makes it easier for Abs to passively cross the mucosal surfaces. From this point of view, effective transport of IgG into mucosal secretions may not occur under natural conditions. These findings indicate that indeed IgA Abs are most effective for prevention of virus invasion at mucosal sites; however, IgG Abs under some circumstances may also be protective for the host [Fig. 3(4), (5)]. Further investigations are expected to clarify the roles of mucosal IgA vs serum IgG Abs against viral infections.

## 6 Strategies for Mucosal Vaccination for Viral Infections

Over 200 years have passed since Jenner used vaccinia virus to prevent smallpox, and vaccination to prevent infectious diseases was initiated. A major goal in medical science is prevention rather than cure, and therefore the development of vaccination strategies is indispensable for eradication of infectious diseases. Currently, ten different vaccines are given to children in the USA from birth to 18 months of age. Many more are in clinical trials or earlier stages of research. The majority of infectious pathogens including viruses invade through mucosal surfaces and these initially replicate in this compartment. It is also well known that Abs are a highly effective host defense mechanism against both extracellular and intracellular phases of intracellular pathogens. Thus, pre-existing Abs at the origin of the infection are able to prevent infections or at least to greatly reduce the levels of virus which invade the host. However, almost all vaccines currently used are given by the parenteral route, despite our knowledge that the systemic and mucosal immune systems are regulated independently. Further, systemic immunization is largely ineffective for induction of protective humoral and cell-mediated immunity in mucosal effector sites. The clear advantage of mucosal vaccination is that this results in the induction of immune responses in both mucosal and systemic compartments. As described above, the mucosal immune system, represented in part by S-IgA Ab responses, protects from infectious pathogens as a modest first line of defense that prevents virus invasion with only minimal inflammatory side effects. By contrast, systemic immunity aggressively attacks the virus and infected cells in systemic tissue compartments. Therefore, mucosal immunization strategies that elicit both mucosal and systemic immune responses, and which result in a dual type of protection against infectious diseases with less inflammatory reactions, are considered to be an ideal strategy to induce immunity with minimal side effects. Further, another merit of this type of vaccination is that because of the CMIS mucosal network, vaccination through an easy route (e.g., oral or nasal route) can induce effective immune responses in the remote mucosal tissues where primary viral infection occurs.

The first successful mucosal vaccine was the live attenuated poliovirus vaccine of Sabin (Ogra 1984). In 1988, a worldwide poliomyelitis eradication program was started and either parenteral or oral polio vaccine was administered to people in all parts of the world. To date, the most widely used and effective of the two is the oral polio vaccine. The oral polio vaccine induces mucosal immune responses including

IgA Ab responses as well as serum IgG Ab responses (Ogra 1984; Faden et al. 1990). As the result of these efforts, an over 90% reduction in the incidence of polio has been achieved, with a corresponding decrease in virus genomic heterogeneity, and the WHO now suggests that polio may be eradicated from the world by the end of year 2002 (Smith et al. 1998; Dowdle et al. 1999).

## 6.1 Mucosal Compartments for Antiviral Immunity

Recently, mucosal vaccination strategies against viral infections have involved the induction of protective immunity in mucosal compartments. Initially, attenuated or inactivated viruses were used; however, viral peptides, recombinant proteins, and plasmid DNAs have also been used as vaccine antigens (Fynan et al. 1993; Sha et al. 1999). In early studies, mucosal vaccination was performed through various routes (e.g., oral, nasal, vaginal, and rectal routes) to obtain optimal immune responses in both mucosal and systemic compartments. However, at present, oral or nasal administration are the major routes of mucosal vaccination, since these routes are both convenient and induce optimal immune responses in the host. In most cases, nasal immunization is the more effective of the two and requires smaller amounts of vaccine antigens. One of the possible reasons is that the acid and digestive enzymes degrade a portion of vaccines when administered orally.

Currently, most efforts are mainly focused on the development of vaccine delivery systems to increase the efficiency of antigen uptake into inductive tissues and the development of adjuvants to enhance specific immune responses, especially S-IgA Ab responses. Liposomes, which are composed of phospholipids and other polar amphipathics, and lipophilic immune-stimulating complexes (ISCOMs) are also effective mucosal vehicles (Manesis et al. 1979; Gregoriadis 1990; Mowat et al. 1991, 1993). They are nonimmunogenic and nontoxic themselves, and protect vaccines from degradation by acidic gastric pH and proteolytic enzymes. In addition, these vehicles have controlled release characteristics. The effectiveness of several viral vaccines such as respiratory syncytial virus, influenza virus, rotavirus, and HIV, incorporated into liposomes or ISCOMs was tested by oral or nasal routes. The induction of mucosal as well as systemic immune responses including mucosal S-IgA Ab responses was shown (Ben Ahmeida et al. 1992; Scheepers and Becht 1994; De Haan et al. 1995, 1996; Van Pinxteren et al. 1997; Hu et al. 1998, 1999; VanCott et al. 1998; Lian et al. 1999). Despite this success, co-administration of cytokines or adjuvants was required to obtain optimal mucosal and systemic immune responses (Okada et al. 1997; Babai et al. 1999a,b; Gluck et al. 1999).

## 6.2 Live Attenuated Delivery Systems

Antigen delivery systems using live attenuated bacteria and viruses expressing foreign antigens by molecular engineering techniques are being actively investigated

(Killeen et al. 1999; Morrow et al. 1999). There are many advantages for using live vectors as well as significant drawbacks. For example, a single administration elicits strong immune responses in the host, since these live vectors can cross mucosal barriers efficiently and replicate in vivo. Further, live vectors are processed and presented by APCs to immune T cells, and immune responses are elicited in a manner similar to that which occurs during a natural infection. Among these systems, attenuated *Salmonella* expressing foreign antigens is the most studied example and has been reported to be safe and useful for delivering foreign antigens to mucosal surfaces and to elicit antigen-specific immune responses when administered either orally or nasally (VanCott et al. 1996). A large number of *Salmonella* vaccines expressing viral proteins from rotavirus, SIV, or HIV have been shown to effectively induce S-IgA Ab responses as well as cell-mediated immune responses in both mucosal and systemic immune compartments (Berggren et al. 1995; Hone et al. 1996; Valentine et al. 1996; Steger et al. 1999; Wang et al. 1999). More recently, attenuated *Salmonella* was also shown to be a useful vehicle for genetic immunization. *Salmonella* carrying eukaryotic expression plasmid vectors was shown to induce protective immune responses in the host, and these plasmids from bacteria were transferred into the nucleus of host cells in vivo (Darji et al. 1997).

## 6.3 Mucosal Adjuvants

Another approach has focused on the development of more suitable mucosal adjuvants. In general, mucosal administration of protein antigens induces antigen-specific systemic immune unresponsiveness, termed mucosally induced tolerance (Weiner 1997; Strobel and Mowat 1998). However, specific immune responses can be elicited in both mucosal and systemic compartments when co-administered with a mucosal adjuvant (Elson and Ealding 1984a,b; Lycke and Holmgren 1986; Clements et al. 1988). The representative and best-studied mucosal adjuvants include cholera toxin (CT), the major enterotoxin produced by *Vibrio cholerae*, and the *Escherichia coli* heat-labile toxin (LT). Both CT and LT are composed of two structurally, functionally, and immunologically separate A and B subunits (Spangler 1992). The B subunit consists of a pentamer and plays a role in binding and invasion to epithelial cells (Van Heyningen 1976; Fukuta et al. 1988), while the A subunit has ADP-ribosyltransferase activity and is associated with toxicity (Field et al. 1989). CT has been shown to induce $CD4^+$ Th2 cell responses (Xu-Amano et al. 1993), while LT enhances both Th1 and Th2 cell responses (Takahashi et al. 1996). The adjuvanticity of CT was first reported to abrogate oral tolerance to co-administered protein and stimulated antigen-specific S-IgA responses in mucosal secretions (Elson and Ealding 1984a,b). The effect of CT as an adjuvant for vaccines against viral infections such as influenza virus and HIV has been shown by a large number of studies (Bukawa et al. 1995; Staats et al. 1996, 1997; Tamura et al. 1996; Belyakov et al. 1998; O'Hagan 1998; Hagiwara et al. 1999). Recent studies also showed that oral or nasal immunization of nonhuman primates with $p55^{gag}$ protein of SIV and CT induced antigen-specific immune responses at mucosal and systemic compartments (KUBOTA et al. 1997; IMAOKA et al. 1998). Further, nasal immunization induced more effective S-IgA Ab responses in the genital tract, which is the major primary infection site of HIV in humans (IMAOKA et al. 1998). Thus, a large number of studies support the usefulness of CT as an adjuvant for mucosal vaccines, and potential clinical applications are being explored. However, CT is quite toxic for humans and is not suitable in its native form. Therefore, mutants of CT (mCTs), which lack ADP-ribosyltransferase activity and diarrhoeagenicity but retain adjuvanticity, have been developed by using molecular biological techniques for the purpose of clinical applications in the near future (TSUJI et al. 1990; YAMAMOTO et al. 1997a,b). Mutants of LT (mLTs) are also well studied as adjuvants for viral vaccines (DICKINSON and CLEMENTS 1995; DOUCE et al. 1995), especially for nasal influenza vaccine, and their effectiveness to induce both serum IgG and mucosal IgA Ab responses in the absence of toxicity were well defined (KOMASE et al. 1998; BARACKMAN et al. 1999). Thus, mLTs, together with mCTs, are considered to be excellent candidates for clinical trials of mucosal vaccines.

## 6.4 Cytokines as Mucosal Adjuvants

Cytokines are also candidates as mucosal adjuvants. It is well known that cytokines play a principal role in the induction and regulation of immune responses. Moreover, some Th2 cytokines are associated with terminal differentiation of B cells to IgA-secreting plasma cells and enhance IgA production in vitro and in vivo (MURRAY et al. 1987; BEAGLEY et al. 1988, 1989; MCGHEE et al. 1989; SONODA et al. 1989; RAMSAY et al. 1994a). Therefore, several groups have used cytokines as mucosal adjuvants for the purpose of enhancing antigen-specific S-IgA Ab responses more effectively at mucosal sites. In fact, nasal administration of viral vaccines encoded by the recombinant vaccinia virus or fowlpox virus vector in the presence of co-expressed IL-5 or IL-6 was shown to markedly enhance mucosal S-IgA Ab responses (RAMSAY and KOHONEN-CORISH 1993; RAMSAY et al. 1994b). Further, IL-12 as mucosal adjuvant has been studied. It is well known that IL-12 is the major Th1-type response-inducing cytokine produced by macrophages and dendritic cells. However, when co-administered orally with soluble antigen and CT, systemic Th1 responses – and interestingly, strong mucosal S-IgA Ab responses – as well as serum Ab levels were induced which were comparable to those in the absence of IL-12 (MARINARO et al. 1997). Based on this observation, IL-12 was used as a mucosal adjuvant for stimulating protective mucosal immunity (BOYAKA et al. 1999; MARINARO et al. 1999). In fact, it was reported that mice immunized nasally with influenza vaccine and IL-12 exhibited enhanced systemic Th1 and mucosal S-IgA Ab responses and were dramatically protected against a lethal dose of influenza virus challenge (ARULANANDAM et al. 1999). Thus, IL-12 itself can work as a powerful mucosal adjuvant. Presently, further efforts are being made to obtain optimal protective antiviral immunity by combination with several vaccine strategies, such as liposomes with mucosal adjuvants or cytokines (Okada et al. 1997; Babai et al. 1999a,b; Gluck et al. 1999).

## 7 Concluding Remarks

Most clinically serious pathogens, including viruses, infect through our vast mucosal surfaces of the upper respiratory, GI, and reproductive tracts. Recent advances in mucosal immunology have revealed that the mucosal immune system is largely independent of the systemic system. Mucosal Ab responses, especially S-IgA Abs, are effective against mucosal viral infections and occur through at least three distinct mechanisms. In addition, and more recently, systemic IgG Abs were also found to play essential roles in host mucosal defense against viral infections in mucosal compartments. Thus, to better understand the regulation and function of the mucosal immune system as well as the relationship between mucosal and systemic immunity, it is essential to apply both basic science and practical immunology, such as development of vaccination strategies. Further, in order to optimally generate mucosal IgA Ab responses more effectively, as well as systemic immune responses, several vaccination strategies are being considered. These include the use of nasal vs oral routes of administration and specialized delivery systems and adjuvants for induction of both systemic and mucosal immunity. Several of these are being clinically tested, and their effectiveness in the absence of any serious side effects has been demonstrated. Clinical applications of these vaccines are expected in the near future.

*Acknowledgements.* We thank Drs. Prosper N. Boyaka and Raymond J. Jackson for critical comments and corrections. Studies from the authors laboratories are supported by U.S. Public Health Service grants AI 35932, DE 09837, AI 18958, DK 44240, AI 43197, P30DK54781 DE 12242, and NIAID contracts AI 65298 and AI 65299.

## References

Ackermann LW, Wollenweber LA, Denning GM (1999) IL-4 and IFN-γ increase steady state levels of polymeric Ig receptor mRNA in human airway and intestinal epithelial cells. J Immunol 162:5112–5118

Ali HM, Scott R, Toms GL (1989) The effect of foster feeding and bottle feeding expressed breast-milk on the susceptibility of guinea-pig infants to influenza virus. Br J Exp Pathol 70:183–191

Arbones ML, Ord DC, Ley K, Ratech H, Maynard-Curry C, Otten G, Capon DJ, Tedder TF (1994) Lymphocyte homing and leukocyte rolling and migration are impaired in L-selectin-deficient mice. Immunity 1:247–260

Arulanandam BP, O'Toole M, Metzger DW (1999) Intranasal interleukin-12 is a powerful adjuvant for protective mucosal immunity. J Infect Dis 180:940–949

Asanuma H, Inaba Y, Aizawa C, Kurata T, Tamura S (1995) Characterization of mouse nasal lymphocytes isolated by enzymatic extraction with collagenase. J Immunol Methods 187:41–51

Asanuma H, Thompson AH, Iwasaki T, Sato Y, Inaba Y, Aizawa C, Kurata T, Tamura S (1997) Isolation and characterization of mouse nasal-associated lymphoid tissue. J Immunol Methods 202:123–131

Baba TW, Liska V, Hofmann-Lehmann R, Vlasak J, Xu W, Ayehunie S, Cavacini LA, Posner MR, Katinger H, Stiegler G, Bernacky BJ, Rizvi TA, Schmidt R, Hill LR, Keeling ME, Lu Y, Wright JE, Chou TC, Ruprecht RM (2000) Human neutralizing monoclonal antibodies of the IgG1 subtype protect against mucosal simian-human immunodeficiency virus infection. Nat Med 6:200–206

Babai I, Samira S, Barenholz Y, Zakay-Rones Z, Kedar E (1999a) A novel influenza subunit vaccine composed of liposome-encapsulated haemagglutinin/neuraminidase and IL-2 or GM-CSF. I. Vaccine characterization and efficacy studies in mice. Vaccine 17:1223–1238

Babai I, Samira S, Barenholz Y, Zakay-Rones Z, Kedar E (1999b) A novel influenza subunit vaccine composed of liposome-encapsulated haemagglutinin/neuraminidase and IL-2 or GM-CSF. II. Induction of TH1 and TH2 responses in mice. Vaccine 17:1239–1250

Barackman JD, Ott G, O'Hagan DT (1999) Intranasal immunization of mice with influenza vaccine in combination with the adjuvant LT-R72 induces potent mucosal and serum immunity which is stronger than that with traditional intramuscular immunization. Infect Immun 67:4276–4279

Beagley KW, Eldridge JH, Kiyono H, Everson MP, Koopman WJ, Honjo T, McGhee JR (1988) Recombinant murine IL-5 induces high rate IgA synthesis in cycling IgA-positive Peyer's patch B cells. J Immunol 141:2035–2042

Beagley KW, Eldridge JH, Lee F, Kiyono H, Everson MP, Koopman WJ, Hirano T, Kishimoto T, McGhee JR (1989) Interleukins and IgA synthesis. Human and murine interleukin 6 induce high rate IgA secretion in IgA-committed B cells. J Exp Med 169:2133–2148

Belai AA, El-Gohery Y, Talaat M (1977) Nasal and paranasal pathology in experimental bilharziasis. J Laryngol Otol 91:391–400

Bell RG, Issekutz T (1993) Expression of a protective intestinal immune response can be inhibited at three distinct sites by treatment with anti-α4 integrin. J Immunol 151:4790–4802

Belyakov IM, Derby MA, Ahlers JD, Kelsall BL, Earl P, Moss B, Strober W, Berzofsky JA (1998) Mucosal immunization with HIV-1 peptide vaccine induces mucosal and systemic cytotoxic T lymphocytes and protective immunity in mice against intrarectal recombinant HIV-vaccinia challenge. Proc Natl Acad Sci USA 95:1709–1714

Ben Ahmeida ET, Jennings R, Erturk M, Potter CW (1992) The IgA and subclass IgG responses and protection in mice immunised with influenza antigens administered as ISCOMS, with FCA, ALH or as infectious virus. Arch Virol 125:71–86

Berggren RE, Wunderlich A, Ziegler E, Schleicher M, Duke RC, Looney D, Fang FC (1995) HIV gp120-specific cell-mediated immune responses in mice after oral immunization with recombinant *Salmonella*. J Acquir Immune Defic Syndr Hum Retrovirol 10:489–495

Berlin C, Berg EL, Briskin MJ, Andrew DP, Kilshaw PJ, Holzmann B, Weissman IL, Hamann A, Butcher EC (1993) α4β7 integrin mediates lymphocyte binding to the mucosal vascular addressin MAdCAM-1. Cell 74:185–5

Bevins CL, Martin-Porter E, Ganz T (1999) Defensins and innate host defence of the gastrointestinal tract. Gut 45:911–915

Bishop RF, Bugg HC, Masendycz PJ, Lund JS, Gorrell RJ, Barnes GL (1996) Serum, fecal, and breast milk rotavirus antibodies as indices of infection in mother-infant pairs. J Infect Dis 174:S22–S29

Blanch VJ, Piskurich JF, Kaetzel CS (1999) Cutting edge: coordinate regulation of IFN regulatory factor-1 and the polymeric Ig receptor by proinflammatory cytokines. J Immunol 162:1232–1235

Bockman DE, Stevens W (1977) Gut-associated lymphoepithelial tissue: bidirectional transport of tracer by specialized epithelial cells associated with lymphoid follicles. J Reticuloendothel Soc 21:245–254

Bohl EH, Gupta RK, Olquin MV, Saif LJ (1972) Antibody responses in serum, colostrum, and milk of swine after infection or vaccination with transmissible gastroenteritis virus. Infect Immun 6:289–301

Bomsel M (1997) Transcytosis of infectious human immunodeficiency virus across a tight human epithelial cell line barrier. Nat Med 3:42–47

Bomsel M, Heyman M, Hocini H, Lagaye S, Belec L, Dupont C, Desgranges C (1998) Intracellular neutralization of HIV transcytosis across tight epithelial barriers by anti-HIV envelope protein dIgA or IgM. Immunity 9:277–287

Bos NA, Bun JC, Popma SH, Cebra ER, Deenen GJ, van der Cammen MJ, Kroese FG, Cebra JJ (1996) Monoclonal immunoglobulin A derived from peritoneal B cells is encoded by both germ line and somatically mutated VH genes and is reactive with commensal bacteria. Infect Immun 64:616–623

Boyaka PN, Marinaro M, Jackson RJ, Menon S, Kiyono H, Jirillo E, McGhee JR (1999) IL-12 is an effective adjuvant for induction of mucosal immunity. J Immunol 162:122–128

Briere F, Bridon JM, Chevet D, Souillet G, Bienvenu F, Guret C, Martinez-Valdez H, Banchereau J (1994) Interleukin 10 induces B lymphocytes from IgA-deficient patients to secrete IgA. J Clin Invest 94:97–104

Bukawa H, Sekigawa K, Hamajima K, Fukushima J, Yamada Y, Kiyono H, Okuda K (1995) Neutralization of HIV-1 by secretory IgA induced by oral immunization with a new macromolecular multicomponent peptide vaccine candidate. Nat Med 1:681–685

Burns JW, Siadat-Pajouh M, Krishnaney AA, Greenberg HB (1996) Protective effect of rotavirus VP6-specific IgA monoclonal antibodies that lack neutralizing activity. Science 272:104–107

Burrows PD, Cooper MD (1997) IgA deficiency. Adv Immunol 65:245–276

Butcher EC, Rouse RV, Coffman RL, Nottenburg CN, Hardy RR, Weissman IL (1982) Surface phenotype of Peyer's patch germinal center cells: implications for the role of germinal centers in B cell differentiation. J Immunol 129:2698–2707

Clements JD, Hartzog NM, Lyon FL (1988) Adjuvant activity of *Escherichia coli* heat-labile enterotoxin and effect on the induction of oral tolerance in mice to unrelated protein antigens. Vaccine 6:269–277

Coffman RL, Lebman DA, Shrader B (1989) Transforming growth factor β specifically enhances IgA production by lipopolysaccharide-stimulated murine B lymphocytes. J Exp Med 170:1039–1044

Coffman RL, Weissman IL (1981) B220: a B cell-specific member of th T200 glycoprotein family. Nature 289:681–683

Craig SW, Cebra JJ (1971) Peyer's patches: an enriched source of precursors for IgA-producing immunocytes in the rabbit. J Exp Med 134:188–200

Craig SW, Cebra JJ (1975) Rabbit Peyer's patches, appendix, and popliteal lymph node B lymphocytes: a comparative analysis of their membrane immunoglobulin components and plasma cell precursor potential. J Immunol 114:492–502

Csencsits KL, Jutila MA, Pascual DW (1999) Nasal-associated lymphoid tissue: phenotypic and functional evidence for the primary role of peripheral node addressin in naive lymphocyte adhesion to high endothelial venules in a mucosal site. J Immunol 163:1382–1389

Cuvelier CA, Quatacker J, Mielants H, De Vos M, Veys E, Roels H (1993) M cells are damaged and increased in number in inflamed human ileal mucosa. Eur J Morphol 31:87–91

Darji A, Guzman CA, Gerstel B, Wachholz P, Timmis KN, Wehland J, Chakraborty T, Weiss S (1997) Oral somatic transgene vaccination using attenuated *S. typhimurium*. Cell 91:765–775

de Haan A, Groen G, Prop J, van Rooijen N, Wilschut J (1996) Mucosal immunoadjuvant activity of liposomes: role of alveolar macrophages. Immunology 89:488–493

de Haan A, Tomee JF, Huchshorn JP, Wilschut J (1995) Liposomes as an immunoadjuvant system for stimulation of mucosal and systemic antibody responses against inactivated measles virus administered intranasally to mice. Vaccine 13:1320–1324

de Martino M, Resti M, Appendino C, Vierucci A (1987) Different degree of antibody response to hepatitis B virus vaccine in breast- and formula-fed infants born to HBsAg-positive mothers. J Pediatr Gastroenterol Nutr 6:208–211

Defrance T, Vanbervliet B, Briere F, Durand I, Rousset F, Banchereau J (1992) Interleukin 10 and transforming growth factor β cooperate to induce anti-CD40-activated naive human B cells to secrete immunoglobulin A. J Exp Med 175:671–682

Denning GM (1996) IL-4 and IFN-γ synergistically increase total polymeric IgA receptor levels in human intestinal epithelial cells: role of protein tyrosine kinase. J Immunol 156:4807–4814

Dickinson BL, Clements JD (1995) Dissociation of *Escherichia coli* heat-labile enterotoxin adjuvanticity from ADP-ribosyltransferase activity. Infect Immun 63:1617–1623

Dogan A, Du M, Koulis A, Briskin MJ, Isaacson PG (1997) Expression of lymphocyte homing receptors and vascular addressins in low-grade gastric B-cell lymphomas of mucosa-associated lymphoid tissue. Am J Pathol 151:1361–1369

Douce G, Turcotte C, Cropley I, Roberts M, Pizza M, Domenghini M, Rappuoli R, Dougan G (1995) Mutants of *Escherichia coli* heat-labile toxin lacking ADP-ribosyltransferase activity act as nontoxic, mucosal adjuvants. Proc Natl Acad Sci USA 92:1644–1648

Dowdle WR, Featherstone DA, Birmingham ME, Hull HF, Aylward RB (1999) Poliomyelitis eradication. Virus Res 62:185–192

Downham MA, Scott R, Sims DG, Webb JK, Gardner PS (1976) Breast-feeding protects against respiratory syncytial virus infections. Br Med J 2:274–276

Elson CO, Ealding W (1984a) Cholera toxin feeding did not induce oral tolerance in mice and abrogated oral tolerance to an unrelated protein antigen. J Immunol 133:2892–2897

Elson CO, Ealding W (1984b) Generalized systemic and mucosal immunity in mice after mucosal stimulation with cholera toxin. J Immunol 132:2736–2741

Estes MK, Cohen J (1989) Rotavirus gene structure and function. Microbiol Rev 53:410–449

Faden H, Modlin JF, Thoms ML, McBean AM, Ferdon MB, Ogra PL (1990) Comparative evaluation of immunization with live attenuated and enhanced-potency inactivated trivalent poliovirus vaccines in childhood: systemic and local immune responses. J Infect Dis 162:1291–1297

Field M, Rao MC, Chang EB (1989) Intestinal electrolyte transport and diarrheal disease (1). N Engl J Med 321:800–806

Franco MA, Greenberg HB (1997) Immunity to rotavirus in T cell deficient mice. Virology 238:169–179

Fubara ES, Freter R (1973) Protection against enteric bacterial infection by secretory IgA antibodies. J Immunol 111:395–403

Fujihashi K, McGhee JR, Kweon MN, Cooper MD, Tonegawa S, Takahashi I, Hiroi T, Mestecky J, Kiyono H (1996a) $\gamma/\delta$ T cell-deficient mice have impaired mucosal immunoglobulin A responses. J Exp Med 183:1929–1935

Fujihashi K, McGhee JR, Yamamoto M, Hiroi T, Kiyono H (1996b) Role of $\gamma\delta$ T cells in the regulation of mucosal IgA response and oral tolerance. Ann N Y Acad Sci 778:55–63

Fujioka H, Emancipator SN, Aikawa M, Huang DS, Blatnik F, Karban T, DeFife K, Mazanec MB (1998) Immunocytochemical colocalization of specific immunoglobulin A with sendai virus protein in infected polarized epithelium. J Exp Med 188:1223–1229

Fukuta S, Magnani JL, Twiddy EM, Holmes RK, Ginsburg V (1988) Comparison of the carbohydrate-binding specificities of cholera toxin and *Escherichia coli* heat-labile enterotoxins LTh-I, LT-IIa, and LT-IIb. Infect Immun 56:1748–1753

Fynan EF, Webster RG, Fuller DH, Haynes JR, Santoro JC, Robinson HL (1993) DNA vaccines: protective immunizations by parenteral, mucosal, and gene-gun inoculations. Proc Natl Acad Sci USA 90:11478–11482

Gallatin WM, Weissman IL, Butcher EC (1983) A cell-surface molecule involved in organ-specific homing of lymphocytes. Nature 304:30–34

Gebert A, Rothkotter HJ, Pabst R (1996) M cells in Peyer's patches of the intestine. Int Rev Cytol 167:91–159

George A, Cebra JJ (1991) Responses of single germinal-center B cells in T-cell-dependent microculture. Proc Natl Acad Sci USA 88:11–15

Glass RI, Gentsch J, Smith JC (1994) Rotavirus vaccines: success by reassortment? Science 265:1389–1391

Gluck U, Gebbers JO, Gluck R (1999) Phase 1 evaluation of intranasal virosomal influenza vaccine with and without *Escherichia coli* heat-labile toxin in adult volunteers. J Virol 73:7780–7786

Goeringer GC, Vidic B (1987) The embryogenesis and anatomy of Waldeyer's ring. Otolaryngol Clin North Am 20:207–217

Gregoriadis G (1990) Immunological adjuvants: a role for liposomes. Immunol Today 11:89–97

Grutzkau A, Hanski C, Hahn H, Riecken EO (1990) Involvement of M cells in the bacterial invasion of Peyer's patches: a common mechanism shared by *Yersinia enterocolitica* and other enteroinvasive bacteria. Gut 31:1011–1015

Hagiwara Y, Komase K, Chen Z, Matsuo K, Suzuki Y, Aizawa C, Kurata T, Tamura S (1999) Mutants of cholera toxin as an effective and safe adjuvant for nasal influenza vaccine. Vaccine 17:2918–2926

Hamann A, Andrew DP, Jablonski-Westrich D, Holzmann B, Butcher EC (1994) Role of $\alpha$4-integrins in lymphocyte homing to mucosal tissues in vivo. J Immunol 152:3282–3293

Hanson LA, Soderstrom T (1981) Human milk: defense against infection. Prog Clin Biol Res 61:147–159

Hecht G (1999) Innate mechanisms of epithelial host defense: spotlight on intestine. Am J Physiol 277:C351–C358

Hendrickson BA, Rindisbacher L, Corthesy B, Kendall D, Waltz DA, Neutra MR, Seidman JG (1996) Lack of association of secretory component with IgA in J chain-deficient mice. J Immunol 157: 750–754

Heritage PL, Underdown BJ, Arsenault AL, Snider DP, McDermott MR (1997) Comparison of murine nasal-associated lymphoid tissue and Peyer's patches. Am J Respir Crit Care Med 156:1256–1262

Hiroi T, Fujihashi K, McGhee JR, Kiyono H (1994) Characterization of cytokine-producing cells in mucosal effector sites: $CD3^+$ T cells of Th1 and Th2 type in salivary gland-associated tissues. Eur J Immunol 24:2653–2658

Hiroi T, Iwatani K, Iijima H, Kodama S, Yanagita M, Kiyono H (1998) Nasal immune system: distinctive Th0 and Th1/Th2 type environments in murine nasal-associated lymphoid tissues and nasal passage, respectively. Eur J Immunol 28:3346–3353

Hiroi T, Yanagita M, Iijima H, Iwatani K, Yoshida T, Takatsu K, Kiyono H (1999) Deficiency of IL-5 receptor α-chain selectively influences the development of the common mucosal immune system independent IgA-producing B-1 cell in mucosa-associated tissues. J Immunol 162:821–828

Hodgins DC, Kang SY, deArriba L, Parreno V, Ward LA, Yuan L, To T, Saif LJ (1999) Effects of maternal antibodies on protection and development of antibody responses to human rotavirus in gnotobiotic pigs. J Virol 73:186–197

Holzmann B, McIntyre BW, Weissman IL (1989) Identification of a murine Peyer's patch-specific lymphocyte homing receptor as an integrin molecule with an α chain homologous to human VLA-4α. Cell 56:37–46

Holzmann B, Weissman IL (1989) Integrin molecules involved in lymphocyte homing to Peyer's patches. Immunol Rev 108:45–61

Hone DM, Wu S, Powell RJ, Pascual DW, Van Cott J, McGhee JR, Fouts TR, Tuskan RG, Lewis GK (1996) Optimization of live oral *Salmonella*-HIV-1 vaccine vectors for the induction of HIV-specific mucosal and systemic immune responses. J Biotechnol 44:203–207

Hopkins SA, Kraehenbuhl JP (1997) Dendritic cells of the murine Peyer's patches colocalize with *Salmonella typhimurium* avirulent mutants in the subepithelial dome. Adv Exp Med Biol 417:105–109

Hu KF, Ekstrom J, Merza M, Lovgren-Bengtsson K, Morein B (1999) Induction of antibody responses in the common mucosal immune system by respiratory syncytical virus immunostimulating complexes. Med Microbiol Immunol (Berl) 187:191–198

Hu KF, Elvander M, Merza M, Akerblom L, Brandenburg A, Morein B (1998) The immunostimulating complex (ISCOM) is an efficient mucosal delivery system for respiratory syncytial virus (RSV) envelope antigens inducing high local and systemic antibody responses. Clin Exp Immunol 113:235–243

Hu MC, Crowe DT, Weissman IL, Holzmann B (1992) Cloning and expression of mouse integrin β p(β7): a functional role in Peyer's patch-specific lymphocyte homing. Proc Natl Acad Sci USA 89:8254–8258

Husband AJ, Gowans JL (1978) The origin and antigen-dependent distribution of IgA-containing cells in the intestine. J Exp Med 148:1146–1160

Imaoka K, Miller CJ, Kubota M, McChesney MB, Lohman B, Yamamoto M, Fujihashi K, Someya K, Honda M, McGhee JR, Kiyono H (1998) Nasal immunization of nonhuman primates with simian immunodeficiency virus p55gag and cholera toxin adjuvant induces Th1/Th2 help for virus-specific immune responses in reproductive tissues. J Immunol 161:5952–5958

Joel DD, Laissue JA, LeFevre ME (1978) Distribution and fate of ingested carbon particles in mice. J Reticuloendothel Soc 24:477–487

Johansen FE, Pekna M, Norderhaug IN, Haneberg B, Hietala MA, Krajci P, Betsholtz C, Brandtzaeg P (1999) Absence of epithelial immunoglobulin A transport, with increased mucosal leakiness, in polymeric immunoglobulin receptor/secretory component-deficient mice. J Exp Med 190:915–922

Jones BD, Ghori N, Falkow S (1994) *Salmonella typhimurium* initiates murine infection by penetrating and destroying the specialized epithelial M cells of the Peyer's patches. J Exp Med 180:15–23

Jones DE, Bevins CL (1992) Paneth cells of the human small intestine express an antimicrobial peptide gene. J Biol Chem 267:23216–23225

Kaetzel CS, Robinson JK, Lamm ME (1994) Epithelial transcytosis of monomeric IgA and IgG cross-linked through antigen to polymeric IgA. A role for monomeric antibodies in the mucosal immune system. J Immunol 152:72–76

Kantor AB, Herzenberg LA (1993) Origin of murine B cell lineages. Annu Rev Immunol 11:501–538

Katz JM, Lu X, Young SA, Galphin JC (1997) Adjuvant activity of the heat-labile enterotoxin from enterotoxigenic *Escherichia coli* for oral administration of inactivated influenza virus vaccine. J Infect Dis 175:352–363

Kelsall BL, Strober W (1996a) Distinct populations of dendritic cells are present in the subepithelial dome and T cell regions of the murine Peyer's patch. J Exp Med 183:237–247

Kelsall BL, Strober W (1996b) The role of dendritic cells in antigen processing in the Peyer's patch. Ann N Y Acad Sci 778:47–54

Kerneis S, Bogdanova A, Kraehenbuhl JP, Pringault E (1997) Conversion by Peyer's patch lymphocytes of human enterocytes into M cells that transport bacteria. Science 277:949–952

Killeen K, Spriggs D, Mekalanos J (1999) Bacterial mucosal vaccines: *Vibrio cholerae* as a live attenuated vaccine/vector paradigm. Curr Top Microbiol Immunol 236:237–254

Komase K, Tamura S, Matsuo K, Watanabe K, Hattori N, Odaka A, Suzuki Y, Kurata T, Aizawa C (1998) Mutants of *Escherichia coli* heat-labile enterotoxin as an adjuvant for nasal influenza vaccine. Vaccine 16:248–254

Kramer DR, Cebra JJ (1995) Early appearance of "natural" mucosal IgA responses and germinal centers in suckling mice developing in the absence of maternal antibodies. J Immunol 154:2051–2062

Kris RM, Asofsky R, Evans CB, Small PA Jr (1985) Protection and recovery in influenza virus-infected mice immunosuppressed with anti-IgM. J Immunol 134:1230–1235

Kubota M, Miller CJ, Imaoka K, Kawabata S, Fujihashi K, McGhee JR, Kiyono H (1997) Oral immunization with simian immunodeficiency virus p55gag and cholera toxin elicits both mucosal IgA and systemic IgG immune responses in nonhuman primates. J Immunol 158:5321–5329

Kuper CF, Hameleers DM, Bruijntjes JP, van der Ven I, Biewenga J, Sminia T (1990) Lymphoid and non-lymphoid cells in nasal-associated lymphoid tissue (NALT) in the rat. An immuno- and enzyme-histochemical study. Cell Tissue Res 259:371–377

Kvale D, Lovhaug D, Sollid LM, Brandtzaeg P (1988) Tumor necrosis factor-α up-regulates expression of secretory component, the epithelial receptor for polymeric Ig. J Immunol 140:3086–3089

Lamm ME (1997) Interaction of antigens and antibodies at mucosal surfaces. Annu Rev Microbiol 51:311–340

Lebman DA, Griffin PM, Cebra JJ (1987) Relationship between expression of IgA by Peyer's patch cells and functional IgA memory cells. J Exp Med 166:1405–1418

Lebman DA, Lee FD, Coffman RL (1990) Mechanism for transforming growth factor β and IL-2 enhancement of IgA expression in lipopolysaccharide-stimulated B cell cultures. J Immunol 144: 952–959

Lebman DA, Park MJ, Hansen-Bundy S, Pandya A (1994) Mechanism for transforming growth factor β regulation of α mRNA in lipopolysaccharide-stimulated B cells. Int Immunol 6:113–119

Lian T, Bui T, Ho RJ (1999) Formulation of HIV-envelope protein with lipid vesicles expressing ganglioside GM1 associated to cholera toxin B enhances mucosal immune responses. Vaccine 18:604–611

Liew FY, Russell SM, Appleyard G, Brand CM, Beale J (1984) Cross-protection in mice infected with influenza A virus by the respiratory route is correlated with local IgA antibody rather than serum antibody or cytotoxic T cell reactivity. Eur J Immunol 14:350–356

Lillard JW Jr, Boyaka PN, Chertov O, Oppenheim JJ, McGhee JR (1999) Mechanisms for induction of acquired host immunity by neutrophil peptide defensins. Proc Natl Acad Sci USA 96:651–656

Linhares AC, Gabbay YB, Freitas RB, da Rosa ES, Mascarenhas JD, Loureiro EC (1989) Longitudinal study of rotavirus infections among children from Belem, Brazil. Epidemiol Infect 102:129–145

Lycke N, Erlandsson L, Ekman L, Schon K, Leanderson T (1999) Lack of J chain inhibits the transport of gut IgA and abrogates the development of intestinal antitoxic protection. J Immunol 163:913–919

Lycke N, Holmgren J (1986) Strong adjuvant properties of cholera toxin on gut mucosal immune responses to orally presented antigens. Immunology 59:301–308

Macpherson AJ, Gatto D, Sainsbury E, Harriman GR, Hengartner H, Zinkernagel RM (2000) A primitive T cell-independent mechanism of intestinal mucosal IgA responses to commensal bacteria. Science 228:2222–2226

Manesis EK, Cameron CH, Gregoriadis G (1979) Hepatitis B surface antigen-containing liposomes enhance humoral and cell-mediated immunity to the antigen. FEBS Lett 102:107–111

Marinaro M, Boyaka PN, Finkelman FD, Kiyono H, Jackson RJ, Jirillo E, McGhee JR (1997) Oral but not parenteral interleukin (IL)-12 redirects T helper 2 (Th2)-type responses to an oral vaccine without altering mucosal IgA responses. J Exp Med 185:415–427

Marinaro M, Boyaka PN, Jackson RJ, Finkelman FD, Kiyono H, Jirillo E, McGhee JR (1999) Use of intranasal IL-12 to target predominantly Th1 responses to nasal and Th2 responses to oral vaccines given with cholera toxin. J Immunol 162:114–121

Mascola JR, Stiegler G, VanCott TC, Katinger H, Carpenter CB, Hanson CE, Beary H, Hayes D, Frankel SS, Birx DL, Lewis MG (2000) Protection of macaques against vaginal transmission of a pathogenic HIV-1/SIV chimeric virus by passive infusion of neutralizing antibodies. Nat Med 6: 207–210

Mata LJ, Wyatt RG (1971) The uniqueness of human milk. Host resistance to infection. Am J Clin Nutr 24:976–986

Mathews M, Jia HP, Guthmiller JM, Losh G, Graham S, Johnson GK, Tack BF, McCray PB Jr (1999) Production of β-defensin antimicrobial peptides by the oral mucosa and salivary glands. Infect Immun 67:2740–2745

Mazanec MB, Coudret CL, Fletcher DR (1995) Intracellular neutralization of influenza virus by immunoglobulin A anti-hemagglutinin monoclonal antibodies. J Virol 69:1339–1343

Mazanec MB, Lamm ME, Lyn D, Portner A, Nedrud JG (1992) Comparison of IgA versus IgG monoclonal antibodies for passive immunization of the murine respiratory tract. Virus Res 23:1–12

Mazanec MB, Nedrud JG, Kaetzel CS, Lamm ME (1993) A three-tiered view of the role of IgA in mucosal defense. Immunol Today 14:430–435

Mazanec MB, Nedrud JG, Lamm ME (1987) Immunoglobulin A monoclonal antibodies protect against Sendai virus. J Virol 61:2624–2626

Mbawuike IN, Pacheco S, Acuna CL, Switzer KC, Zhang Y, Harriman GR (1999) Mucosal immunity to influenza without IgA: an IgA knockout mouse model. J Immunol 162:2530–2537

McDermott MR, Bienenstock J (1979) Evidence for a common mucosal immunologic system. I. Migration of B immunoblasts into intestinal, respiratory, and genital tissues. J Immunol 122:1892–1898

McGhee JR, Kiyono H (1999) The Mucosal Immune System. Fundamental Immunology, Fourth Edition. edited by William E. Paul 909–945

McGhee JR, Mestecky J, Elson CO, Kiyono H (1989) Regulation of IgA synthesis and immune response by T cells and interleukins. J Clin Immunol 9:175–199

McLean B, Holmes IH (1980) Transfer of antirotaviral antibodies from mothers to their infants. J Clin Microbiol 12:320–325

McWilliams M, Phillips-Quagliata JM, Lamm ME (1975) Characteristics of mesenteric lymph node cells homing to gut-associated lymphoid tissue in syngeneic mice. J Immunol 115:54–58

Mestecky J (1987) The common mucosal immune system and current strategies for induction of immune responses in external secretions. J Clin Immunol 7:265–276

Mestecky J, McGhee JR (1987) Immunoglobulin A (IgA): molecular and cellular interactions involved in IgA biosynthesis and immune response. Adv Immunol 40:153–245

Mills J 5th, Van Kirk JE, Wright PF, Chanock RM (1971) Experimental respiratory syncytial virus infection of adults. Possible mechanisms of resistance to infection and illness. J Immunol 107:123–130

Minamishima I, Ueda K, Minematsu T, Minamishima Y, Umemoto M, Take H, Kuraya K (1994) Role of breast milk in acquisition of cytomegalovirus infection. Microbiol Immunol 38:549–552

Mizukawa N, Sugiyama K, Ueno T, Mishima K, Takagi S, Sugahara T (1999) Defensin-1, an antimicrobial peptide present in the saliva of patients with oral diseases. Oral Dis 5:139–142

Morrow CD, Novak MJ, Ansardi DC, Porter DC, Moldoveanu Z (1999) Recombinant viruses as vectors for mucosal immunity. Curr Top Microbiol Immunol 236:255–273

Mowat AM, Donachie AM, Reid G, Jarrett O (1991) Immune-stimulating complexes containing Quil A and protein antigen prime class I MHC-restricted T lymphocytes in vivo and are immunogenic by the oral route. Immunology 72:317–322

Mowat AM, Maloy KJ, Donachie AM (1993) Immune-stimulating complexes as adjuvants for inducing local and systemic immunity after oral immunization with protein antigens. Immunology 80: 527–534

Murakami M, Yoshioka H, Shirai T, Tsubata T, Honjo T (1995) Prevention of autoimmune symptoms in autoimmune-prone mice by elimination of B-1 cells. Int Immunol 7:877–882

Murray PD, McKenzie DT, Swain SL, Kagnoff MF (1987) Interleukin 5 and interleukin 4 produced by Peyer's patch T cells selectively enhance immunoglobulin A expression. J Immunol 139:2669–2674

Naficy AB, Abu-Elyazeed R, Holmes JL, Rao MR, Savarino SJ, Kim Y, Wierzba TF, Peruski L, Lee YJ, Gentsch JR, Glass RI, Clemens JD (1999) Epidemiology of rotavirus diarrhea in Egyptian children and implications for disease control. Am J Epidemiol 150:770–777

Neutra MR, Frey A, Kraehenbuhl JP (1996a) Epithelial M cells: gateways for mucosal infection and immunization. Cell 86:345–348

Neutra MR, Pringault E, Kraehenbuhl JP (1996b) Antigen sampling across epithelial barriers and induction of mucosal immune responses. Annu Rev Immunol 14:275–300

O'Hagan DT (1998) Recent advances in vaccine adjuvants for systemic and mucosal administration. J Pharm Pharmacol 50:1–10

Offit PA, Clark HF (1985) Protection against rotavirus-induced gastroenteritis in a murine model by passively acquired gastrointestinal but not circulating antibodies. J Virol 54:58–64

Ogra PL (1984) Mucosal immune response to poliovirus vaccines in childhood. Rev Infect Dis 6: S361–S368

Ogra PL, Losonsky GA, Fishaut M (1983) Colostrum-derived immunity and maternal-neonatal interaction. Ann N Y Acad Sci 409:82–95

Okada E, Sasaki S, Ishii N, Aoki I, Yasuda T, Nishioka K, Fukushima J, Miyazaki J, Wahren B, Okuda K (1997) Intranasal immunization of a DNA vaccine with IL-12- and granulocyte-macrophage colony-stimulating factor (GM-CSF)-expressing plasmids in liposomes induces strong mucosal and cell-mediated immune responses against HIV-1 antigens. J Immunol 159:3638–3647

Ouellette AJ (1999) IV. Paneth cell antimicrobial peptides and the biology of the mucosal barrier. Am J Physiol 277:G257–G261

Outlaw MC, Dimmock NJ (1990) Mechanisms of neutralization of influenza virus on mouse tracheal epithelial cells by mouse monoclonal polymeric IgA and polyclonal IgM directed against the viral haemagglutinin. J Gen Virol 71:69–76

Owen RL (1977) Sequential uptake of horseradish peroxidase by lymphoid follicle epithelium of Peyer's patches in the normal unobstructed mouse intestine: an ultrastructural study. Gastroenterology 72:440–451

Panyutich AV, Szold O, Poon PH, Tseng Y, Ganz T (1994) Identification of defensin binding to C1 complement. FEBS Lett 356:169–173

Peters MG, Secrist H, Anders KR, Nash GS, Rich SR, MacDermott RP (1989) Normal human intestinal B lymphocytes. Increased activation compared with peripheral blood. J Clin Invest 83:1827–1833

Phillips JO, Everson MP, Moldoveanu Z, Lue C, Mestecky J (1990) Synergistic effect of IL-4 and IFN-γ on the expression of polymeric Ig receptor (secretory component) and IgA binding by human epithelial cells. J Immunol 145:1740–1744

Pickering LK, Morrow AL, Herrera I, O'Ryan M, Estes MK, Guilliams SE, Jackson L, Carter-Campbell S, Matson DO (1995) Effect of maternal rotavirus immunization on milk and serum antibody titers. J Infect Dis 172:723–728

Quayle AJ, Porter EM, Nussbaum AA, Wang YM, Brabec C, Yip KP, Mok SC (1998) Gene expression, immunolocalization, and secretion of human defensin-5 in human female reproductive tract. Am J Pathol 152:1247–1258

Ramsay AJ, Husband AJ, Ramshaw IA, Bao S, Matthaei KI, Koehler G, Kopf M (1994a) The role of interleukin-6 in mucosal IgA antibody responses in vivo. Science 264:561–563

Ramsay AJ, Kohonen-Corish M (1993) Interleukin-5 expressed by a recombinant virus vector enhances specific mucosal IgA responses in vivo. Eur J Immunol 23:3141–3145

Ramsay AJ, Leong KH, Boyle D, Ruby J, Ramshaw IA (1994b) Enhancement of mucosal IgA responses by interleukins 5 and 6 encoded in recombinant vaccine vectors. Reprod Fertil Dev 6:389–392

Renegar KB, Jackson GD, Mestecky J (1998) In vitro comparison of the biologic activities of monoclonal monomeric IgA, polymeric IgA, and secretory IgA. J Immunol 160:1219–1223

Renegar KB, Small PA Jr (1991a) Immunoglobulin A mediation of murine nasal anti-influenza virus immunity. J Virol 65:2146–2148

Renegar KB, Small PA Jr (1991b) Passive transfer of local immunity to influenza virus infection by IgA antibody. J Immunol 146:1972–1978

Reuman PD, Keely SP, Schiff GM (1989) Isolation and characterization of mouse nasal lymphocytes. J Immunol Methods 125:49–55

Rosen SD, Chi SI, True DD, Singer MS, Yednock TA (1989) Intravenously injected sialidase inactivates attachment sites for lymphocytes on high endothelial venules. J Immunol 142:1895–1902

Rosen SD, Singer MS, Yednock TA, Stoolman LM (1985) Involvement of sialic acid on endothelial cells in organ-specific lymphocyte recirculation. Science 228:1005–1007

Rott LS, Rose JR, Bass D, Williams MB, Greenberg HB, Butcher EC (1997) Expression of mucosal homing receptor α4β7 by circulating $CD4^+$ cells with memory for intestinal rotavirus. J Clin Invest 100:1204–1208

Roux ME, McWilliams M, Phillips-Quagliata JM, Weisz-Carrington P, Lamm ME (1977) Origin of IgA-secreting plasma cells in the mammary gland. J Exp Med 146:1311–1322

Ruggeri FM, Johansen K, Basile G, Kraehenbuhl JP, Svensson L (1998) Antirotavirus immunoglobulin A neutralizes virus in vitro after transcytosis through epithelial cells and protects infant mice from diarrhea. J Virol 72:2708–2714

Saitoh-Inagawa W, Hiroi T, Yanagita M, Iijima H, Uchio E, Ohno S, Aoki K, Kiyono H (2000) Unique characteristics of lacrimal glands as a part of mucosal immune network: high frequency of IgA-committed B-1 cells and $NK1.1^+$ αbeta T cells. Invest Ophthalmol Vis Sci 41:138–144

Savidge TC, Smith MW (1995) Evidence that membranous (M) cell genesis is immuno-regulated. Adv Exp Med Biol 371A:239–241

Savidge TC, Smith MW, James PS, Aldred P (1991) Salmonella-induced M-cell formation in germ-free mouse Peyer's patch tissue. Am J Pathol 139:177–184

Scheepers K, Becht H (1994) Protection of mice against an influenza virus infection by oral vaccination with viral nucleoprotein incorporated into immunostimulating complexes. Med Microbiol Immunol (Berl) 183:265–278

Schonwetter BS, Stolzenberg ED, Zasloff MA (1995) Epithelial antibiotics induced at sites of inflammation. Science 267:1645–1648

Schoub BD, Prozesky OW, Lecatsas G, Oosthuizen R (1978) The role of breast-feeding in the prevention of rotavirus infection. J Med Microbiol 11:25–31

Scott GH, Sydiskis RJ (1976) Responses of mice immunized with influenza virus by serosol and parenteral routes. Infect Immun 13:696–703

Sha Z, Vincent MJ, Compans RW (1999) Enhancement of mucosal immune responses to the influenza virus HA protein by alternative approaches to DNA immunization. Immunobiology 200:21–30

Sheridan JF, Smith CC, Manak MM, Aurelian L (1984) Prevention of rotavirus-induced diarrhea in neonatal mice born to dams immunized with empty capsids of simian rotavirus SA-11. J Infect Dis 149:434–438

Shimada S, Kawaguchi-Miyashita M, Kushiro A, Sato T, Nanno M, Sako T, Matsuoka Y, Sudo K, Tagawa Y, Iwakura Y, Ohwaki M (1999) Generation of polymeric immunoglobulin receptor-deficient mouse with marked reduction of secretory IgA. J Immunol 163:5367–5373

Singh PK, Jia HP, Wiles K, Hesselberth J, Liu L, Conway BA, Greenberg EP, Valore EV, Welsh MJ, Ganz T, Tack BF, McCray PB Jr (1998) Production of β-defensins by human airway epithelia. Proc Natl Acad Sci USA 95:14961–14966

Smith CB, Bellanti JA, Chanock RM (1967) Immunoglobulins in serum and nasal secretions following infection with type 1 parainfluenza virus and injection of inactivated vaccines. J Immunol 99: 133–141

Smith J, Aylward RB, Salisbury D, Wassilak S, Oblapenko G (1998) Certifying the elimination of poliomyelitis from Europe: advancing towards global eradication. Eur J Epidemiol 14:769–773

Sollid LM, Kvale D, Brandtzaeg P, Markussen G, Thorsby E (1987) Interferon-γ enhances expression of secretory component, the epithelial receptor for polymeric immunoglobulins. J Immunol 138:4303–4306

Sonoda E, Hitoshi Y, Yamaguchi N, Ishii T, Tominaga A, Araki S, Takatsu K (1992) Differential regulation of IgA production by TGF-β and IL-5: TGF-β induces surface IgA-positive cells bearing IL-5 receptor, whereas IL-5 promotes their survival and maturation into IgA-secreting cells. Cell Immunol 140:158–172

Sonoda E, Matsumoto R, Hitoshi Y, Ishii T, Sugimoto M, Araki S, Tominaga A, Yamaguchi N, Takatsu K (1989) Transforming growth factor β induces IgA production and acts additively with interleukin 5 for IgA production. J Exp Med 170:1415–1420

Spangler BD (1992) Structure and function of cholera toxin and the related *Escherichia coli* heat-labile enterotoxin. Microbiol Rev 56:622–647

Spencer J, MacDonald TT, Finn T, Isaacson PG (1986) The development of gut associated lymphoid tissue in the terminal ileum of fetal human intestine. Clin Exp Immunol 64:536–543

Spit BJ, Hendriksen EG, Bruijntjes JP, Kuper CF (1989) Nasal lymphoid tissue in the rat. Cell Tissue Res 255:193–198

Staats HF, Montgomery SP, Palker TJ (1997) Intranasal immunization is superior to vaginal, gastric, or rectal immunization for the induction of systemic and mucosal anti-HIV antibody responses. AIDS Res Hum Retroviruses 13:945–952

Staats HF, Nichols WG, Palker TJ (1996) Mucosal immunity to HIV-1: systemic and vaginal antibody responses after intranasal immunization with the HIV-1 C4/V3 peptide T1SP10 MN(A). J Immunol 157:462–472

Steger KK, Valentine PJ, Heffron F, So M, Pauza CD (1999) Recombinant, attenuated *Salmonella typhimurium* stimulate lymphoproliferative responses to SIV capsid antigen in rhesus macaques. Vaccine 17:923–932

Stolzenberg ED, Anderson GM, Ackermann MR, Whitlock RH, Zasloff M (1997) Epithelial antibiotic induced in states of disease. Proc Natl Acad Sci USA 94:8686–8690

Streeter PR, Rouse BT, Butcher EC (1988) Immunohistologic and functional characterization of a vascular addressin involved in lymphocyte homing into peripheral lymph nodes. J Cell Biol 107:1853–1862

Strobel S, Mowat AM (1998) Immune responses to dietary antigens: oral tolerance. Immunol Today 19:173–181

Svinarich DM, Gomez R, Romero R (1997a) Detection of human defensins in the placenta. Am J Reprod Immunol 38:252–255

Svinarich DM, Wolf NA, Gomez R, Gonik B, Romero R (1997b) Detection of human defensin 5 in reproductive tissues. Am J Obstet Gynecol 176:470–475

Szabo MC, Butcher EC, McEvoy LM (1997) Specialization of mucosal follicular dendritic cells revealed by mucosal addressin-cell adhesion molecule-1 display. J Immunol 158:5584–5588

Takahashi I, Marinaro M, Kiyono H, Jackson RJ, Nakagawa I, Fujihashi K, Hamada S, Clements JD, Bost KL, McGhee JR (1996) Mechanisms for mucosal immunogenicity and adjuvancy of *Escherichia coli* labile enterotoxin. J Infect Dis 173:627–635

Tamura S, Miyata K, Matsuo K, Asanuma H, Takahashi H, Nakajima K, Suzuki Y, Aizawa C, Kurata T (1996) Acceleration of influenza virus clearance by Th1 cells in the nasal site of mice immunized intranasally with adjuvant-combined recombinant nucleoprotein. J Immunol 156:3892–3900

Taylor HP, Dimmock NJ (1985) Mechanism of neutralization of influenza virus by secretory IgA is different from that of monomeric IgA or IgG. J Exp Med 161:198–209

Tsuji T, Inoue T, Miyama A, Okamoto K, Honda T, Miwatani T (1990) A single amino acid substitution in the A subunit of *Escherichia coli* enterotoxin results in a loss of its toxic activity. J Biol Chem 265:22520–22525

Valentine PJ, Meyer K, Rivera MM, Lipps C, Pauza D, Maziarz RT, So M, Heffron F (1996) Induction of SIV capsid-specific CTL and mucosal sIgA in mice immunized with a recombinant *S. typhimurium* aroA mutant. Vaccine 14:138–146

van der Ven I, Sminia T (1993) The development and structure of mouse nasal-associated lymphoid tissue: an immuno- and enzyme-histochemical study. Reg Immunol 5:69–75

van Ginkel FW, Wahl SM, Kearney JF, Kweon MN, Fujihashi K, Burrows PD, Kiyono H, McGhee JR (1999) Partial IgA-deficiency with increased Th2-type cytokines in TGF-β1 knockout mice. J Immunol 163:1951–1957

van Heyningen S (1976) The subunits of cholera toxin: structure, stoichiometry, and function. J Infect Dis 133:5–13

Van Pinxteren LA, Campbell I, Clarke CJ, Snodgrass DR, Bruce MG (1997) A single oral dose of inactivated rotavirus and ISCOM matrices induces partial protection in lambs. Biochem Soc Trans 25:340S

Van Wetering S, Mannesse-Lazeroms SP, Van Sterkenburg MA, Daha MR, Dijkman JH, Hiemstra PS (1997) Effect of defensins on interleukin-8 synthesis in airway epithelial cells. Am J Physiol 272:L888–L896

VanCott JL, Staats HF, Pascual DW, Roberts M, Chatfield SN, Yamamoto M, Coste M, Carter PB, Kiyono H, McGhee JR (1996) Regulation of mucosal and systemic antibody responses by T helper cell subsets, macrophages, and derived cytokines following oral immunization with live recombinant *Salmonella*. J Immunol 156:1504–1514

VanCott TC, Kaminski RW, Mascola JR, Kalyanaraman VS, Wassef NM, Alving CR, Ulrich JT, Lowell GH, Birx DL (1998) HIV-1 neutralizing antibodies in the genital and respiratory tracts of mice intranasally immunized with oligomeric gp160. J Immunol 160:2000–2012

Wagner N, Lohler J, Kunkel EJ, Ley K, Leung E, Krissansen G, Rajewsky K, Muller W (1996) Critical role for β7 integrins in formation of the gut-associated lymphoid tissue. Nature 382:366–370

Wang L, Huang JA, Nagesha HS, Smith SC, Phelps A, Holmes I, Martyn JC, Coloe PJ, Reeves PR (1999) Bacterial expression of the major antigenic regions of porcine rotavirus VP7 induces a neutralizing immune response in mice. Vaccine 17:2636–2645

Warnock RA, Campbell JJ, Dorf ME, Matsuzawa A, McEvoy LM, Butcher EC (2000) The role of chemokines in the microenvironmental control of T versus B cell arrest in Peyer's patch high endothelial venules. J Exp Med 191:77–88

Watson SR, Imai Y, Fennie C, Geoffroy JS, Rosen SD, Lasky LA (1990) A homing receptor-IgG chimera as a probe for adhesive ligands of lymph node high endothelial venules. J Cell Biol 110: 2221–2229

Weiner HL (1997) Oral tolerance: immune mechanisms and treatment of autoimmune diseases. Immunol Today 18:335–343

Weinstein PD, Cebra JJ (1991) The preference for switching to IgA expression by Peyer's patch germinal center B cells is likely due to the intrinsic influence of their microenvironment. J Immunol 147:4126–4135

Weinstein PD, Schweitzer PA, Cebra-Thomas JA, Cebra JJ (1991) Molecular genetic features reflecting the preference for isotype switching to IgA expression by Peyer's patch germinal center B cells. Int Immunol 3:1253–1263

Weltzin R, Monath TP (1999) Intranasal antibody prophylaxis for protection against viral disease. Clin Microbiol Rev 12:383–393

Weltzin R, Traina-Dorge V, Soike K, Zhang JY, Mack P, Soman G, Drabik G, Monath TP (1996) Intranasal monoclonal IgA antibody to respiratory syncytial virus protects rhesus monkeys against upper and lower respiratory tract infection. J Infect Dis 174:256–261

Williams MB, Butcher EC (1997) Homing of naive and memory T lymphocyte subsets to Peyer's patches, lymph nodes, and spleen. J Immunol 159:1746–1752

Williams RC, Gibbons RJ (1972) Inhibition of bacterial adherence by secretory immunoglobulin A: a mechanism of antigen disposal. Science 177:697–699

Wilson CL, Ouellette AJ, Satchell DP, Ayabe T, Lopez-Boado YS, Stratman JL, Hultgren SJ, Matrisian LM, Parks WC (1999) Regulation of intestinal α-defensin activation by the metalloproteinase matrilysin in innate host defense. Science 286:113–117

Wolf JL, Bye WA (1984) The membranous epithelial (M) cell and the mucosal immune system. Annu Rev Med 35:95–112

Wolf JL, Rubin DH, Finberg R, Kauffman RS, Sharpe AH, Trier JS, Fields BN (1981) Intestinal M cells: a pathway for entry of reovirus into the host. Science 212:471–472

Wu HY, Russell MW (1993) Induction of mucosal immunity by intranasal application of a streptococcal surface protein antigen with the cholera toxin B subunit. Infect Immun 61:314–322

Xu-Amano J, Kiyono H, Jackson RJ, Staats HF, Fujihashi K, Burrows PD, Elson CO, Pillai S, McGhee JR (1993) Helper T cell subsets for immunoglobulin A responses: oral immunization with tetanus toxoid and cholera toxin as adjuvant selectively induces Th2 cells in mucosa associated tissues. J Exp Med 178:1309–1320

Yamamoto S, Kiyono H, Yamamoto M, Imaoka K, Fujihashi K, Van Ginkel FW, Noda M, Takeda Y, McGhee JR (1997a) A nontoxic mutant of cholera toxin elicits Th2-type responses for enhanced mucosal immunity. Proc Natl Acad Sci USA 94:5267–5272

Yamamoto S, Takeda Y, Yamamoto M, Kurazono H, Imaoka K, Fujihashi K, Noda M, Kiyono H, McGhee JR (1997b) Mutants in the ADP-ribosyltransferase cleft of cholera toxin lack diarrheagenicity but retain adjuvanticity. J Exp Med 185:1203–1210

Yarus S, Rosen JM, Cole AM, Diamond G (1996) Production of active bovine tracheal antimicrobial peptide in milk of transgenic mice. Proc Natl Acad Sci USA 93:14118–14121

Youngman KR, Fiocchi C, Kaetzel CS (1994) Inhibition of IFN-γ activity in supernatants from stimulated human intestinal mononuclear cells prevents up-regulation of the polymeric Ig receptor in an intestinal epithelial cell line. J Immunol 153:675–681

Zhao C, Nguyen T, Liu L, Shamova O, Brogden K, Lehrer RI (1999) Differential expression of caprine β-defensins in digestive and respiratory tissues. Infect Immun 67:6221–6224

Zhao C, Wang I, Lehrer RI (1996) Widespread expression of β-defensin hBD-1 in human secretory glands and epithelial cells. FEBS Lett 396:319–322

# The Role of the Complement System in Virus Infections

G.T. Spear, M. Hart, G.G. Olinger, F.B. Hashemi, and M. Saifuddin



## 1 Introduction

The complement system plays important roles in the humoral immune response to microorganisms. This chapter reviews literature regarding the interaction of the complement system with viruses, with an emphasis on recent studies. While complement can inactivate viruses, a number of viruses have evolved mechanisms to avoid or subvert complement activation (Table 1), including directing expression of proteins that inactivate complement or incorporating host-cell-complement regulatory proteins. Some viruses also utilize cell-surface complement proteins as receptors. The interaction of human complement with animal viruses has importance for gene therapy with virus vectors. Complement is also involved in the development of antibody responses to viruses.

Department of Immunology/Microbiology, 1653 W, Congress Parkway, Rush-Presbyterian-St. Luke's Medical Center, Chicago, IL 60612, USA

**Table 1.** Interactions of viruses with the complement system

| |
|---|
| Antiviral effects mediated by complement |
| Antibody independent virus inactivation |
| Antibody-dependent virus inactivation |
| Induction of antiviral antibody responses |
| Avoidance and subversion of the complement system by viruses |
| Inhibition of complement via virus-encoded proteins |
| Virus incorporation of host cell complement control proteins |
| Utilization of cell surface complement proteins as virus receptors |
| Enhanced binding of viruses to complement receptor positive cells |

## 2 The Complement System

The complement system acts in the recognition and destruction of microorganisms. It is composed of close to 30 proteins which are mostly present in an inactive form in serum (Liszewski and Atkinson 1993; Morgan 1999). Inflammatory stimuli activate the complement cascade by triggering serine proteases. Complement components have been identified in all vertebrate classes, and a precursor to the complement system has even been found in sea urchins (Smith et al. 1999).

There are three pathways that can induce activation of complement: the classical pathway, the alternative pathway, and the lectin pathway. The classical pathway is activated upon binding of at least one IgM antibody or at least two closely spaced IgG antibodies to an antigen followed by binding and activation of complement component C1. The lectin pathway is activated by binding of mannose binding lectin (MBL) to carbohydrates on microorganisms. Both C1 and MBL in turn activate components C4 and C2, which leads to cleavage of C3 to C3b. In the alternative pathway, C3 spontaneously undergoes hydrolysis resulting in interaction with factor B and cleavage of C3 to C3b. Activation of any one of the three pathways can lead to activation of components C5, C6, C7, C8, and C9. The C5–C9 components of complement form what has been termed the membrane attack complex (MAC), since they combine to form a pore in membranes.

The complement system has three main functions. The first is opsonization of microorganisms for phagocytosis. This is mediated by interaction of complement receptors on the surface of phagocytic cells with C3b and its breakdown products, which become covalently deposited on the surface of microorganisms due to complement activation. The second function of complement is release of peptide factors that stimulate various cell types, including phagocytes. The third function of complement is lysis of microorganisms, which is mediated by the MAC.

The complement system is tightly regulated in several ways to protect host cells and tissues from damage at sites where complement is activated. Activated components have short half-lives due to spontaneous decay. Factor I, factor H, C1 inhibitor and C4-binding protein are all soluble plasma proteins that limit early component activation. Two proteins present on the surface of most cells, CD46 (membrane cofactor protein) and CD55 (decay-accelerating factor), act at the level of C3 to regulate complement activation. Formation of the MAC is also highly regulated. First, formation of the MAC is relatively inefficient, due to a short half-life of activated intermediates. Second, the protein CD59 is present on the surface of cells and prevents recruitment of C9 to form the full lytic pore. Third, nucleated cells have the ability to recover from complement attack by shedding or endocytosis of membrane vesicles that are enriched in MAC components.

## 3 Virus Inactivation by Complement

When viruses are exposed to a source of complement, complement activation can either directly mediate neutralization of viruses or augment antibody-dependent neutralization (reviewed in Sissons 1987). Examples of the former are Newcastle disease virus, in which inactivation is dependent on properdin of the alternative pathway (Wedgewood et al. 1956). In a recent example of the latter mechanism, vesicular stomatitis virus (VSV) titers in spleens were reduced by 20- to 80-fold in mice injected with antiviral antibodies with intact complement systems when compared to antibody-injected C3-knockout mice (Ochsenbein et al. 1999).

There are several mechanisms by which complement mediates neutralization (Table 2). First, a classic study by Berry and Almeida (1968) with avian infectious bronchitis virus showed that, as complement activation proceeds at the virus surface, there is a buildup of complement components. This coating of virus with complement components interferes with virus binding to its receptor on cells; components C1q, C4b, and C3b are most important for this effect. Second, in vivo, complement activation at the surface of virions can lead to deposition of complement components (opsonization) followed by uptake and destruction of the virus by phagocytic cells that express complement receptors (Van Strijp et al. 1991). Further, studies indicate that clearance of viruses from plasma can be significantly altered by activation of complement (Davies et al. 1993; Madi et al. 1992). Third, as the membrane attack complex of complement is activated, components C5b, C6, C7, C8, and C9 form pores in the membrane of enveloped viruses (reviewed in Cooper and Welsh 1979) resulting in lysis of the virus, or virolysis. Virolysis can be observed in several ways. By electron microscopy, the lytic pores can be observed in the virus membrane, and this is accompanied by changes in virus morphology.

**Table 2.** Mechanisms of virus inactivation by the complement system

| Type of inactivation | Complement components critical for inactivation | Mechanism of virus inactivation |
|---|---|---|
| Virus phagocytosis and/or clearance | C3b attachment to virus | Uptake and destruction by phagocytic cells |
| Virus coating | C1q, C4b, C3b | Coating virions prevents interaction with its receptor |
| Virus lysis | C5b-9 (membrane attack complex) | Pore formation and disruption of virus membrane |

Also, during virolysis of retroviruses, release of internal components, including reverse transcriptase, can be detected.

It is likely that several of these neutralization mechanisms act simultaneously to inactivate viruses. For example, studies show that exposure of human immunodeficiency virus type-1 (HIV) to antibody and complement leads to virolysis, releasing reverse transcriptase and core protein from virions (Saifuddin et al. 1994; Sullivan et al. 1996). However, complement-dependent neutralization of HIV is observed in the absence of a MAC, indicating that coating of virus with complement components also contributes to the inactivation of virus (Sullivan and Spear 1998).

# 4 Mechanisms Used by Viruses to Evade Complement-Mediated Destruction

## 4.1 Virus-Encoded Complement Resistance Proteins

A number of proteins are encoded by viruses that, when expressed, inhibit complement function and are therefore believed to contribute to survival and pathogenesis of viruses in vivo. One of the most well characterized of these is the gC glycoprotein of herpes simplex virus type-1 (HSV-1). Complement binding by gC was first demonstrated by showing that infection of cells with HSV-1 resulted in the newly acquired ability to bind C3b due to expression of gC on the cell surface (Friedman et al. 1984). gC is also expressed at the virion surface and protects HSV-1 from destruction by complement in vitro by inhibiting the alternative pathway (McNearney et al. 1987). Studies to localize the binding site on C3 for gC showed binding of C3 and C3 breakdown products C3b and C3c, but not C3d (Kostavasili et al. 1997). gC has been shown to inhibit complement by sterically blocking the interaction of C3 with C5 and properdin, and by accelerating the decay of the C3bBb alternative pathway convertase (Kostavasili et al. 1997).

Expression of gC by HSV-1 has been shown in several in vivo studies to increase virus survival and pathogenesis. Virus lacking gC replicated at lower levels and caused less disease than gC wild-type virus in guinea pigs that had intact complement, while, in contrast, the wild-type and gC mutant viruses behaved similarly in C3-deficient guinea pigs (Lubinski et al. 1998). In a mouse model, mutation of gC in either the C3b-binding domain or the C5/properdin-binding domain reduced virus-induced pathology, although mutation of the C3b-binding domain reduced pathogenesis to a much greater extent (Lubinski et al. 1999). In C3-knockout mice, the gC mutations had little effect. That study concluded that gC was very important for complement evasion due to the large difference in virulence between wild-type and mutant viruses.

Many other herpesviruses encode complement control proteins. Herpes simplex virus type-2 (HSV-2) expresses a gC that is structurally similar to the HSV-1 gC, but, while it binds C3 and increases the resistance of virions to complement-mediated destruction (McNearny et al. 1987), it does not function to accelerate the decay of C3bBb (Kostavasili et al. 1997). Herpesvirus saimiri encodes two proteins with complement regulatory activity. One is called complement control protein homologue (CCPH), since it has structural homology to several human complement control proteins and inhibits C3 convertase activity (Fodor et al. 1995). The other is called HVSCD59, since it has functional and structural homology with human CD59 (Rother et al. 1994). Murine γ-herpesvirus 68, which is closely related to herpesvirus saimiri, encodes a protein with homology to membrane cofactor protein and decay-accelerating factor that inhibits in vitro complement activity (Kapadia et al. 1999). Human herpesvirus 8 (HHV8), the Kaposi-sarcoma associated virus, encodes several proteins with homology to complement control proteins (Russo et al. 1996). A protein from pseudorabies virus was found to bind to C3 (Huemer et al. 1992), while a gene in equine herpesvirus type-1 has been shown to have homology with HSV-1 gC (Allen and Coogle 1998). Purified Epstein-Barr virus (EBV) was shown to down-regulate complement activation through cofactor activity for the factor-I-mediated cleavage of C3b and C4b to inactive breakdown products and also accelerated the decay of the alternative pathway C3 convertase. While it was not clear if these complement regulatory functions were mediated by EBV-encoded protein(s), it was established that the activities were not mediated by host-cell-complement receptor-1, which has similar complement regulatory activities (Mold et al. 1988).

Vaccinia virus also produces a protein, called vaccinia complement control protein (VCP), that inhibits complement activation. VCP is secreted from infected cells instead of being expressed on the virus membrane, as is the case with complement control proteins made by herpesviruses (Kotwal et al. 1990). VCP functions to inhibit complement by accelerating decay of the C3 convertase and acting as a cofactor for factor-I-mediated cleavage of C3b and C4b (McKenzie et al. 1992). The anti-complement activity of VCP appears to be important for survival of virus and inducing pathology, since it prevents antibody-dependent alternative-complement-pathway-mediated destruction of vaccinia virus in vitro, and a VCP-deficient strain of vaccinia induced less pathology in animals (Isaacs et al. 1992).

## 4.2 Incorporation of Host Cell Complement Control Proteins by Viruses

Incorporation of complement control proteins utilized by the host cell is another strategy used by some enveloped viruses to escape the effects of complement. As described above, mammalian cells have several ways to protect themselves from the bystander effects of complement activation that can occur during inflammation. An important mechanism is expression of the complement regulatory proteins CD46, CD55, and CD59 on the cell surface. There is evidence that a number of enveloped viruses, including human cytomegalovirus (HCMV), HIV, human T cell leukemia/lymphoma virus type-I (HTLV-I) and vaccinia virus, incorporate one or more of these proteins into the virus membrane as they bud from infected cells.

HCMV has been shown to incorporate host cell complement components in several ways. COOPER (1994) found that purified HCMV virions had complement regulatory activities that were functionally similar to the regulatory effects of CD46 and CD55. Further, specific CD46 and CD55 monoclonal antibodies (mAbs) blocked the complement regulatory activity of the purified virus. Western blot studies of purified virus also showed that HCMV incorporated CD46 and CD55 as well as CD59 (SPILLER et al. 1997; SPEAR et al. 1995). Treatment of virions with antibody to CD55 and complement resulted in HCMV neutralization, providing further evidence that CD55 is expressed on the surface of virions (SPEAR et al. 1995). However, treatment of virus with anti-CD59 and complement did not alter the HCMV titer, suggesting the possibility that CD59 is not displayed on the virus surface.

Interestingly, HCMV infection of fibroblasts increases expression of CD46 and CD55 by three to eight times that seen in uninfected cells (SPILLER et al. 1996). While this increased expression provides enhanced protection of the infected cells from complement-mediated destruction, it also likely results in incorporation of higher levels of these proteins into virions.

HIV has also been shown to incorporate complement control proteins. Studies in our lab showed that virus produced in seven different cell lines had widely variable sensitivity to complement-mediated virolysis even though all were genetically identical since they were produced by transfection of the infectious pNL4-3 plasmid (SAIFUDDIN et al. 1994, 1995b). While this indicated that host cell factors substantially affected the complement sensitivity of virus, further studies showed that treatment of virus with highly specific antisera to CD55 and CD59 in the presence of complement induced virolysis of both T-cell-line-adapted strains of HIV and HIV primary isolates, providing evidence that both of these proteins were on the surface of virions. MONTEFIORI et al. (1994) and MARSCHANG et al. (1995) showed that a T-cell-line-adapted HIV strain as well as HIV primary isolates could be immunocaptured by antibodies to CD46, CD55, and CD59, indicating the presence of all three proteins on the surface of virions. Virus in the plasma of infected persons was also immunocaptured by antibody to CD46 and CD59 (SULLIVAN et al. 1996). Further studies showed that the incorporated complement control proteins functioned to protect HIV virions from complement. Thus, antibodies that blocked the function of CD55 (MARSCHANG et al. 1995; STOIBER et al. 1996) increased HIV sensitivity to complement. Also, treatment of virions with phosphatidylinositol-specific phopholipase C (PI-PLC), which removes glycosylphophatidylinositol-anchored CD55 and CD59 (SAIFUDDIN et al. 1995b), or production of virus in cells which lack CD55 and CD59 (SAIFUDDIN 1994), increased the sensitivity of HIV to complement-mediated destruction. Reconstitution of PI-PLC-treated virus with purified CD55 or CD59 increased the virus resistance to lysis (SAIFUDDIN et al. 1995b). Finally, transfection of either CD46, CD55, or CD59 into virus-producing cells increased the resistance of virus to complement lysis, indicating that all three proteins can functionally protect virus (SAIFUDDIN et al. 1997).

Other retroviruses have also been shown to incorporate complement control proteins. Simian immunodeficiency virus contained CD46, CD55, and CD59 proteins as determined by immunocapture (MONTEFIORI et al. 1994). HTLV-I also incorporated CD55 and CD59, as determined by Western blot of purified virus, and these proteins were shown to protect virions from lysis (SPEAR et al. 1995).

A recent study by VANDERPLASSCHEN et al. (1998) showed that the extracellular enveloped virus (EEV) form but not the intracellular mature virus (IMV) form of vaccinia virus incorporated complement control proteins from the host HeLa cells. The EEV were neutralized by human complement when produced in rat or rabbit cells but were not neutralized when produced in human cells, showing a species restriction in complement sensitivity that is common for mammalian cell-surface complement regulatory proteins. Western blots of purified EEV showed immunoreactivity for CD46, CD55, and CD59, while immunogold staining localized those antigens to the surface of EEV virions. Mutation or deletion of virus-encoded proteins that may have participated in EEV complement resistance had no affect on complement sensitivity, but transfection of host rat cells with human CD55 and CD59 substantially increased the resistance of EEV to human complement.

## 5 Reactivity of Human Complement with Animal Viruses: Relevance to Gene Therapy and Transplantation

The ability of human complement to inactivate animal retroviruses was first described by Welsh et al. in 1975. Many viruses, including murine leukemia viruses (MLV), are inactivated by the classical pathway and the mechanism was originally described as an antibody-independent ability of the viruses to bind and activate complement component C1 (COOPER et al. 1976).

Animal retroviruses are currently being investigated for use as vectors in human gene therapy. Thus, the anti-retroviral activity of human complement provides a barrier for use of this mode of gene therapy in vivo. As described above, some enveloped viruses incorporate complement control proteins from host cells, which provides a degree of protection from complement inactivation. Therefore, several groups have expressed human complement control proteins in murine cells used to package retrovirus gene vectors in order to protect the viruses from destruction. Some degree of protection of MLV from human complement was shown when human CD55 (HIASA et al. 1999) and CD59 (BREUN et al. 1999) were expressed in the packaging lines, while a fusion protein composed of CD55 and the envelope protein of MLV also provided protection from human complement (SPITZER et al. 1999). Alternatively, production of MLV in human HOS cells also provided protection from complement destruction (TAKEUCHI et al. 1994), presumably by leading to incorporation of HOS-derived human complement control proteins into MLV.

Expression of complement regulatory proteins by virus vectors is being investigated as a means of gene therapy, since complement activation itself has been shown to contribute to the pathogenesis of many inflammatory diseases. McGrath et al. (1999) have developed an adenovirus vector to deliver and express rat CD59, decay accelerating factor, and Crry, a rat complement control protein that has functions of both decay accelerating factor and membrane cofactor protein. All of these proteins were expressed in infected cells and inhibited complement-mediated lysis of the cells and can be tested in rat models of inflammatory diseases.

Recently, antibodies that bind to carbohydrate epitopes on animal retroviruses and other animal enveloped viruses have been recognized as being responsible for much of the human complement activation on viruses produced in animal cells (Rother et al. 1995a). These antibodies are reactive with Gal-α1–3Gal-β1–4GlcNAc-R, which is also called αgal. Influenza viruses, murine retroviruses, and lymphocytic choriomeningitis virus, when produced in animal cells, all bind antibody to αgal and are attacked by human complement (Gallili et al. 1999). Thus, it has been conjectured that complement along with αgal-specific antibody protects humans from some animal virus infections.

The αgal epitopes on viruses are present on glycolipids and N-linked glycoproteins produced in mammalian cells, and up to 1% of circulating human antibodies react with this epitope. Most of this naturally occurring antibody is of the IgG class (Gallili et al. 1995) and is thought to be produced in humans in response to normal gastrointestinal bacterial flora (Gallili 1993). Old World primates, apes, and humans lack these epitopes due to inactivating mutations of the α1–3galactosyl transferase gene. Therefore, while viruses produced in human cells do not react with αgal antibody, viruses produced in most animals react with these antibodies. Expression of the enzyme in human cells leads to expression of the αgal epitope and makes the cells and the retroviruses they produce sensitive to destruction by human serum (Takeuchi et al. 1996). Complement appears important for inactivation of viruses after binding of αgal antibodies (Rother et al. 1995b), since human sera that are depleted of even a single classical-pathway complement component do not inactivate murine retroviruses efficiently. Changing the enzymes that result in αgal glycosylation in animal cells could also lead to production of virus vectors for gene therapy that are resistant to inactivation in human serum (Gallili et al. 1999). Alternatively, utilization of human packaging cell lines instead of animal cells substantially increases the half-life of retroviral vectors in chimpanzees (DePolo et al. 1999).

There has been much interest in the use of pig organs for transplantation into humans. However, expression of αgal on pig organs leads to hyperacute rejection, which consists of rapid destruction of the vascular endothelium of the transplanted tissue, due largely to complement activation after binding of αgal antibody. Several approaches that are being tried to prevent hyperacute rejection include developing pigs that lack the αgal epitopes (either by transgenically expressing alternative glycosylating enzymes or by knocking out the α1–3galactosyl transferase gene) or transgenic expression of human complement control proteins CD46, CD55, and CD59 in pigs (Weiss 1999). There have been concerns, however, that "humanizing" of pig tissues would also lead to a humanizing of any pig viruses in transplanted tissue and an increase in susceptibility of humans to infection. In fact, several studies show that an endogenous retrovirus in pigs (PERV) can infect human cells in tissue culture (Patience et al. 1997; Wilson et al. 1998). PERV was lysed after incubation in human serum, suggesting that complement and αgal antibody can prevent infection of human cells (Patience et al. 1997). Humanizing of pigs would likely lead to resistance of PERV to the neutralizing effects of human serum.

# 6 Virus Utilization of Complement Receptors for Infection of Cells

## 6.1 Direct Binding of Viruses to Complement Receptors

There are a number of viruses that utilize cell-surface complement proteins as receptors. Two of the most studied are EBV and measles virus. EBV was found to utilize complement receptor-2 (CR2) as a receptor for infection of B lymphocytes (Frade et al. 1985), while measles virus uses membrane cofactor protein CD46 as a receptor (Dorig et al. 1993; Naniche et al. 1993). Decay accelerating factor, CD55, was shown to serve as a receptor for some echoviruses (Bergelson et al. 1994). More recently, CD55 was also shown to be a receptor for enterovirus 70 (Karnauchow et al. 1996), but while attachment of some coxsackieviruses was mediated by CD55, CD55 by itself did not lead to infection of cells (Shafren et al. 1995).

## 6.2 Complement Enhancement of Virus Infection

While treatment of many viruses with complement results in at least partial neutralization, treatment of either West Nile virus, HTLV-I, or HIV with complement under certain conditions can actually increase, or enhance, virus infectivity. Cardosa et al. (1983, 1986) showed that treatment of West Nile virus with specific IgM antibody and complement leads to significantly higher virus replication in macrophages. This was probably mediated by an increase in virus infection after binding to complement receptor-3 (CR3) on the macrophages since the increased infection was blocked by antibody to CR3. Treatment of HTLV-I with complement also led to increased virus replication and increased provirus formation in cells that express complement receptors (Saifuddin 1995a).

Robinson, Montefiori, and Mitchell first reported that treatment of HIV with antiviral antibody and human complement increased virus replication in the MT2 T cell line (Robinson et al. 1988). While some enhancement of HIV infection can also occur in the absence of antibody, antibody-dependent or antibody-independent complement-enhanced infection of MT2 cells appears to be the result of deposition of C3b-breakdown products on the surface of virions, leading to higher amounts of virus binding to cells via CR2 expressed on MT2 cells, which in turn leads to an increase in CD4-dependent entry and infection by the virus (ROBINSON et al. 1990; JUNE et al. 1991; TACNET-DELORME et al. 1999). This augmented virus-binding mechanism for complement-mediated enhancement of HIV infection is analogous to the mechanism proposed for antibody-mediated enhancement of infection of cells by dengue virus, which probably occurs after interaction of non-neutralizing antibodies with virions resulting in virus binding at higher levels to cells that express Fc receptors (ROTHMAN and ENNIS 1999). While complement mediates enhanced infection of the MT2 T cell line by HIV, it has not been entirely clear what cells would be the target of this type of infection in vivo, since T cells do not express either appreciable levels of CR2 or other complement receptors that are necessary for binding of opsonized virus. However, it is possible that complement could enhance virus binding and infection of monocytes and macrophages that express complement receptors, and in fact, complement has been shown to enhance infection of monocytic cells in vitro (REISINGER et al. 1990).

Several recent studies have introduced alternative models for how complement-mediated enhancement of HIV infection could occur in vivo (Table 3). First, several studies indicate that complement-opsonized HIV can bind to uninfectable cells via complement receptors and then get efficiently transferred to T cells. For example, studies in our lab showed that incubation of HIV primary isolates with antibody and complement increased virus binding to CR2 on primary B lymphocytes derived from peripheral blood or tonsils (JAKUBIK et al. 1999, 2000). HIV immune complexes bound to B cells by this mechanism are not appreciably internalized, but remain on the cell surface and are much more infectious for primary CD4 T cells than cell-free HIV immune complexes. This efficient transfer of virus from B cells to T cells appears to occur during B cell-T cell adhesion and is blocked by anti-LFA antibodies (Olinger, unpublished studies). Second, complement has also been shown to be important for the in vitro binding of immune complexes containing infectious HIV to CR2 on follicular dendritic cells (FDC) (JOLING et al. 1993), and FDC have also been shown to efficiently pass HIV immune complexes to T cells during cell-cell interactions (HEATH et al. 1995). HIV has, in fact, been

**Table 3.** Possible cellular targets for complement-mediated enhancement of HIV infection

| Infected cell | Mechanism of enhanced infection |
|---|---|
| T cells | Binding of opsonized HIV to T cells that express complement receptors and CD4 (MT2 cell line) |
| T cells | Binding of opsonized HIV to follicular dendritic cells via complement receptors and subsequent transfer of virus to T cells |
| T cells | Binding of opsonized HIV to B cells via complement receptors and subsequent transfer of virus to T cells |
| Monocytes/macrophages | Binding of opsonized HIV to monocytes or macrophages that express complement receptors and CD4 |
| B cells | Binding of opsonized virus to B cells that express complement receptors and CD4 due to CD40 stimulation |

observed to bind at high levels to the surface of FDC in the lymph nodes of HIV-infected persons, where CR2 is highly expressed, suggesting that HIV that have activated complement bind to FDC via CR2 in infected persons (Armstrong and Horne 1984; Spiegel et al. 1992). Finally, stimulation of primary B cells with CD40 induces expression of CD4 and the HIV-coreceptor CXCR4, thus allowing HIV infection (Moir et al. 1999). Whether CR2 continues to be expressed on CD40-stimulated B cells is not known, but that study suggests that, under some circumstances, B cells could be infected in an enhanced fashion similar to MT2 cells, in which CR2 and CD4 are expressed on the same cell.

Although there is no direct evidence that complement enhances HIV infection in vivo, a portion of HIV isolated from the plasma of infected persons is apparently opsonized with C3 breakdown products (Sullivan et al. 1996, 1998). Further, a recent study by Szabo et al. (1999) found a strong correlation between HIV plasma virus load and the level of complement-mediated infection induced by patient sera.

## 7 Recent Studies on the Interaction of HIV with Complement

Recent studies have provided important insights on the interaction of HIV with complement. A number of these have looked at the role of complement during neutralization of virus, and one provided evidence that complement can play an in vivo role for antibody-mediated protection from HIV infection. Gauduin et al. (1998) found that administration to PBL-SCID mice of a mAb to the HIV envelope protein, gp120, protected the mice from HIV infection. Complement appeared to be important for this protective effect, since mice were not protected if they were first given cobra venom factor, which was shown to deplete most of the complement activity in the mice.

Another study (Spruth et al. 1999) used an interesting approach to neutralizing HIV which took advantage of the fact that, as HIV buds from infected cells, it takes along with it a portion of the cell membrane and a number of host cell proteins, including high levels of class I and class II MHC antigens (Arthur et al. 1992). HIV was neutralized when incubated with sera from persons that had high levels of anti-MHC antibodies due to multiple transfusions. Complement was shown to be necessary for full neutralizing activity. The authors suggested that alloimmunization of individuals could provide some protection against HIV infection.

Our laboratory has also recently investigated the ability of anti-MHC antibodies to neutralize HIV. We found that an HIV primary isolate was fully neutralized after incubation with serum from a person that had high levels of anti-MHC antibodies along with a source of complement (Fig. 1). Active complement was necessary for full neutralizing activity since heat inactivation of the complement source resulted in only partial neutralization. This is of some interest, since

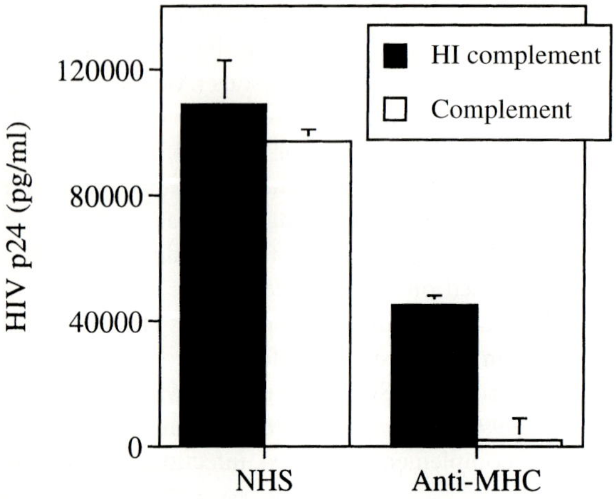


**Fig. 1.** Complement is necessary for efficient neutralization of an HIV primary isolate by anti-MHC antibodies. An HIV primary isolate was produced in phytohemagglutinin (PHA)-stimulated peripheral blood mononuclear cells (PBMC) (Takefman et al. 1998) and incubated with sera as sources of antibody and complement for 1h. After treatment, virus was cultured with PHA-stimulated PBMC, cells were cultured for 7 days, and the level of virus replication was assessed by measuring the virus core protein (p24) in culture supernatants by ELISA (mean ± standard deviation is shown). The complement source was normal human serum from an $AB^{+}$ donor and was used at a final dilution of 1:3. As a control, complement activity was destroyed by heat-inactivation (*HI Complement*). The antibody sources were heat-inactivated serum diluted 1:3 either from a person that had high levels of antibodies to class I and class II MHC due to multiple transfusions (*Anti-MHC*), or from a normal donor that had no anti-MHC antibodies (*NHS*). Untreated virus replicated to 121,000 ± 7,000pg p24/ml

HIV primary isolates are relatively difficult to neutralize with antiviral antibodies either in the presence or absence of complement (Takefman et al. 1998).

Wu et al. (1999) showed that other host cell components can be targets for neutralization mediated by complement and antibody. Antibodies to ganglioside GM2 were shown to activate complement and induce virolysis of HIV. Interestingly, GM2 is expressed at higher levels in cells due to HIV infection.

Previous studies had shown that complement from non-immune rabbits and guinea pigs neutralized HIV (Hosoi et al. 1990; Spear et al. 1991), and it had been hypothesized that this was due to antibody-independent direct activation of complement. However, Gerencer et al. (1998) showed that natural IgM antibodies in baby rabbit serum bind to high-mannose glycans on HIV type 1 glycoprotein 120/160 and activate the classical complement pathway. Thus, natural IgM in animal sera that are reactive with carbohydrates on virus produced in human cells appear to be responsible for complement activation.

Several studies have also looked at antibody-independent mechanisms for complement activation by HIV. Tacnet-Delorme et al. (1999) purified C1q, C1r, C1s, C4, C2, C3, factor H, and factor I and showed that the purified components, when combined at physiologic levels, resulted in complement activation on HIV. This indicates that HIV can directly activate the classical complement pathway. MBL has also been shown to bind to the high-mannose glycans on HIV gp120/

gp160 and activate complement independent of antibody (Ezekowitz et al. 1989; Saifuddin et al. 2000).

## 8 The Role of Complement in Development of Antiviral Immune Responses

Complement is generally regarded as an immune effector mechanism involved directly in destruction and opsonization of pathogens. However, the complement system is also important for the development of antibody responses. Evidence for this effect of complement was originally obtained by showing that animals either depleted of C3 or injected with antibodies that blocked complement receptors had defective humoral immune responses to antigens (Pepys 1972; Heyman et al. 1990). There appears to be two ways that complement can be involved in the induction of humoral immunity, with both mechanisms involving covalent attachment of C3b to antigens (Carroll 2000). Attachment of C3b to antigens leads to binding of antigen to FDC in lymph nodes via complement receptors, which in turn leads to the enhanced presentation of that antigen to B cells. Second, deposition of C3b on antigens induces cross-linking of the B cell antigen receptor and CR2 on the B cell surface, which results in a dramatically more efficient stimulation of B cells. Consequently, much less antigen is required for B cell stimulation.

Several recent studies showed that these mechanisms are active in inducing humoral immune responses against viruses. Mice deficient in either C3, C4, or CR2 (CD21) and complement receptor-1 (CD35) did not develop normal humoral memory responses to herpes simplex virus (Da Costa et al. 1999). Mice deficient in complement components also had delayed IgM titers, but not IgG titers during the early antibody responses to infection with VSV, poliomyelitis virus, and vaccinia virus (Ochsenbein et al. 1999). In that study, IgG responses to vaccination with VSV proteins were also delayed in complement-deficient mice. These results suggest that complement can play an important role in development of humoral responses either after vaccination with virus antigens or during slow-growing virus infection. In contrast, complement may play less of a role during situations of rapid virus replication in which overwhelming antigen stimulation of B cells occurs.

An in vitro study with influenza virus also suggests a role for complement in stimulating influenza-specific T cells (Boackle et al. 1998). Incubation of influenza virus with serum from influenza-immune donors formed immune complexes (IC) that bound to B cells via CR2. The IC-pulsed B cells were then capable of stimulating specific T cells, indicating that antigen processing by the B cells took place. Interestingly, in that study, the uptake of influenza by the B cells was not antigen-specific.

## References


Allen GP, Coogle LD (1988) Characterization of an equine herpesvirus type 1 gene encoding a glycoprotein (gp13) with homology to herpes simplex virus glycoprotein C. J Virol 62:2850–2858

Armstrong JA, Horne R (1984) Follicular dendritic cells and virus-like particles in AIDS-related lymphadenopathy. Lancet 2:370–372

Arthur LO, Bess JW Jr, Sowder RC 2d, Benveniste RE, Mann DL, Chermann JC, Henderson LE (1992) Cellular proteins bound to immunodeficiency viruses: implications for pathogenesis and vaccines. Science 258:1935–1938

Bergelson JM, Chan M, Solomon KR, St John NF, Lin H, Finberg RW (1994) Decay-accelerating factor (CD55), a glycosylphosphatidylinositol-anchored complement regulatory protein, is a receptor for several echoviruses. Proc Natl Acad Sci USA 91:6245–6249

Berry DM, Almeida JD (1968) The morphological and biological effects of various antisera on avian infectious bronchitis virus. J Gen Virol 3:97–102

Boackle SA, Morris MA, Holers VM, Karp DR (1998) Complement opsonization is required for presentation of immune complexes by resting peripheral blood B cells. J Immunol 161:6537–6543

Breun S, Salmons B, Gunzburg WH, Baumann JG (1999) Protection of MLV vector particles from human complement. Biochem Biophys Res Commun 264:1–5

Cardosa MJ, Porterfield JS, Gordon S (1983) Complement receptor mediates enhanced flavivirus replication in macrophages. J Exp Med 158:258–263

Cardosa MJ, Gordon S, Hirsch S, Springer TA, Porterfield JS (1986) Interaction of West Nile virus with primary murine macrophages: role of cell activation and receptors for antibody and complement. J Virol 57:952–959

Carroll MC (2000) The role of complement in B cell activation and tolerance. Adv Immunol 74:61–88

Cooper NR, Jensen FC, Welsh RM Jr, Oldstone MB (1976) Lysis of RNA tumor viruses by human serum: direct antibody-independent triggering of the classical complement pathway. J Exp Med 144:970–984

Cooper NR, Welsh RM (1979) Antibody and Complement-dependent viral neutralization. Springer Sem Immunopathol 2:285–310

Cooper NR (1994) Interactions of the complement system with microorganisms. In: Erdei A (ed) New aspects of complement structure and function. RG Landes, Austin, pp 133–149

Da Costa XJ, Brockman MA, Alicot E, Ma M, Fischer MB, Zhou X, Knipe DM, Carroll MC (1999) Humoral response to herpes simplex virus is complement-dependent. Proc Natl Acad Sci USA 96:12708–12712

Davies KA, Erlendsson K, Beynon HL, Peters AM, Steinsson K, Valdimarsson H, Walport MJ (1993) Splenic uptake of immune complexes in man is complement-dependent. J Immunol 151:3866–3873

DePolo NJ, Harkleroad CE, Bodner M, Watt AT, Anderson CG, Greengard JS, Murthy KK, Dubensky TW Jr, Jolly DJ (1999) The resistance of retroviral vectors produced from human cells to serum inactivation in vivo and in vitro is primate species dependent. J Virol 73:6708–6714

Dorig RE, Marcil A, Chopra A, Richardson CD (1993) The human CD46 molecule is a receptor for measles virus (Edmonston strain). Cell 75:295–305

Ezekowitz RAB, Kuhlman M, Groopman JE, Byrn A (1989) A human serum mannose-binding protein inhibits in vitro infection by the human immunodeficiency virus. J Exp Med 169:185–196

Fodor WL, Rollins SA, Bianco-Caron S, Rother RP, Guilmette ER, Burton WV, Albrecht JC, Fleckenstein B, Squinto SP (1995) The complement control protein homolog of herpesvirus saimiri regulates serum complement by inhibiting C3 convertase activity. J Virol 69:3889–3892

Frade R, Barel M, Ehlin-Henriksson B, Klein G (1985) gp140, the C3d receptor of human B lymphocytes, is also the Epstein-Barr virus receptor. Proc Natl Acad Sci USA 82:1490–1493

Friedman HM, Cohen GH, Eisenberg RJ, Seidel CA, Cines DB (1984) Glycoprotein C of herpes simplex virus 1 acts as a receptor for the C3b complement component on infected cells. Nature 309: 633–635

Galili U (1993) Evolution and pathophysiology of the human natural anti-alpha-galactosyl IgG (anti-Gal) antibody. Springer Semin Immunopathol 15:155–171

Galili U, Tibell A, Samuelsson B, Rydberg L, Groth CG (1995) Increased anti-Gal activity in diabetic patients transplanted with fetal porcine islet cell clusters. Transplantation 59:1549–1556

Gallili U, Wang L, LaTemple DC, Radic MZ (1999) The natural anti-gal antibody. In: Gallili U, Avila JL (eds) AlphaGal and anti-Gal, subcellular biochemistry, vol 32. Kulwer, Dordrecht, pp 79–106

Gauduin MC, Weir R, Fung MS, Koup RA (1998) Involvement of the complement system in antibody-mediated post-exposure protection against human immunodeficiency virus type 1. AIDS Res Hum Retroviruses 14:205–211

Gerencer M, Barrett PN, Kistner O, Mitterer A, Dorner F (1998) Natural IgM antibodies in baby rabbit serum bind high-mannose glycans on HIV type 1 glycoprotein 120/160 and activate classic complement pathway. AIDS Res Hum Retroviruses 14:599–605

Heath SL, Tew JG, Tew JG, Szakal AK, Burton GF (1995) Follicular dendritic cells and human immunodeficiency virus infectivity. Nature 377:740–744

Heyman B, Wiersma EJ, Kinoshita T (1990) In vivo inhibition of the antibody response by a complement receptor-specific monoclonal antibody. J Exp Med 172:665–668

Hiasa A, Watanabe M, Okada H, Ikenaka K, Fujita T, Yoshimatsu T, Kanematsu T, Shiku H (1999) Retroviruses prepared from human DAF expressing murine packaging cells acquire resistance against human serum. Int J Oncol 14:1091–1096

Hosoi S, Borsos T, Dunlop N, Nara PL (1990) Heat-labile, complement-like factor(s) of animal sera prevent(s) HIV-1 infectivity in vitro. J Acquir Immune Defic Syndr 3:366–371

Huemer HP, Larcher C, Coe NE (1992) Pseudorabies virus glycoprotein III derived from virions and infected cells binds to the third component of complement. Virus Res 23:271–280

Isaacs SN, Kotwal GJ, Moss B (1992) Vaccinia virus complement-control protein prevents antibody-dependent complement-enhanced neutralization of infectivity and contributes to virulence. Proc Natl Acad Sci USA 89:628–632

Jakubik JJ, Saifuddin M, Takefman DM, Spear GT (1999) B lymphocytes in lymph nodes and peripheral blood are important for binding immune complexes containing HIV-1. Immunology 96:612–619

Jakubik JJ, Saifuddin M, Takefman DM, Spear GT (2000) Immune complexes containing HIV-1 primary isolates bind to lymphoid tissue B lymphocytes and are infectious for T lymphocytes. J Virol 74: 552–555

Joling P, Bakker LJ, Van Strijp JA, Meerloo T, de Graaf L, Dekker ME, Goudsmit J, Verhoef J, Schuurman HJ (1993) Binding of human immunodeficiency virus type-1 to follicular dendritic cells in vitro is complement dependent. J Immunol 150:1065–1073

June RA, Schade SZ, Bankowski MJ, Kuhns M, McNamara A, Lint TF, Landay AL, Spear GT (1991) Complement and antibody mediate enhancement of HIV infection by increasing virus binding and provirus formation. AIDS 5:269–274

Kapadia SB, Molina H, van Berkel V, Speck SH, Virgin HW 4th (1999) Murine gammaherpesvirus 68 encodes a functional regulator of complement activation. J Virol 73:7658–7670

Karnauchow TM, Tolson DL, Harrison BA, Altman E, Lublin DM, Dimock K (1996) The HeLa cell receptor for enterovirus 70 is decay-accelerating factor (CD55). J Virol 70:5143–5152

Kostavasili I, Sahu A, Friedman HM, Eisenberg RJ, Cohen GH, Lambris JD (1997) Mechanism of complement inactivation by glycoprotein C of herpes simplex virus. J Immunol 158:1763–1771

Kotwal GJ, Isaacs SN, McKenzie R, Frank MM, Moss B (1990) Inhibition of the complement cascade by the major secretory protein of vaccinia virus. Science 250:827–830

Liszewski MK, Atkinson JP (1993) The complement system. In: Paul WE (ed) Fundamental immunology, 3rd edn. Raven, New York, pp 917–939

Lubinski JM, Wang L, Soulika AM, Burger R, Wetsel RA, Colten H, Cohen GH, Eisenberg RJ, Lambris JD, Friedman HM (1998) Herpes simplex virus type 1 glycoprotein gC mediates immune evasion in vivo. J Virol 72:8257–8263

Lubinski J, Wang L, Mastellos D, Sahu A, Lambris JD, Friedman HM (1999) In vivo role of complement-interacting domains of herpes simplex virus type 1 glycoprotein gC. J Exp Med 190:1637–1646

Madi N, Steiger G, Estreicher J, Schifferli JA (1992) Abnormal immune adherence and elimination of hepatitis B surface antigen/antibody complexes in patients with AIDS. J Immunol 148:723–728

Marschang P, Sodroski J, Wurzner R, Dierich MP (1995) Decay-accelerating factor (CD55) protects human immunodeficiency virus type 1 from inactivation by human complement. Eur J Immunol 25:285–290

McGrath Y, Wilkinson GW, Spiller OB, Morgan BP (1999) Development of adenovirus vectors encoding rat complement regulators for use in therapy in rodent models of inflammatory diseases. J Immunol 163:6834–6840

McKenzie R, Kotwal GJ, Moss B, Hammer CH, Frank MM (1992) Regulation of complement activity by vaccinia virus complement-control protein. J Infect Dis 166:1245–1250

McNearney TA, Odell C, Holers VM, Spear PG, Atkinson JP (1987) Herpes simplex virus glycoproteins gC-1 and gC-2 bind to the third component of complement and provide protection against complement-mediated neutralization of viral infectivity. J Exp Med 166:1525–1535

Moir S, Lapointe R, Malaspina A, Ostrowski M, Cole CE, Chun TW, Adelsberger J, Baseler M, Hwu P, Fauci AS (1999) CD40-Mediated induction of CD4 and CXCR4 on B lymphocytes correlates with restricted susceptibility to human immunodeficiency virus type 1 infection: potential role of B lymphocytes as a viral reservoir. J Virol 73:7972–7980

Mold C, Bradt BM, Nemerow GR, Cooper NR (1988) Epstein-Barr virus regulates activation and processing of the third component of complement. J Exp Med 168:949–969

Montefiori DC, Cornell RJ, Zhou JY, Zhou JT, Hirsch VM, Johnson PR (1994) Complement control proteins, CD46, CD55, and CD59, as common surface constituents of human and simian immunodeficiency viruses and possible targets for vaccine protection. Virology 205:82–92

Morgan BP (1999) Regulation of the complement membrane attack pathway. Crit Rev Immunol 19: 173–198

Naniche D, Varior-Krishnan G, Cervoni F, Wild TF, Rossi B, Rabourdin-Combe C, Gerlier D (1993) Human membrane cofactor protein (CD46) acts as a cellular receptor for measles virus. J Virol 67:6025–6032

Ochsenbein AF, Pinschewer DD, Odermatt B, Carroll MC, Hengartner H, Zinkernagel RM (1999) Protective T cell-independent antiviral antibody responses are dependent on complement. J Exp Med 190:1165–1174

Patience C, Takeuchi Y, Weiss RA (1997) Infection of human cells by an endogenous retrovirus of pigs. Nat Med 3:282–286

Pepys MB (1972) Role of complement in induction of the allergic response. Nat New Biol 237:157–159

Reisinger EC, Vogetseder W, Berzow D, Kofler D, Bitterlich G, Lehr HA, Wachter H, Dierich MP (1990) Complement-mediated enhancement of HIV-1 infection of the monoblastoid cell line U937. AIDS 4:961–965

Robinson WE Jr, Montefiori DC, Mitchell WM (1988) Antibody-dependent enhancement of human immunodeficiency virus type 1 infection. Lancet 1:790–794

Robinson WE Jr, Montefiori DC, Mitchell WM (1990) Complement-mediated antibody-dependent enhancement of HIV-1 infection requires CD4 and complement receptors. Virology 175:600–604

Rother RP, Rollins SA, Fodor WL, Albrecht JC, Setter E, Fleckenstein B, Squinto SP (1994) Inhibition of complement-mediated cytolysis by the terminal complement inhibitor of herpesvirus saimiri. J Virol 68:730–737

Rother RP, Fodor WL, Springhorn JP, Birks CW, Setter E, Sandrin MS, Squinto SP, Rollins SA (1995a) A novel mechanism of retrovirus inactivation in human serum mediated by anti-α-galactosyl natural antibody. J Exp Med 182:1345–1355

Rother RP, Squinto SP, Mason JM, Rollins SA (1995b) Protection of retroviral vector particles in human blood through complement inhibition. Hum Gene Ther 6:429–435

Rothman AL, Ennis FA (1999) Immunopathogenesis of dengue hemorrhagic fever. Virology 257:1–6

Russo JJ, Bohenzky RA, Chien MC, Chen J, Yan M, Maddalena D, Parry JP, Peruzzi D, Edelman IS, Chang Y, Moore PS (1996) Nucleotide sequence of the Kaposi sarcoma-associated herpesvirus (HHV8). Proc Natl Acad Sci USA 93:14862–14867

Saifuddin M, Ghassemi M, Patki C, Parker CJ, Spear GT (1994) Host cell components affect the sensitivity of HIV-1 to complement-mediated virolysis. AIDS Res Human Retroviruses 10: 829–837

Saifuddin M, Landay AL, Ghassemi M, Patki C, and Spear GT (1995a) HTLV-1 activates complement leading to increased binding to complement receptor positive cells. AIDS Res Hum Retrovir 11:1115–1122

Saifuddin M, Parker CJ, Peeples ME, Zolla-Pazner S, Ghassemi M, Rooney IA, Atkinson JP, Spear GT (1995b) Role of virion-associated phosphatidylinositol-linked proteins CD55 and CD59 in complement resistance of cell line-derived and primary isolates of HIV-1. J Exp Med 182:501–509

Saifuddin M, Hedayati T, Atkinson JP, Holguin MH, Parker CJ, Spear GT (1997) HIV-1 incorporates both glycosyl phosphatidylinositol anchored CD55 and CD59 and integral membrane CD46 at levels that protect from complement-mediated destruction. J Gen Virol 78:1907–1911

Saifuddin M, Hart ML, Gewurz H, Zhang Y, Jakubik JJ, Spear GT (2000) Interaction of mannose-binding lectin with primary isolates of HIV-1. J Gen Virol 81:949–955

Shafren DR, Bates RC, Agrez MV, Herd RL, Burns GF, Barry RD (1995) Coxsackieviruses B1, B3, and B5 use decay accelerating factor as a receptor for cell attachment. J Virol 69:3873–3877

Sissons JGP (1987) Complement and viruses. In: Whaley K (ed) Complement in health and disease. MTP, Lancaster, pp 255–266

Smith LC, Azumi K, Nonaka M (1999) Complement systems in invertebrates. The ancient alternative and lectin pathways. Immunopharmacology 42:107–120

Spear GT, Sullivan BL, Takefman DM, Landay AL, Lint TF (1991) Human Immunodeficiency Virus (HIV)-infected cells and free virus directly activate the classical complement pathway in rabbit, mouse and guinea pig sera; activation results in virus neutralization by virolysis. Immunology 73:377–382

Spear GT, Lurain NS, Parker CJ, Ghassemi M, Payne GH, Saifuddin M (1995) Host cell-derived complement control proteins CD55 and CD59 are incorporated into the virions of two unrelated enveloped viruses; HTLV-I and human cytomegalovirus. J Immunol 155:4376–4381

Spiegel H, Herbst H, Niedobitek G, Foss HD, Stein H (1992) Follicular dendritic cells are a major reservoir for human immunodeficiency virus type 1 in lymphoid tissues facilitating infection of CD4+ T-helper cells. Am J Pathol 140:15–22

Spiller OB, Morgan BP, Tufaro F, Devine DV (1996) Altered expression of host-encoded complement regulators on human cytomegalovirus-infected cells. Eur J Immunol 26:1532–1538

Spiller OB, Hanna SM, Devine DV, Tufaro F (1997) Neutralization of cytomegalovirus virions: the role of complement. J Infect Dis 176:339–347

Spitzer D, Hauser H, Wirth D (1999) Complement-protected amphotropic retroviruses from murine packaging cells. Hum Gene Ther 10:1893–1902

Spruth M, Stoiber H, Kacani L, Schonitzer D, Dierich MP (1999) Neutralization of HIV type 1 by alloimmune sera derived from polytransfused patients. AIDS Res Hum Retroviruses 15:533–543

Stoiber H, Pinter C, Siccardi AG, Clivio A, Dierich MP (1996) Efficient destruction of human immunodeficiency virus in human serum by inhibiting the protective action of complement factor H and decay accelerating factor (DAF, CD55). J Exp Med 183:307–310

Sullivan BL, Knopoff EJ, Takefman DM, Sha BE, Spear GT (1996) Susceptibility of HIV-1 plasma virus to complement-mediated lysis: evidence for a role in clearance of virus in vivo. J Immunol 157:1791–1798

Sullivan BL, Spear GT (1998) Complement can neutralize HIV-1 plasma virus by a C5-independent mechanism. Virology 248:173–181

Szabo J, Prohaszka Z, Toth FD, Gyuris A, Segesdi J, Banhegyi D, Ujhelyi E, Minarovits J, Fust G (1999) Strong correlation between the complement-mediated antibody-dependent enhancement of HIV-1 infection and plasma viral load. AIDS 13:1841–1849

Tacnet-Delorme P, Boyer V, Thielens NM, Hernandez JF, Bally I, Sim RB, Desgranges C, Arlaud GJ (1999) In vitro analysis of complement-dependent HIV-1 cell infection using a model system. J Immunol 162:4088–4093

Takefman DM, Sullivan BL, Sha BE, Spear GT (1998) Mechanisms of resistance of HIV-1 primary isolates to complement mediated lysis. Virology 246:370–378

Takeuchi Y, Cosset FL, Lachmann PJ, Okada H, Weiss RA, Collins MK (1994) Type C retrovirus inactivation by human complement is determined by both the viral genome and the producer cell. J Virol 68:8001–8007

Takeuchi Y, Porter CD, Strahan KM, Preece AF, Gustafsson K, Cosset FL, Weiss RA, Collins MK (1996) Sensitization of cells and retroviruses to human serum by (alpha 1–3) galactosyltransferase. Nature 379:85–88

Vanderplasschen A, Mathew E, Hollinshead M, Sim RB, Smith GL (1998) Extracellular enveloped vaccinia virus is resistant to complement because of incorporation of host complement control proteins into its envelope. Proc Natl Acad Sci USA 95:7544–7549

van Strijp JA, van der Tol ME, Miltenburg LA, van Kessel KP, Verhoef J (1991) Tumour necrosis factor triggers granulocytes to internalize complement-coated virus particles. Immunology 73:77–82

Wedgewood RJ, Ginsberg HS, Pillemer H (1956) The properdin system and immunity. VI. The inactivation of Newcastle disease virus by the properdin system. J Exp Med 104:707–725

Weiss RA (1999) Xenografts and retroviruses. Science 285:1221–1222

Welsh RM Jr, Cooper NR, Jensen FC, Oldstone MB (1975) Human serum lyses RNA tumour viruses. Nature 257:612–614

Wilson CA, Wong S, Muller J, Davidson CE, Rose TM, Burd P (1998) Type C retrovirus released from porcine primary peripheral blood mononuclear cells infects human cells. J Virol 72:3082–3087

Wu X, Okada N, Momota H, Irie RF, Okada H (1999) Complement-mediated anti-HIV-1 effect induced by human IgM monoclonal antibody against ganglioside GM2. J Immunol 162:533–539

# Intrabodies as Antiviral Agents

W.A. Marasco



## 1 Introduction

Antibodies are nature's paradigm of evolutionary engineering, being incredibly diverse in their ability to be generated against virtually any target molecule, whether self or foreign, protein or nucleic acid, carbohydrate or lipid. However, in their natural function against disease-causing microorganisms, antibodies function outside the cell, where they aid in blocking cellular entry and in clearing these microbes from the host. Expanding the ability of antibodies to function inside of cells as intrabodies has greatly increased their utility and has opened up a number of important and exciting basic research and clinical opportunities.

The basic principle behind intrabodies is a marriage between an antibody binding site and the molecular instructions to direct this binding site to a precise

---

Department of Cancer Immunology and AIDS, Dana-Farber Cancer Institute, Harvard Medical School, 44 Binney Street, Jimmy Fund Building, Room 824, Boston, MA 02115, USA intracellular location. By harvesting the power and diversity of the immune system in the form of rearranged antibody genes, the immune system itself accomplishes the binding site design, thereby bypassing some of the technical hurdles involved in other methods of gene inactivation such as targeted gene disruption, dominant negative mutants, and RNA-based strategies that use antisense, ribozymes, and RNA decoys.

At least four mechanisms of inhibition by intrabodies have been described. First, intrabodies can bind to the active site of an intracellular target molecule and block its activity. Second, intrabodies can disrupt intracellular protein–protein interactions. Third, intrabodies can interfere with intracellular target protein transport. Fourth, the binding of intrabodies to their target can accelerate target protein degradation (reviewed in Marasco 1998). Intrabodies have also been demonstrated to activate enzymes (Martineau et al. 1998), act as transcriptional activators (Mary et al. 1999), and restore mutant protein function (Caron de Fromentel et al. 1999). Indeed, the list of potential new applications of intrabodies in fields as diverse as functional genomics has been constantly growing (Visintin et al. 1999).

There have been two recent comprehensive reviews on the design and function of intrabodies (Marasco 1998; Cattaneo 1999). In this chapter, I will focus on the work primarily from my laboratory on using intrabodies as antiviral agents for the gene therapy of HIV-1-infection and AIDS. However, the use of intrabodies in basic research and in the therapy of other viral diseases, including hepatitis B (Zu Putlitz et al. 1999), murine leukemia virus (MuLV) (Gargano 1997; Marin et al. 2000), and tick-borne flaviviruses (Jiang et al. 1995), has been reported.

## 2 Starting Materials for Intrabody Construction

The single-chain antibody, also known as single-chain variable region fragments, or sFv, is a ~28-kDa recombinant single polypeptide in which the heavy- and light-chain variable domains are separated by a flexible linker peptide, generally $(Gly_4\text{-}Ser)_3$. The sFv format has been the most extensively used form of rearranged antibodies for the construction of "sFv intrabodies". The majority of sFv intrabodies that have been reported to date are derived from murine monoclonal antibodies (mAbs). These hybridomas have provided a reliable source of well-characterized mAb reagents for the construction of intrabodies and are particularly useful when their epitope reactivity and affinity have been previously characterized.

Recently, large naive or unimmunized human sFv phage-display libraries have been created and offer a rich source of rearranged antibody genes against an unlimited plethora of target molecules. Smaller immunoglobulin libraries from individuals with autoimmune (Portolano et al. 1993; Barbas et al. 1995) or infectious diseases (Barbas et al. 1992a; Zebedee et al. 1992) have allowed more disease-specific antibodies to be readily isolated. Starting materials for these recombinant-DNA-based strategies have included RNA from mouse spleens (CLACKSON et al. 1991) and human peripheral-blood lymphocytes (PBLs) (PORTOLANO et al. 1993; BARBAS et al. 1991, 1995; MARKS et al. 1991), and lymphoid organs and bone marrow from HIV-1-infected donors (BURTON et al. 1991; BARBAS et al. 1992b).

We have recently constructed a large, unimmunized, human, single-chain-antibody (sFv) phage-display library that can be used to directly isolate human sFv genes for the construction of intrabodies (MEHTA et al. 2000). To maximize the genetic diversity of the unimmunized antibody library, B cells were isolated from the peripheral blood of 57 volunteers at Dana-Farber Cancer Institute. We felt that using a large number of donors would increase the polymorphism of the V region gene segments. We prepared RNA, made cDNA, and PCR-amplified all classes of rearranged human $V_H$, $V_{kappa}$ and $V_{lambda}$ genes separately from each individual. This translates to the isolation of human $V_H$, $V_{kappa}$, and $V_{lambda}$ genes from $2.75 \times 10^8$ B cells (25% of $1.1 \times 10^9$ PBLs), giving us a theoretical heavy-chain diversity of 275 million members. The rearranged $V_H$ genes were separately cloned into a pUC19-based vector to obtain $3.5 \times 10^8$ transformants. The $V_{kappa}$ and $V_{lambda}$ genes were cloned into the pFARBER phagemid vector as $V_L$–gene-III fusion proteins to yield light-chain libraries with $120 \times 10^6 V_{kappa}$ and $80 \times 10^6 V_{lambda}$ members, respectively.

The final human sFv phage library was constructed by performing 80 electroporations of >275 million human $V_H$ genes randomly combined with 1.6 million each of $V_{kappa}$– and $V_{lambda}$–gene-III fusions in the pFARBER phagemid vector. These ratios were chosen to maintain maximal $V_H$ diversity, since the majority of binding energy is contributed by $V_H$ CDR3. A total of $1.63 \times 10^{10}$ transformants were isolated. Analysis by restriction enzyme digestion demonstrated an sFv insert efficiency of >92%, yielding a library of 15 billion members. This library is now supplying us with an unlimited source of human sFv genes.

## 3 Intrabody Expression Vectors

The selective trafficking and compartmentalization of specific molecules throughout the cell require the recognition of sorting determinants that are most often present in the primary structure of those molecules. The identification of some of these determinants has been one of the major accomplishments of molecular cell biology, and this information has been used extensively in the design of intrabodies that can be targeted to precise subcellular locations. Figure 1 shows a partial list of a series of plasmid-based intrabody expression vectors that contain different intracellular trafficking signals. The vectors have been constructed so that sFv genes from our human sFv phage-display library can be directly cloned into these vectors through identical *Sfi*I/*Not*I cloning sites.

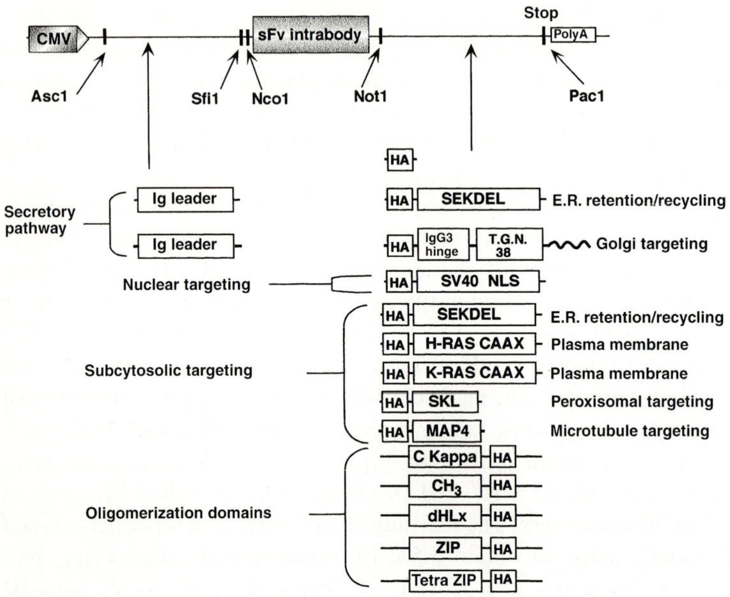


**Fig. 1.** Intrabody expression vectors. Using pcDNA3.1 backbone (Invitrogen), a series of intracellular trafficking vectors have been constructed that accommodate sFv genes from the pFARBER vector through compatible *Sfi*I/*Not*I cloning sites. A hemagglutinin (HA) tag has been included in the vector design to normalize intrabody expression levels as well as for immunoprecipitation, pulse-chase, and Western blot experiments

# 4 Pharmacologic and Genetic Therapies for HIV-1 Infection and AIDS

Human immunodeficiency virus productively infects human CD4$^+$ cells that express appropriate coreceptors (reviewed in Luster 1998), including CD4$^+$ T cells, cells of the monocyte/macrophage lineage, and other cell types, such as neuronal cells (Gallo et al. 1984; Gartner et al. 1986). In the absence of antiretroviral therapy, infection with HIV-1 leads, in most cases, to a progressive decline in the number and functions of CD4$^+$ T cells, with the eventual appearance of clinical manifestations of cellular immunodeficiency, such as opportunistic infections and malignancies – i.e., AIDS (Gottlieb et al. 1981; Fauci et al. 1984; Schroff et al. 1983).

Recent improvements in antiretroviral drug treatments, particularly highly active antiretroviral therapies (HAART), have resulted in remarkable reductions of viral loads in HIV-1-seropositive patients. These treatment advances have led to an equally remarkable improvement in quality of life and a decrease in mortality in HIV-1-infected individuals. However, there are limitations to the therapy, particularly in patients that have been on multiple monotherapies in the past and/or in patients that have started HAART later in the course of their HIV-1 infection. In these patients, the antiretroviral responses have been poor. In addition, strict compliance with HAART therapy is required to maintain long-term suppression of HIV-1 replication, with viral rebound occurring even with short drug holidays (DAAR et al. 1998; MARKOWITZ et al. 1999). Furthermore, a reservoir of latently infected cells in peripheral blood, lymph nodes, and GALT persists in these HIV-1-infected individuals (CHUN et al. 1997; WONG et al. 1997; FINZI et al. 1997, 1999; MARKOWITZ et al. 1999; DYBUL et al. 2000). Thus, unless virus can be completely eliminated, there is concern that low levels of viral replication will lead to continuous depletion and damage to the immune system. In fact, the selective pressure of chemotherapy may generate drug-resistant strains of virus that retain the pathogenic virulence of wild-type strains (EMERY and LANE 1996). These limitations of HAART, combined with the lifelong adherence that is probably necessary to maintain suppression of viral replication (HAVLIR et al. 1998; PIALOUX et al. 1998), have underscored the need and intensified the search for additional therapeutic approaches to enhance immune function against HIV-1.

Gene therapy for the treatment of HIV-1 infection and AIDS has captured the renewed interest of a number of investigators as an attractive addition to HAART, because alteration of the host cell could potentially confer permanent suppression of viral replication after infection, or perhaps even lasting protection against viral infection. Several strategies have been investigated that have different and sometimes overlapping modes of action. One approach is based on enhancement of the immune response against the virus by using genetically modified cells that express viral gene products to induce antiviral cellular immune responses. Another approach uses an extracellular mode of action by secretion from the transduced cell of a factor directly affecting the HIV-1 life cycle (e.g., sCD4-IgG, anti-gp120 Fab antibody fragments) or of a factor affecting the host's defense mechanism (e.g., a cytokine). "Intracellular immunization" is a third strategy and is aimed at achieving inhibition of HIV-1 replication by interference with essential structural or regulatory viral genes that are involved in the molecular pathogenesis of HIV-1-infection. The stable transfer of genetic elements that inhibit viral replication, so-called resistance genes, into those cells of a patient that are potential targets for viral infection may lead to resistance to viral infection in vivo (BALTIMORE 1988; GILBOA and SMITH 1994). Furthermore, protection from HIV-1 of even a small percentage of transduced, peripheral-blood $CD4^+$ mononuclear cells may lead to clinical benefit if in vivo expansion of the protected cells occurs (WEI et al. 1995; HO et al. 1995; PERELSON et al. 1996). Indeed, recent studies have provided evidence that, if virus replication can be inhibited, the thymus function may be restored, at least partially, in some HIV-1-infected individuals (ARNO et al. 1999; SMITH et al. 2000; SULIS et al. 2000). Gene therapy with transduced $CD34^+$ stem cells will be a logical extension of this approach if transduced $CD4^+$ T lymphocytes are protected in vivo.

Intracellular immunization strategies that are aimed at inhibiting target gene expression can be RNA (antisense, ribozymes, RNA decoys) or protein (intrabodies, dominant-negative mutants) based, and each group of inhibitors has distinct advantages and limitations. While RNA-based strategies are often limited by the inability to achieve high levels of inhibitor expression or to allow accurate subcellular localization, protein-based strategies may be limited by their potential immunogenicity, particularly when the genetic strategy is to treat a chronic infection such as AIDS. Like its normal cellular-protein counterparts, the intracellularly expressed protein transgene will be degraded by the 20S and 26S proteasome and presented by MHC-I to antigen-presenting cells (Goldberg 1995; Rock 1996). When the MHC-I-presented peptides are recognized as foreign, a subsequent cellular immune response can be elicited against the transduced cells. Indeed, while results of several cancer gene-therapy marking studies (Brenner et al. 1993a,b) and gene-replacement studies (Bordignon et al. 1995; Blaese et al. 1995) that have used the neomycin selection marker have shown persistence of the marker gene, and a HIV-1 revM10-based intracellular immunization study has shown preferential survival of the revM10-expressing cells compared to the non-expressing frameshift revM10-transduced cells (Woffendin et al. 1996; Ranga et al. 1998), there is growing evidence that a cytotoxic T lymphocyte (CTL) response can limit long-term protein transgene expression, particularly in the clinical setting where multiple administrations of gene-modified cells are required (Riddell et al. 1996).

## 5 Intrabodies as Anti-HIV-1 Agents

As an anti-HIV-1 treatment strategy, genes encoding anti-HIV-1 intrabodies can be stably introduced into mature $CD4^+$ T cells and/or $CD34^+$ hematopoietic stem cells using appropriate gene-transfer vectors, including MuLV and lentivirus vectors. The intrabodies are then synthesized by the cell and targeted to specific cellular compartments where they bind to their target HIV-1 protein and inhibit its function (Marasco 1995, 1997; Marasco et al. 1999). Previous studies have demonstrated intrabody-mediated inhibition of HIV-1 replication in vitro by blocking gp120 processing and incorporation into virions (Marasco et al. 1993; Chen et al. 1994a,b, 1996; Zhou et al. 1998), or by inhibiting critical HIV-1 regulatory protein functions such as Tat-mediated viral transcriptional transactivation (Mhashilkar et al. 1995, 1997, 1999; Poznansky et al. 1998, 1999), or Rev-mediated nuclear export of singly spliced or genomic viral RNA (Duan et al. 1994a,b). In addition, intrabodies against HIV-1 reverse transcriptase (RT) (Maciejewski et al. 1995; Shaheen et al. 1996) and matrix (MA, p17) (Levin et al. 1997) have been shown to inhibit earlier events of the viral life cycle, including reverse transcription of viral RNA to DNA and integration into cellular chromosomal DNA (Maciejewski et al. 1995; Shaheen et al. 1996; Levy-Mintz et al. 1996; Levin et al. 1997). Anti-integrase intrabodies (Levy-Mintz et al. 1996, Kitamura et al. 1999) have also been reported to cause pre-integration blockade of HIV-1 infection, and in a more recent advance in the field, anti-integrase intrabodies have been incorporated into HIV-1 virions as Vpr–sFv or WXXF–sFv fusion proteins (BouHamdan et al. 1998; Okui et al. 1998, 2000; Kulksoky et al. 1999). Vpr is known to be quantitatively incorporated into HIV-1 virions, and the WXXF motif has been shown to bind to Vpr and thus can be used to incorporate WXXF-motif-bearing sFvs into wild-type, *vpr*-encoding HIV-1 virions. In both strategies, the incorporation of anti-integrase sFvs into the virions has resulted in decreased infectivity of the virus particles.

Despite these numerous in vitro gene-therapy studies showing the efficacy of anti-HIV-1 intrabodies as molecular inhibitors of HIV-1 replication, at present, studies that examine the protection of transduced cells expressing anti-HIV-1 intrabodies in vivo have not been reported. This has in part been due to the fact that most of the anti-HIV-1 intrabodies reported to date have been derived from murine mAbs. Thus, the likelihood of eliciting a CTL response against the transduced cells expressing the murine anti-HIV-1 intrabodies that would severely limit the lifespan of the transduced cells is considerable, particularly if transduction of mature $CD4^+$ T-cells is planned. Because of these considerations and our desire to conduct an anti-Tat intrabody clinical gene-therapy trial, humanization of a potent anti-Tat intrabody sFvtat1Ck was considered a prerequisite before the in vivo protective effects of the anti-Tat sFv intrabody gene-therapy approach could be examined. We have recently reviewed our pre-clinical studies on the humanized anti-Tat intrabody sFvhutat2 and the outline of our human clinical trial (Marasco et al. 1999), and therefore this work will not be presented in this review. However, in the text that follows, I will present recent data on the continued progress we have made with anti-HIV-1 intrabodies and our efforts to evaluate their in vivo clinical efficacy.

## 5.1 Post-Integration Blockade of HIV-1 Infection

### 5.1.1 The SHIV-Macaque Gene Therapy Model

Although the initial clinical gene-therapy trials in HIV-1-infected individuals using intracellular-immunization-based strategies have demonstrated in vivo protection of the transduced $CD4^+$ T cells, the studies have also highlighted the limitations of these approaches, including duration of in vivo survival of transduced cells, loss of detectable antiviral transgene expression, and rapid clearing of transduced cells due to immune responses against the marker gene (Woffendin et al. 1996; Ranga et al. 1998; Riddell et al. 1996). These clinical results strongly suggest that pre-clinical evaluation of new anti-HIV-1 gene therapies in an animal model is required to address many of these current limitations.

Chimeric simian-human immunodeficiency viruses (SHIV) containing the *tat*, *rev*, *vpu* and *env* genes of HIV-1 have proven useful in the analysis of the role of discrete HIV genes in viral pathogenesis in macaque monkeys. Non-pathogenic SHIVs establish chronic infection (Reimann et al. 1996a) and pathogenic SHIVs, arising from point mutations in the envelope glycoproteins, cause high levels of viremia, rapid and profound $CD4^+$ T cell depletion, and AIDS (Reimann et al.

1996b; KARLSSON et al. 1998; CAYABYAB et al. 1999). Thus, the SHIV-macaque model offers a unique and important experimental system to evaluate the effects of anti-HIV-1 gene therapy. In this model, the efficacy of anti-HIV-1 intrabody genes can be examined without the concomitant use of antiretroviral drugs. Moreover, we are simultaneously able to control input virus, to evaluate for development of resistance, and to analyze for immune responses against the antiviral gene products.

This model is being initially explored with the humanized anti-Tat intrabody, sFvhutat2 (MHASHILKAR et al. 1999; MARASCO et al. 1999). Importantly, the Tat epitope to which the sFvhutat2 intrabody is directed is identical in the non-pathogenic and pathogenic isogenic virus pairs SHIV-HXBc2 and SHIV-HXBc2 32.3 and SHIV-89.9 and SHIV-89.6 KB9, respectively. The retroviral vectors being used for these studies as well as our clinical trials are shown in Fig. 2. The truncated human nerve growth factor receptor (ΔNGFR) is used as a phenotypic selection marker (RUDOLL et al. 1996; BONINI et al. 1997; FEHSE et al. 1997; VERZELETTI et al. 1998). Standard separation techniques are used to deplete $CD8^+$ T cells from Ficoll density-gradient-separated macaque PBMCs. The cells are transduced using retrovirus-containing supernatants from PG13 packaging cell lines and then enriched for NGFR expression after treatment with anti-NFGR Mab using goat anti-mouse-IgG-coated microbeads.

In a typical experiment, shown in Fig. 3, transduction efficiencies of 3.18% and 8.86% were achieved with retroviral supernatants containing pLNCΔNGFR (empty vector) (Fig. 3A) or pLNCΔNGFRsFvhutat2 (sFvhutat2-expressing vector) (Fig. 3C), respectively, whereas mock-transduced cells were barely detectable (data not shown). Following selection, these two cell populations were enriched to 80.4% and 89.82% (Fig. 3B,D), respectively, whereas there was no enrichment for NGFR expression in the mock-transduced cells (data not shown).

The bulk populations of $CD8^+$ T-cell-depleted, transduced macaque PBMCs were infected with 10,000 RT units/$10^6$ cells with SHIV-HXBc2, SHIV-HXBc2 32.3 or $SIV_{mac}239$. As shown in Fig. 4, the cell-free RT values were significantly lower in cultures of sFvhutat2-transduced cells than in vector-transduced cells for SHIV-HXBc2 and SHIV-HXBc2 32.3, but not for $SIV_{mac}239$, which does not carry the sFvhutat2 epitope in its genome.

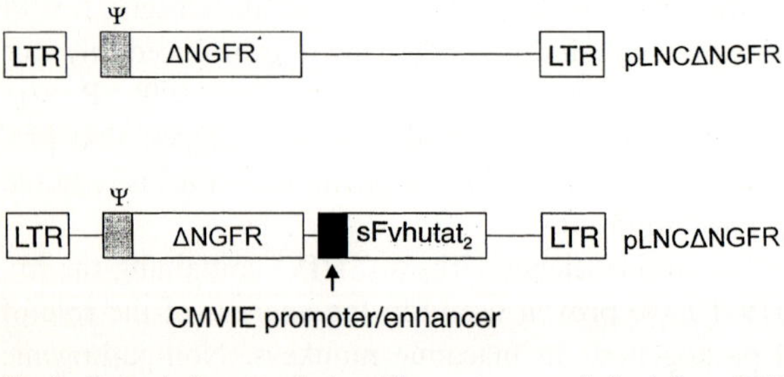


**Fig. 2.** Retroviral vector design. The vectors are derived from LNCX (MILLER 1993). The neomycin selection marker was replaced with truncated human nerve growth factor receptor (Δ*NGFR*) to produce LNCΔNGFR; sFvhutat2 expression is driven from an internal CMV promoter in LNCΔNGFRsFvhutat2

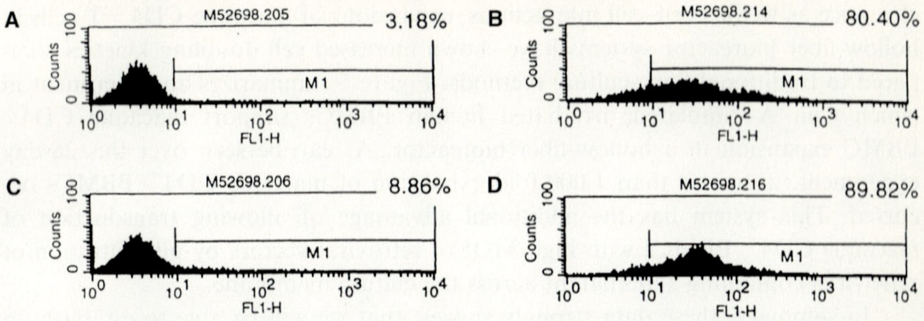


**Fig. 3A–D.** Transduction of rhesus monkey PBMCs with pLNCΔNGFR and pLNCΔNGFRsFvhutat2 vectors. Cells were mock-transduced (data not shown) or transduced with retroviral supernatants containing pLNCΔNGFR (empty vector) (**A**, **B**) or pLNCΔNGFRsFvhutat2 (expression sFvhutat2) (**C**, **D**). Bulk populations of unselected (**A**, **C**) cells and cells selected for NGFR expression by affinity column enrichment (**B**, **D**) are shown, respectively. Following selection, these two cell populations were enriched to 80.4% and 89.82% (**B**, **D**) respectively, whereas there was no enrichment for NGFR expression in the mock-transduced cells (data not shown)

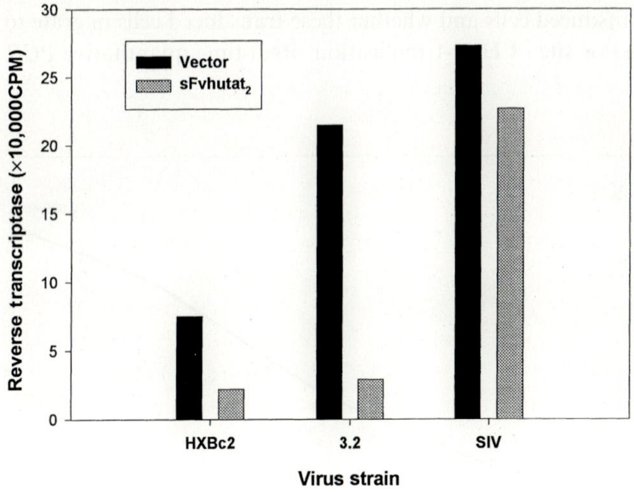


**Fig. 4.** Protection of transduced macaque PBMCs from SHIV infection. $CD8^+$ T-cell-depleted macaque PBMCs were transduced with retroviral supernatants containing pLNCΔNGFR (*solid bars*) or pLNCΔNFGRsFvhutat2 (*hatched bars*). Following ΔNGFR selection, the bulk populations of transduced cells were infected with 10,000 reverse transcriptase (RT) units/$10^6$ cells with SHIV-HXBc2, SHIV-HXBc2 32.2 or $SIV_{mac239}$. The RT values shown were obtained on the day of peak viremia of the vector-transduced cells ($SIV_{mac239}$, day 10; SHIV-HXBc2, day 22; SHIV-HXBc2 32.3, day 24)

#### 5.1.1.1 Optimization of Ex Vivo Expansions of Transduced Macaque $CD4^+$ T Cells

In preparation of our in vivo studies using the SHIV-macaque model, we have spent a considerable effort optimizing the conditions for ex vivo expansions of transduced macaque $CD4^+$ T cells. By maximizing nutrient delivery and waste clearance as well as cell–cell interactions, expansions of macaque CD4$^+$ T cells in hollow-fiber bioreactor systems have shown increased cell doubling kinetics compared to traditional flask-culture methods. Figure 5 summarizes an experiment in which Con A-stimulated, irradiated human PBMCs support macaque CD4$^+$ PBMC expansion in a hollow-fiber bioreactor. As can be seen over this 28-day experiment, a greater than 1,000-fold expansion of macaque CD4$^+$ PBMCs occurred. This system has the additional advantage of allowing transduction of macaque CD4$^+$ PBMCs with high MOI of retroviral vectors by ultrafiltration of retrovirus-containing supernatant across the culture membrane.

In summary, these data strongly suggest that we will be able to establish an in vivo SHIV-macaque gene-therapy model using an isogeneic pair of non-pathogenic and pathogenic SHIVs. Both pre-SHIV and post-SHIV infection models are planned using both non-pathogenic and pathogenic SHIVs in both models. For both models, we will infuse equal numbers of empty-vector-transduced or sFvhutat2-transduced CD4$^+$-enriched T cells into four macaques (gene therapy arm) and mock-transduced CD4$^+$-enriched T cells into four macaques (control arm). The primary goals of these studies are to determine if there is increased survival of sFvhutat2- vs vector-transduced cells and whether these transduced cells migrate to the lymph nodes, the major site of HIV-1 replication. Real-time quantitative PCR

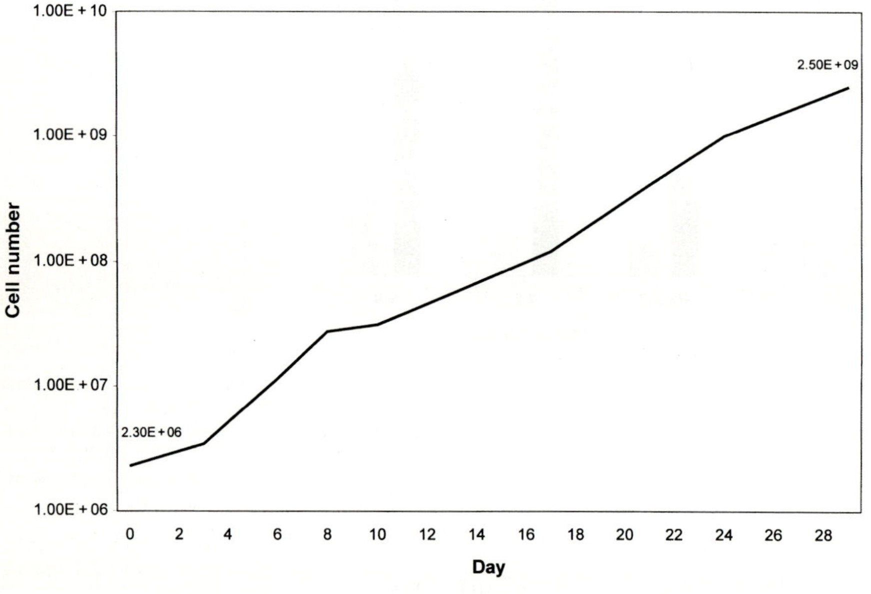


**Fig. 5.** Ex vivo expansion of transduced macaque CD4$^+$ mononuclear cells (PBMCs) in a hollow-fiber bioreactor; 10–15ml of anti-coagulated macaque peripheral blood was enriched by Ficoll density gradient centrifugation for PBMCs. Cells were depleted of CD8$^+$ T cells, activated, transduced, and enriched for NGFR expression as described in Fig. 2. Expansion is supported with Con A-stimulated, irradiated human PBMCs. Under the conditions used in this experiment, greater than 98% of the cells are CD4$^+$, NGFR$^+$

will be used to detect the two populations of transduced cells in peripheral blood (Marasco et al. 1999). The secondary goals of these studies are to determine if in vivo resistance to sFvhutat2 develops and whether a host immune response develops against the transduced cells. These studies will establish a valuable new primate model for the testing of anti-HIV-1 gene therapies. In addition, the results from these studies will substantially advance our understanding of how to apply and improve this promising technology for the treatment of HIV-1 infection and AIDS.

## 5.2 Pre-Integration Blockade of HIV-1 Infection

After 1984, when it was established that CD4 was a receptor for HIV (Dalgleish et al. 1984; Klatzmann et al. 1984), it became clear that other receptors remained to be found. In 1996, the role of chemokine receptors in HIV entry into $CD4^+$ cells was established (Feng et al. 1996; Deng et al. 1996; Dragic et al. 1996, Choe et al. 1996). This discovery shed new light on the pathogenesis of HIV-1 disease. The C-C chemokine receptor CCR5 is of major importance in this respect, since it is used by virus isolates (R5 isolates) in the blood of HIV-infected persons from the beginning of infection until AIDS develops (Connor et al. 1997). Moreover, a 32-bp deletion in the CCR5 gene, which results in a mutated gene, *Δ32CCR5*, encoding a truncated molecule that is not expressed at the cell surface, is correlated with non-infectability of homozygotes (Liu et al. 1996) and with some delay in HIV infection progression for heterozygotes (Mummidi et al. 1998).

Although CD4 cell-surface density determines the efficiency of infection of $CD4^+$ cell lines by primary isolates (Kabat et al. 1994), the membrane density of this receptor is very constant in normal peripheral-blood $CD4^+$ T lymphocytes. In contrast, there is a large degree of interindividual variability at the level of CCR5 expression (Moore 1997; Reynes et al. 2000). This variability also determines cell infectability, as has been shown in vitro with normal PBLs (Wu et al. 1997), cell lines (Platt et al. 1998), and macrophages (Fear et al. 1998). A correlation between HIV RNA plasma level and peripheral-blood, $CD4^+$, T-cell-surface CCR5 density in infected individuals has also been reported (Reynes et al. 2000). In some HIV-exposed, seronegative individuals, autoantibodies against CCR5 have been detected that are able to block MIP-1β binding to CCR5 and chemotaxis and that are capable of neutralizing infection with R5 primary isolates (Lopalco et al. 2000). While the mechanism(s) that are responsible for induction of this autoimmunity after vial infection are not clear, several additional observations suggest that the development of autoantibodies against CCR5 may be a therapeutic vaccine strategy for the treatment of HIV-1 infection (Ditzel et al. 1998; Zuber et al. 1998; Lehner et al. 1999). In addition, although there is evidence that HIV-1 coreceptor usage can change from R5 tropic to X4 tropic viruses within an HIV-1-infected individual during the natural progression from HIV-1 infection to AIDS (Connor et al. 1997), expanded coreceptor repertoire usage by HIV-1 is not a prerequisite for a progressive clinical course of HIV-1 infection (De Roda Husman et al. 1999).

In much the same way as cancer researchers have recently focused on blocking the blood supply to tumors, through anti-angiogenesis drugs that are active on tumor vascular endothelial cells that, unlike the tumors cells, do not develop drug resistance, AIDS investigators have begun targeting the coreceptors as a means to achieve pre-integration blockade of HIV-1 infection, thereby avoiding the develop of resistance to antiretroviral drugs. Whether blockage of CCR5 entry by immunologic, pharmacologic, or genetic means leads to coreceptor adaptation in the clinical setting remains to be determined. Thus, unless there is evidence to the contrary, the ability to use gene therapy for the systemic delivery of human anti-CCR5 antibodies, a strategy that has prevented disease progression from virus infection in other animal models (Pelegrin et al. 2000), remains a viable possibility.

### 5.2.1 Phenotypic Knockout of the HIV-1 Coreceptors

The use of gene therapy for the "phenotypic knockout" of CCR5 as an intracellular immunization strategy for the treatment of HIV-1 infection has been the recent focus of a number of laboratories, including our own. These in vitro strategies have included the use of the ER-directed, KDEL-retained "intrakines" MIP-1$\alpha$ and RANTES (Yang et al. 1997), SDF-1 (Chen et al. 1997), or both RANTES and SDF-1 (Bai et al. 1998) to block R5, X4 or R5X4 viruses, respectively, as well as anti-CCR5 ribozymes (Bai et al. 2000) and an ER-directed, KDEL-retained rabbit anti-CCR5 sFv intrabody to block R5 virus infection of $CD4^+$, $CCR5^+$ human cells (Steinberger et al. 2000). The rationale for the anti-CCR5 intrabody studies are shown in Fig. 6.

Because of our interest in developing anti-CCR5 intrabodies for clinical trials, we have concentrated on isolating human anti-CCR5 sFv antibodies that could be used as intrabodies. For the reasons discussed above, human sFv intrabodies are felt to be a prerequisite for moving into primate gene-therapy studies or into human clinical trials. Two different approaches, discussed below, have been used to produce human anti-CCR5 intrabodies.

#### 5.2.1.1 Use of a Synthetic, N-Terminal CCR5 Polypeptide for Screening a Human sFv Phage-Display Library

In the first approach, performed in collaboration with Drs. James Marks and David Powers (University of California, San Francisco), a 36-amino-acid polypeptide corresponding to the 33-amino-acid, N-terminal extracellular domain of CCR5 was synthesized. An additional carboxy-terminal G-G-K sequence was added to act as a flexible spacer and site for biotinylation of the $\varepsilon$-amino side chain of the terminal lysine. This polypeptide was used for panning with the Sheets library, a naive human sFv library composed of $\sim 5 \times 10^9$ members (Sheets et al. 1998). After the third round of panning, phage were screened by ELISA for binding

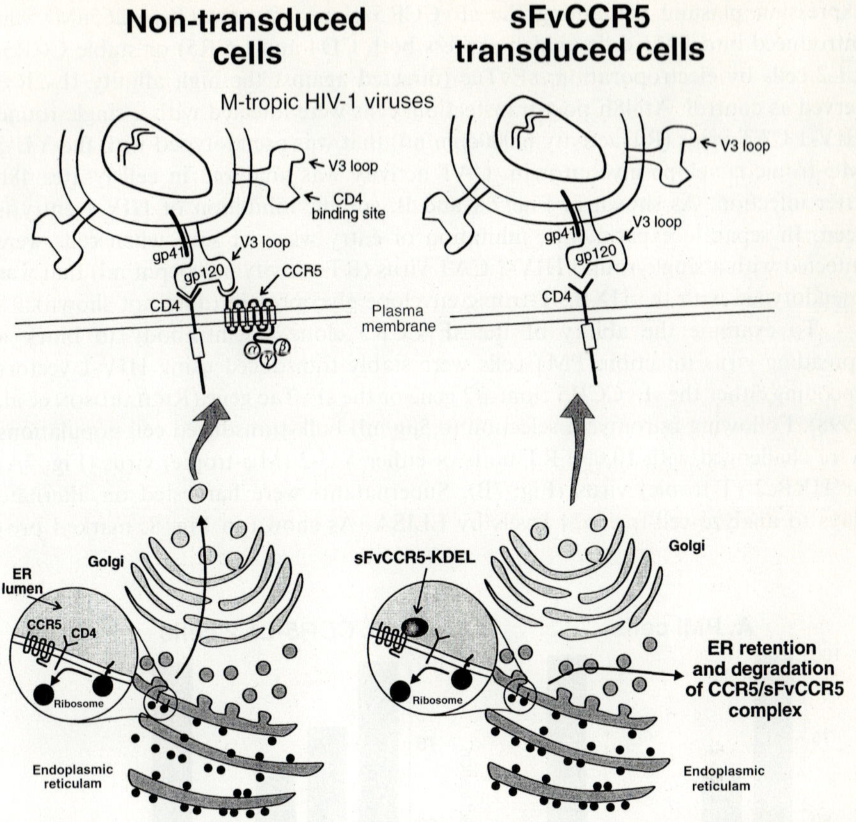


**Fig. 6.** Entry of M-tropic (R5) HIV-1 is mediated by binding of the envelope glycoprotein gp120 to CD4 and CCR5 coreceptors (*left*). CCR5, a seven-transmembrane spanning protein, does not contain a classical leader sequence to direct its transport to the endoplasmic reticulum; however, its N-terminus and three extracellular loops are potentially available in the lumen of the ER for binding to an ER-directed intrabody (*right*). An ER-retained intrabody, by virtue of containing a carboxy-terminal KDEL retention signal, could bind CCR5 in the ER lumen and prevent its transport to the cell surface to achieve "phenotypic knockout" of CCR5

to the biotinylated CCR5 polypeptide bound to streptavidin-coated ELISA plates. Six unique sFvs were identified from the 15/96 randomly selected clones that were found to bind to the CCR5 peptide. The equilibrium dissociation constant ($K_d$) of these sFvs for binding to the biotinylated CCR5 polypeptide, as determined by surface plasmon resonance in a BIACORE, ranged from 265 to 72nM. However, these purified sFvs bound relatively weakly to the surface of CCR5-transfected COS-1 cells.

The anti-CCR5 sFv genes were cloned into an intrabody expression vector containing a chimeric IgG$V_H$ pel B leader sequence and a carboxy-terminal SEKDEL signal for ER retention using identical *Sfi*I/*Not*I cloning sites as the pHEN phage-display vector (Hoogenboom et al. 1991) (Fig. 1). The intrabody expression plasmid containing the sFvCCR5 clone #2 gene ($K_D = 265nM$) was introduced into PM1 cells (which express both CD4 and CCR5) or stable CCR5-L1.2 cells by electroporation; sFvTac (directed against the high affinity IL-2Rα) served as control. At 48h post-transfection, cells were infected with a single-round HIV-1 CAT virus (RT activity 6,000cpm/ml) that was pseudotyped with the YU-2 Mϕ-tropic envelope glycoprotein. CAT activity was analyzed in cell lysates 48h after infection. As shown in Fig. 7A and B, specific inhibition of HIV-1 entry is seen. In separate experiments, inhibition of entry was not seen when cells were infected with a single-round HIV-1 CAT virus (RT activity 6,000cpm/ml) that was pseudotyped with the HXB2 T-tropic envelope glycoprotein (data not shown).

To examine the ability of the sFvCCR5 clone #2 intrabody to block a spreading virus infection, PM1 cells were stably transduced using HIV-1 vectors encoding either the sFvCCR5 clone #2 gene or the sFvTac gene (RICHARDSON et al. 1998). Following puromycin selection (0.5μg/ml) bulk-transduced cell populations were challenged with 10,000 RT units of either YU-2 (Mϕ-tropic) virus (Fig. 7A) or HXBc2 (T-tropic) virus (Fig. 7B). Supernatants were harvested on alternate days to analyze cell-free p24 levels by ELISA. As shown in Fig. 8, marked pro-

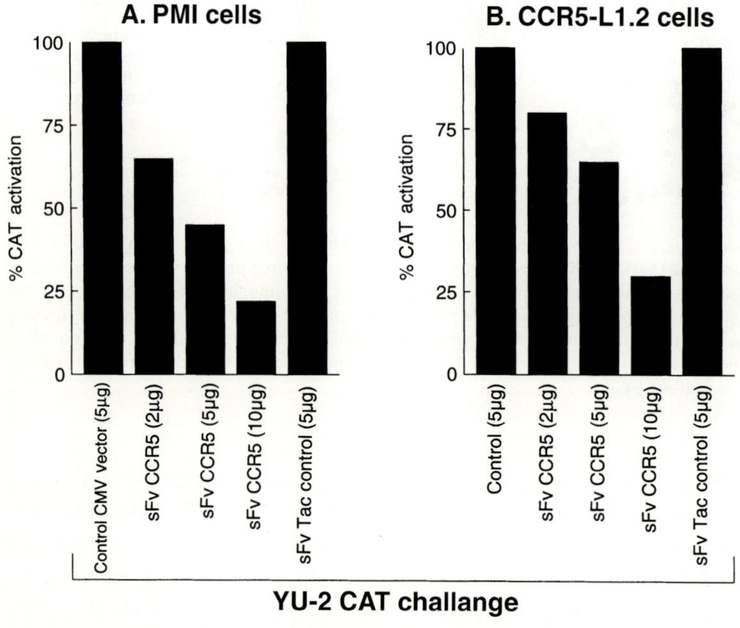


**Fig. 7.** Phenotypic knockout of CCR5 inhibits entry of an HIV-1 CAT reporter virus that has been pseudotyped with Mϕ-tropic Yu-2 envelope glycoprotein. The expression plasmid containing the sFvCCR5 clone #2 gene ($K_D = 265M$) was transiently transfected into PM1 cells (which express both CD4 and CCR5) or stable CCR5-L1.2 cells (cotransfected with CD4 expression plasmid) by electroporation. sFvTac directed against the high-affinity interleukin-2-receptor-α (IL-2R-α) served as control. At 48h post-transfection, cells were infected with single-round HIV-1 reporter virus (6,000cpm/ml RT activity). CAT activity was analyzed in cell lysates 48h after infection

tection from infection against the YU-2 virus is seen with the sFvCCR5 clone #2 expressing cells, and, as expected, the sFvCCR5 clone #2 cells are not protected against the HXBc2 virus. Thus, these data demonstrate that sFvCCR5 intrabodies can be used to block infection of Mϕ-tropic HIV-1 virions.

#### 5.2.1.2 Use of Paramagnetic Proteoliposomes Containing a Pure, Native, and Orientated Human CCR5 for Screening a Human sFv Phage-Display Library

It was recently determined that post-translational tyrosine sulfation modifications occur in the tyrosine-rich N-terminus of CCR5 and that these modifications are important in chemokine binding and in HIV-1 entry (Farzan et al. 1998, 1999). These post-translational modifications were not synthesized into the original CCR5 N-terminal polypeptide that was used for panning the human sFv library described above, and thus may have contributed to the weakened binding of the anti-CCR5 sFv proteins to the modified CCR5 on the cell surface. These observations led us to develop another panning procedure in which we used paramagnetic proteoliposomes containing a pure, native, and orientated human CCR5 (Mirzabekov et al. 2000) for screening our newly constructed, 15-billion-member, unimmunized, human sFv phage-display library (Mehta et al. 2000). After three rounds of panning, the phage were counterselected with control liposomes lacking CCR5. After four rounds of panning, 96 of the selected phage were examined for the ability to bind Cf2Th/synCCR5- and Cf2Th/synCXCR4-expressing cells relative to binding to Cf2Th cells (Mirzabekov et al. 1999). Of the 96 phage, 19 specifically bound CCR5-expressing cells. After five rounds of panning, 49 out of 96 phage tested specifically bound CCR5-expressing cells. Selected single-chain antibodies containing carboxy terminal c-Myc and $His_6$ tags were produced in *Escherichia coli*. Periplasm prepa-

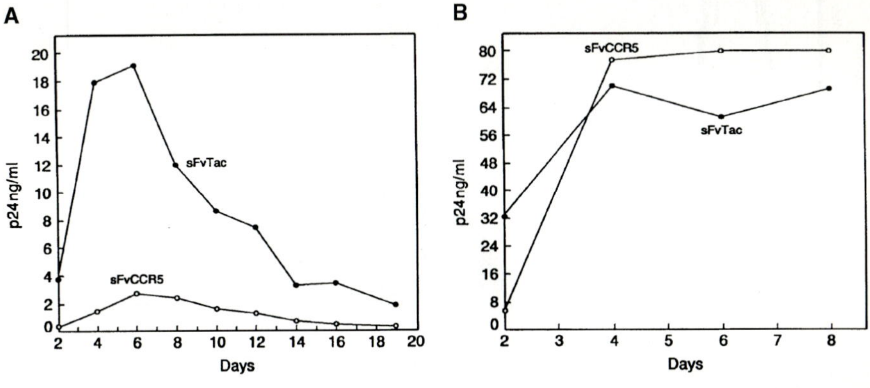


**Fig. 8A,B.** CC4$^+$ PM1 cells were transduced with the lentiviral vectors that contain a puromycin resistance gene and either the RE-directed, KDEL-retained sFvCCR5#2 or sFvTac intrabody. Following puromycin selection (0.5µg/ml), bulk-transduced cell populations were challenged with 10,000 RT units of either YU-2 (Mf-tropic) virus (**A**) or HXBc2 (T-tropic) virus (**B**). Supernatants were harvested on alternative days to analyze cell-free p24 levels by ELISA rations were passed over nickel–nitrilotriacetic-acid agarose columns and eluted with EDTA. Following buffer exchange, the sFvs were further purified over a Sepharose 75 column. Figure 9 shows the binding of one selected single-chain antibody to Cf2Th/synCCR5 cells but not to the parental Cf2Th cells. Several of these human anti-CCR5 sFvs also bind to macaque CCR5 (data not shown).

This and several of the other anti-CCR5 sFvs that have been isolated from the Mehta library (Mehta et al. 2000) were cloned into the same ER-directed, KDEL-retaining intrabody vectors described in Fig. 1. Following transient cotransfection of the anti-CCR5 sFv intrabodies and human CCR5 expressor plasmids into 293 T cells, surface CCR5 surface expression was examined by FACS analysis. As can be seen in Fig. 10, marked reduction in surface CCR5 expression was seen. Further characterization of these human anti-CCR5 sFv antibodies and intrabodies are underway.

## 6 Future Directions

The future clinical use of intrabody gene therapy will obviously depend on the outcome of in vivo animal studies and human clinical trials. A successful study, as

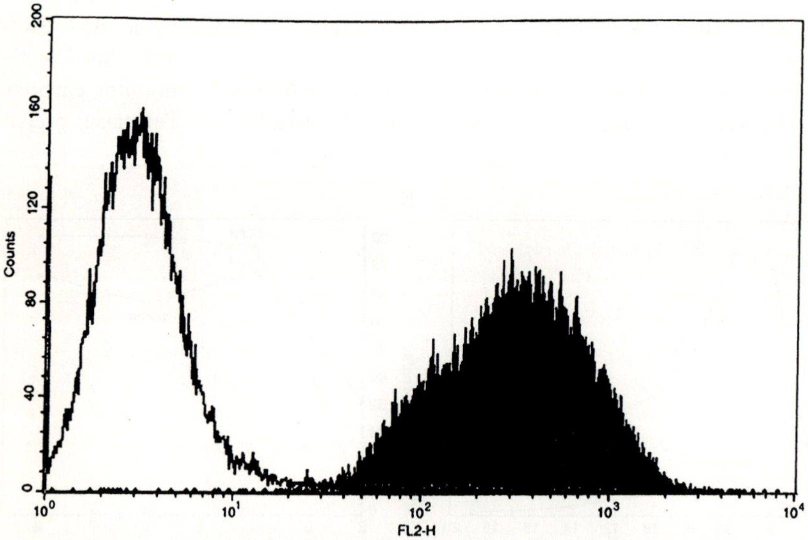


**Fig. 9.** A purified, human anti-CCR5 sFv, selected by panning the Mehta (Mehta et al. 2000) phage-display library with CCR5-protoliposomes, was incubated with either Cf2Th canine thymocytes or Cf2Th/synCCR5 cells, which express human CCR5 (Mirzabekov et al. 1999), for 1h on ice. After washing once with 2% bovine serum albumin in phosphate-buffered saline, the cells were incubated with mouse anti-c-Myc IgG followed by anti-mouse phycoerythrin (PE)-conjugated IgG and analyzed by FACS

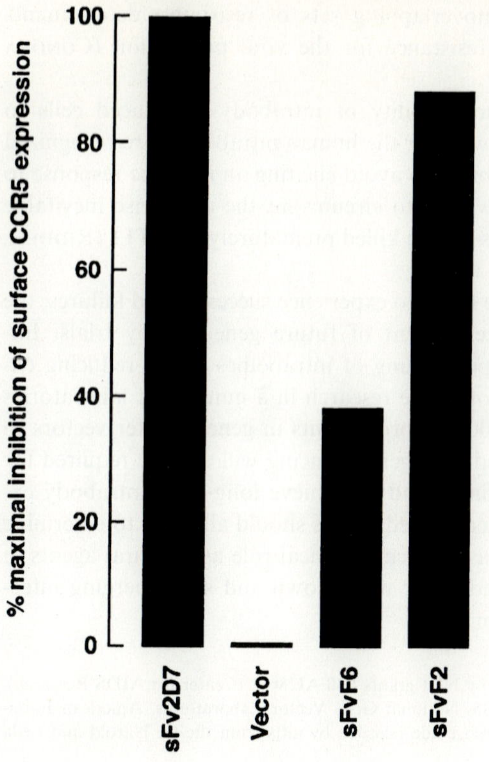


**Fig. 10.** Inhibition of transient CCR5 surface expression upon contransfection with anti-CCR5 intrabodies. 293 T cells were transiently transfected with codon-optimized synCCR5 plasmid (MIRZABEKOV et al. 1999) and cotransfected with the indicated plasmids. Control plasmid, pcDNA3.1; murine CCR5sFv2D7, human CCR5sFvF2, and human CCR5sFvF6 (shown in Fig. 9). The data are presented as % maximal inhibition of CCR5 expression as compared to murine CCR5sFv2D7 (~60% inhibition), a well-characterized, murine anti-CCR5 monoclonal antibody with potent HIV-1 neutralizing activity (RUFFING et al. 1998)

in the case of our sFvhutat2 intrabody approach, would ideally demonstrate that sFvhutat2-transduced CD4$^+$ T cells show a reduced ability to support HIV-1 replication, as evidenced by their extended survival in vivo. For this to occur, the sFvhutat2 intrabody would have to be active against a suspected wide range of quasispecies of HIV-1 with the differing Tat genes that are likely to exist in these patients with advanced HIV-1-infection (MEYERHANS et al. 1989; LEGUERN et al. 1993; NEUVEUT et al. 1996). In addition, for true molecular immune reconstitution to occur, these protected cells would have to be able to support the development of a strong T-helper response to naive and memory antigens. One of the potential limitations of this approach is that resistance to the intrabodies may occur in vivo despite our inability to demonstrate the development of resistance to sFvhutat2 in vitro (MARASCO et al. 1999). Targeting intrabodies to cellular cofactors that are required for HIV-1 replication, such as human cyclin T1, is an example in which resistance is unlikely to occur. Likewise, the CCR5 coreceptor remains an attractive target if intrabody therapy to this cellular protein does not promote virus adaptation to use other chemokine coreceptors. However, as with anti-Tat intrabody gene therapy, the addition of intrabody gene therapies directed against critical cellular proteins that are required for HIV-1 replication could add an important new component to divergent combination therapy; the addition of another distinct therapeutic class of drugs with nonoverlapping sets of resistance determinants should raise the genetic barrier to resistance for the viral population (Condra 1998).

Equally important to the reduced ability of intrabody-transduced cells to support HIV-1 replication in vivo is whether the human intrabodies are recognized as host or as foreign proteins. Our ability to avoid eliciting an immune response to the protein transgenes is critical if we are to circumvent the otherwise inevitable consequence that the transduced cells will be killed prematurely by CTLs (Riddell et al. 1996).

As with all new therapies, we are likely to experience successes and failures; the latter will be a learning tool for refinement of future gene-therapy trials. Improvements in the stability and proper folding of intrabodies in the reducing environment of the cytosol is an area of active research in a number of laboratories (Ohage et al. 1999; Worn et al. 2000). Improvements in gene transfer vectors to allow in vivo targeting and decreased transgene silencing will also be required for this approach to be economically viable and to achieve long-term intrabody expression in vivo, respectively. These combined efforts should allow us to determine if intrabody gene therapy will have a significant clinical role as antiviral agents in the treatment of HIV-1-infection and other well-known and still emerging infectious diseases of this new millennium.

*Acknowledgements.* This work was supported by NIH grants P30 AI28691 (Center for AIDS Research), P30 CA06516 (Cancer Center grant), AI28785, National Gene Vector Laboratories, American Foundation for AIDS Research. This work was also made possible by gifts from the G. Harold and Leila Y. Mathers Charitable Foundation.

## References

Arno A, Ruiz L, Juan M, Jou A, Balague M, Zayat MK, Marfil S, Martinez-Picado J, Martinez MA, Romeu J, Pujol-Borrell R, Lane C, Clotet B (1999) Efficacy of low-dose subcutaneous interleukin-2 to treat advanced human immunodeficiency virus type 1 in persons with 250µl CD4 T cells and undetectable plasma virus load. J Infec Dis 180:56–60

Bai J, Gorantla S, Banda N, Cagnon L, Rossi J, Akkina R (2000) Characterization of anti-CCR5 ribozyme-transduced CD34+ hematopoietic progenitor cells in vitro and in a SCID-hu mouse model in vivo. Mol Ther 1:244–254

Bai X, Chen J-D, Yang A-G, Torti F, Chen S-Y (1998) Genetic co-inactivation of macrophage- and T-tropic HIV-1 chemokine coreceptors CCR-5 and CXCR-4 by intrakines. Gene Ther 5:984–994

Baltimore D (1988) Intracellular immunization. Nature 335:395–396

Barbas CF III, Bjorling E, Chiodi F, et al. (1992) Recombinant human Fab fragments neutralize human type 1 immunodeficiency virus in vitro. Proc Natl Acad Sci USA 89:9339–9343

Barbas CF III, Kang AS, Lerner RA, et al. (1991) Assembly of combinatorial antibody libraries on phage surfaces: the gene III site. Proc Natl Acad Sci USA 88:7978–7982

Barbas SM, Ditzel HJ, Salonen EM, et al. (1995) Human auto antibody recognition of DNA. Proc Natl Acad Sci USA 92:2529–2533

Blaese RM, Culver KW, Miller AD, Carter CS, Fleisher T, Clerici M, Shearer G, Chang L, Chiang Y, Tolstoshev P, Greenblatt JJ, Rosenberg SA, Klein H, Berger M, Mullen CA, Ramsey WJ, Muul L, Morgan RA, Erson WF (1995) T lymphocyte-directed gene therapy for ADA-SCID: Initial trial results after 4 years. Science 270:475–480

Bonini C, Ferrari G, Verzeletti S, Servida P, Zappone E, Ruggieri L, Ponzoni M, Rossini S, Mavilio F, Traversari C, Bordignon C (1997) HSV-TK gene transfer into donor lymphocytes for control of allogenic graft-versus-leukemia. Science 276:1719–1724

Bordignon C, Notarangelo LD, Nobili N, Ferrari G, Casorati G, Panina P, Mazzolari E, Maggioni D, Rossi C, Servida P, Ugazio AG, Mavilio F (1995) Gene therapy in peripheral blood lymphocytes and bone marrow for ADA- immunodeficient patients. Science 270:470–475

BouHamdan M, Xue Y, Baudat Y, Hu B, Sire J, Pomerantz RJ, Duan LX (1998) Diversity of HIV-1 Vpr interactions involves usage of the WXXF motif of host cell proteins. J Biol Chem 273:8009–8016

Brenner MK, Rill DR, Moen RC, Krance RA, Mirro J Jr, Anderson WR, Ihle JN (1993a) Gene-marking to trace origin of relapse after autologous bone-marrow transplantation. The Lancet 341:85–86

Brenner MK, Rill DR, Holladay MS, Heslop HE, Moen RC, Buschle M, Krance RA, Santana VM, Anderson WF, Ihle JN (1993b) Gene marking to determine whether autologous marrow infusion restores long-term haemopoiesis in cancer patients. The Lancet 342:1134–1137

Burton Dr, Barbas CF III, Persson MA, et al. (1991) A large array of human monoclonal antibodies to type 1 human immunodeficiency virus from combinatorial libraries of asymptomatic seropositive individuals. Proc Natl Acad Sci USA 88:10134–10137

Caron de Fromentel C, Gruel N, Venot C, Debussche L, Conseiller E, Dureuil C, Teillaud J-L, Tocque B, Bracco L (1999) Restoration of transcriptional activity of p53 mutants in human tumour cells by intracellular expression of anti-p53 single chain Fv fragments. Oncogene 18:551–557

Cattaneo A, Biocca S (1999) The selection of intracellular antibodies. TIB Tech 17:115–121

Cayabyab M, Karlsson GB, Etemad-Moghadam BA, Hofmann W, Steenbeke T, Halloran M, Fanton JW, Axthelm MK, Letvin NL, Sodroski JG (1999) Changes in human immunodeficiency virus type 1 envelope glycoproteins responsible of the pathogenicity of a multiply passaged simian-human immunodeficiency virus (SHIV-HXBc2). J Virol 73:976–984

Chen J-D, Bai X, An-Gang Y, Cong Y, Chen S-Y (1997) Inactivation of HIV-1 chemokine co-receptor CXCR-4 by a novel intrakine strategy. Nat Med 3:1110–1116

Chen SY, Bagley J, Marasco WA (1994a) Intracellular antibodies as a new class of therapeutic molecules for gene therapy. Hum Gene Ther 5:595–600

Chen SY, Khouri Y, Bagley J, Marasco WA (1994b) Combined intra and extra-cellular immunization against human immunodeficiency virus type 1 infection using as human anti-gp120 antibody. Proc Natl Acad Sci USA 91:5932–5936

Chen J, Yang Q, Yang A-G, Marasco WA, Chen S-Y (1996) Intra and extra-cellular immunization against HIV-1 infection with lymphocytes transduced with an AAV vector expressing a human anti-gp120 antibody gene. Hum Gene Ther. 7:1515–1525

Choe H, Farzan M, Sun Y, Sullivan N, Rollins B, Ponath PD, Wu L, Mackay CR, LaRosa G, Newman W, Gerard N, Gerard C, Sodroski J (1996) The â-chemokine receptors CCR3 and CCR5 facilitate infection by primary HIV-1 isolates. Cell 85:1135–1148

Chun T-W, Carruth L, Finzi D, Shen X, DiGiuseppe JA, Taylor H, Hermankova M, Chadwick K, Margolick J, Quinn TC, Kuo Y-H, Brookmeyer R, Zelger MA, Barditch-Crovo P, Siliciano RF (1997) Quantification of latent tissue reservoirs and total body viral load in HIV-1 infection. Nature 387:183–188

Clackson T, Hoogenboom HR, Griffiths AD, et al. (1991) Making antibody fragments using phage display libraries. Nature 352:624–628

Condra JH (1998) Resisting resistance: maximizing the durability of antiretroviral therapy. Ann Intern Med 128:951–954

Connor RI, Sheridan KE, Ceradini D, Choe S, Landau NR (1997) Change in coreceptor use correlates with disease progression in HIV-1-infected individuals. J Exp Med 185:621–628

Daar ES, Bai J, Hausner MA, Majchrowicz M, Tamaddon M, Giorgi JV (1998) Acute HIV syndrome after discontinuation of antiretroviral therapy in a patient treated before seroconversion. Ann Intern Med 128:827–829

Dalgleish AG, Beverly PCL, Clapham PR, Crawford DH, Greaves MR, Weiss RA (1984) The CD4 (T4) antigen is an essential component of the receptor for the AIDS retrovirus. Nature 312: 763–767

Deng H, Liu R, Ellmeier W, Choe S, Unutmaz D, Burkhart M, Di Marzio P, Marmon S, Sutton RE, Hill CM, Davis CB, Peiper SC, Schall TJ, Littman DR, Landau NR. Identification of a major co-receptor for primary isolates of HIV-1. Nature 381:661–666

de Roda Husman A-M, van Rij RP, Blaak H, Broersen S, Schuitemaker H (1999) Adaptation to promiscuous usage of chemokine receptors is not a prerequisite for human immunodeficiency virus Type 1 disease progression. J Infec Dis 180:1106–1115

Ditzel HJ, Rosenkilde MM, Garred P, Wang M, Koefoed K, Pedersen C, Burton DR, Schwartz TW (1998) The CCR5 receptor acts as an alloantigen in CCR5Ä32 homozygous individuals: identification of chemokine and HIV-1 blocking human antibodies. Proc Natl Acad Sci USA 95:5241–5245

Dragic T, Litwin V, Allaway GP, Martin SR, Huang Y, Nagashima KA, Cayanan C, Maddon PJ, Koup RA, Moore JP, Paxton WA (1996) HIV-1 entry into CD4+ cells is mediated by the chemokine receptor CC-CKR-5. Nature 381:667

Duan L, Bagasra O, Laughlin MA, Oakes JW, Pomerantz RJ (1994a) Potent inhibition of human immunodeficiency virus type 1 replication by an intracellular anti-Rev single chain antibody. Proc Natl Acad Sci USA 91:5075–5079

Duan L, Zhang H, Oakes JW, Bagasra O, Pomerantz RJ (1994b) Molecular and virological effects of intracellular anti-rev single-chain variable fragments on the expression of various human immunodeficiency virus-1 strains. Hum Gene Ther 5:1315–1324

Dybul M, Chun T-W, Ward DJ, Hertogs K, Larder B, Fox CH, Orenstein JM, Baird BF, Li Y, Green LG, Engel D, Liu S, Mican JM, Fauci AS (2000) Evaluation of lymph node virus burden in human immunodeficiency virus-infected patients receiving Efavirenz-based protease inhibitor-sparing highly active antiretroviral therapy. J Infec Dis 181:1273–1279

Emery S, Lane HC (1996) Immune-based therapies in HIV infection: recent developments. AIDS 10:Suppl A S159–S163

Fauci AS, Macher AM, Longo DL, Lane HC, Rook AH, Masur H, Gelmann EP (1984) Acquired immunodeficiency syndrome: Epidemiologic, clinical, immunologic, therapeutic considerations. Ann Int Med 100:92–99

Farzan M, Choe H, Vaca L, Martin K, Sun Y, Desjardins E, Ruffing N, Wu L, Wyatt R, Gerard N, Gerard C, Sodroski J (1998) A tyrosine-rich region in the N terminus of CCR5 is important for human immunodeficiency virus type 1 entry and mediates an association between gp120 and CCR5. J Virol 72:1160–1164

Farzan M, Mirzabekov T, Kolchinsky P, Wyatt R, Cayabyab M, Gerard NP, Gerard C, Sodroski J, Choe H (1999) Tyrosine sulfation of the amino terminus of CCR5 facilitates HIV-1 entry. Cell 96:667–676

Fear WR, Kesson AM, Naif H, Lynch GW (1998) Differential tropism and chemokine receptor expression of human immunodeficiency virus type 1 in neonatal monocytes, monocyte-derived macrophages, and placental macrophages. J Virol 72:1334–1344

Fehse B, Uhde A, Fehse N, Eckert H-G, Clausen J, Ruger R, Koch S, Ostertag W, Zander AR, Stockschlader M (1997) Selective immunoaffinity-based enrichment of $CD34^+$ cells transduced with retroviral vectors containing an intracytoplasmatically truncated version of the human low-affinity nerve growth factor receptor (ÄLNGFR) gene. Hum Gene Ther 8:1815–1824

Feng Y, Broder CC, Kennedy PE, Berger EA (1996) HIV-1 entry cofactor functional cDNA cloning of a seven-transmembrane, G protein-coupled receptor. Science 272:872–877

Finzi D, Hermankova M, Pierson T, Carruth LM, Buck C, Chaisson RE, Quinn TC, Chadwick K, Margolick J, Brookmeyer R, Gallant J, Markowitz M, Ho DD, Richman DD, Siliciano RF (1997) Identification of a reservoir for HIV-1 in patients on highly active antiretroviral therapy. Science 278:1295–1300

Finzi D, Blankson J, Siliciano JD, Margolick JB, Chadwick K, Pierson T, Smith K, Lisziewicz J, Lori F, Flexner C, Quinn TC, Chaisson RE, Rosenberg E, Walker B, Gange S, Gallant J, Siliciano RF (1999) Latent infection of $CD4^+$ T cells provides a mechanism for lifelong persistence of HIV-1, even in patients on effective combination therapy. Nat Med 5:512–517

Gallo RC, Salahuddin SZ, Popovic M, Shearer GM, Kaplan M, Hayes BF, Palker TJ, Redfield R, Oleske J, Safai B, White G, Foster P (1984) Frequent detection and isolation of cytopathic retroviruses (HTLV-III) from patients with AIDS and at risk for AIDS. Science 224:500–503

Gargano N, Cattaneo A (1997) Inhibition of murine leukaemia virus retrotranscription by the intracellular expression of a phage-derived anti-reverse transcriptase antibody fragment. J Gen Virol 78:2592–2599

Gartner S, Markovitz P, Markovitz DM, Kaplan MH, Gallo RC, Popovic M (1986) The role of mononuclear phagocytes in HTLV-III/LAV infection. Science 233:215–219

Gilboa E, Smith C (1994) Gene therapy for infectious diseases: the AIDS model. Trends Genet 10: 139–144

Goldberg AL (1995) Functions of the proteasome: the Lysis at the end of the tunnel. Science 268:522–523

Gottlieb MS, Schroff R, Schanker HM, Weisman JD, Fan PT, Wolf RA, Saxon A (1981) Pneumocystis carinii pneumonia and mucosal candidiasis in previously healthy homosexual men: Evidence of a new acquired cellular immunodeficiency. N Eng J Med 305:1425

Havlir DV, Marschner IC, Hirsch MS, Collier AC, Tebas P, Bassett RL, Ioannidis JPA, Holohan MK, Leavitt R, Boone G, Richman DD, for the AIDS Clinical Trials Group Study 343 Team (1998) Maintenance antiretroviral therapies in HIV-infected subjects with undetectable plasma HIV RNA after triple-drug therapy. New Engl J Med 339:1261–1268

Hoogenboom HR, Griffiths AD, Johnson KS, Chiswell DJ, Hudson P, Winter G (1991) Multi-subunit proteins on the surface of filamentous phage: methodologies for displaying antibody (Fab) heavy and light chains. Nucleic Acids Res 19:4133–4137

Jiang W, Venugopal K, Gould E (1995) Intracellular interference of tick-borne flavivirus infection by using a single-chain antibody fragment delivered by recombinant Sindbis virus. J Virol 69:1044–1049

Kabat D, Kozak SL, Wehrly K, Chesebro B (1994) Differences in CD4 dependence for infectivity of laboratory-adapted and primary patient isolates of human immunodeficiency type 1. J Virol 68:2570–2577

Karlsson GB, Halloran M, Schenten D, Lee J, Racz P, Tenner-Racz K, Manola J, Gelman R, Etemad-Moghadam B, Desjardins E, Wyatt R, Gerard NP, Marcon L, Margolin D, Fanton J, Axthelm MK, Letvin NL, Sodroski J (1998) The envelope glycoprotein ectodomains determine the efficiency of CD4+ T-lymphocyte depletion in simian-human immunodeficiency virus-infected macaques. J Exp Med 188:1159–1171

Kitamura Y, Ishikawa T, Okui N, Kobayashi N, Kanda T, Shimada T, Miyake K, Yoshiike K (1999) Inhibition of replication of HIV-1 at both early and late stages of the viral life cycle by single-chain antibody against viral integrase. J Acquired Immune Def Syndr Hum Retro 20:105–114

Klatzmann D, Champagne E, Chamaret S, Gruest J, Guetard D, Hercend T, Gluckman JC, Montagnier L (1984) T-lymphocyte T4 molecule behaves as the receptor for human retrovirus LAV. Nature 312:767–768

Kulkosky J, BouHamdan M, Geist A, Pomerantz RJ (1999) A novel Vpr peptide interactor fused to integrase (IN) restores integration activity to IN-defective HIV-1 virions. Virol 255:77–85

LeGuern M, Shioda T, Levy JA, Cheng-Mayer C (1993) Single amino acid change in tat determines the different rates of replication of two sequential HIV-1 isolates. Virol 195:441–447

Lehner T, Wang Y, Doyle C, Tao L, Bergmeier LA, Mitchell E, Bogers WMJ, Heeney J, Kelly CG (1999) Induction of inhibitory antibodies to the CCR5 chemokine receptor and their complementary role in preventing SIV infection in macaques. Eur J Immunol 28:2427–2435

Levin R, Mhashilkar AM, Dorfman T, Bukovsky A, Zani C, Bagley J, Hinkula J, Niedrig M, Albert J, Wahren B, Göttlinger HG, Marasco WA (1997) Inhibition of early and late events of the HIV-1 replication cycle by cytoplasmic Fab intrabodies against the matrix protein, p17. Mol Med 3:96–110

Levy-Mintz P, Duan L, Zhang H, Hu B, Domadula G, Zhu M, Kulkosky J, Bizub-Bender D, Skalka AM, Pomerantz RJ (1996) Intracellular expression of single-chain variable fragments to inhibit early stages of the viral life cycle by targeting human immunodeficiency virus type 1 integrase. J Virol 70:8821–8832

Liu R, Paxton WA, Choe S, Ceradini D, Martin SR, Horuk R, MacDonald ME, Stuhlmann H, Koup RA, Landau NR (1996) Homozygous defect in HIV-1 coreceptor accounts for resistance of some multiply-exposed individuals to HIV-1 infection. Cell 86:367–377

Lopalco L, Barassi C, Pastori C, Longhi R, Burastero SE, Tambussi G, Mazzotta F, Lazzarin A, Clerici M, Siccardi AG (2000) CCR5-reactive antibodies in seronegative partners of HIV-seropositive individuals down-modulate surface CCR5 in vivo and neutralize the infectivity of R5 strains of HIV-1 in vitro. J Immunol 164:3426–3433

Luster AD (1998) Chemokines – chemotactic cytokines that mediate inflammation. N Engl J Med 338:436–445

Maciejewski JP, Weichold FF, Young NS, Cara A, Zella D, Reitz MS, Jr, Gallo RC (1995) Intracellular expression of antibody fragments directed against HIV reverse transcriptase prevents HIV infection in vitro. Nat Med 1:667–673

Marasco WA, Haseltine WA, Chen S-Y (1993) Design, intracellular expression and activity of a human anti-human immunodeficiency virus type 1 gp120 single-chain antibody. Proc Natl Acad Sci USA 90:7889–7893

Marasco WA (1995) Intracellular antibodies (intrabodies) as research reagents and therapeutic molecules for gene therapy. Immunotech 1:1–19

Marasco WA (ed) (1998) Intrabodies – basic research and clinical gene therapy applications. Springer, Berlin, Heidelberg, New York

Marasco WA (1997) Intrabodies: turning the humoral immune system outside in for intracellular immunization. Gene Ther 4:11–15

Marasco WA, LaVecchio J, Winkler A (1999) Human anti-HIV-1 tat sFv intrabodies for gene therapy of advanced HIV-1-infection and AIDS. J Immunol Methods 231:223–238

Marin M, Pelegrin-Zurilla M, Bachrach E, Noel D, Brockly F, Piechaczyk M (2000) Antiviral activity of an intracellularly expressed single-chain antibody fragment directed against the murine leukemia virus capsid protein. Hum Gene Ther 11:389–401

Markowitz M, Vesanen M, Tenner-Racz K, Cao Y, Binley JM, Talai A, Hurley A, Ji X, Chaudhry MR, Yaman M, Frankel S, Heath-Chiozzi M, Leonard JM, Moore JP, Racz P, Nixon DF, Ho DD (1999) The effect of commencing combination antiretroviral therapy soon after human immunodeficiency virus type 1 infection on viral replication and antiviral immune responses. J Infec Dis 179:525–537

Marks JD, Hoogenboom HR, Bonnert TP, et al. (1991) By-passing immunization, human antibodies from V-gene libraries displayed on phage. J Mol Biol 222:581–597

Martineau P, Jones P, Winter G (1998) Expression of an antibody fragment at high levels in the bacterial cytoplasm. J Mol Biol 280:117–127

Mary M-N, Venot C, Caron de Fromentel C, Debussche L, Conseiller E, Cochet O, Gruel N, Teillaud J-L, Schweighoffer F, Tocque B, Bracco L (1999) A tumor specific single chain antibody dependent gene expression system. Oncogene 18:559–564

Mehta S, Kontos H, Wang Y, Marasco WA, manuscript in preparation

Meyerhans A, Cheynier R, Albert J, Seth M, Kwok S, Sninsky J, Morfeldt-Mason L, Asjo B, Wain-Hobson S (1989) Temporal fluctuations in HIV quasispecies in vivo are not reflected by sequential HIV isolations. Cell 58:901–910

Mhashilkar AM, Bagley J, Chen SY, Szilvay AM, Helland DG, Marasco WA (1995) Inhibition of HIV-1 Tat-mediated LTR transactivation and HIV-1 infection by anti-Tat single chain intrabodies. EMBO J 14:1542–1551

Mhashilkar AM, Biswas DK, LaVecchio J, Pardee AB, Marasco WA (1997) Inhibition of human immunodeficiency virus type 1 replication in vitro by a novel combination of anti-Tat single-chain intrabodies and NF-êB antagonists. J Virol 71:6486–6494

Mhashilkar AM, LaVecchio J, Eberhardt B, Brooks J, Boisot S, Dove JH, Pumphrey C, Li X, Weissmahr RN, Ring DB, Ramstedt U, Marasco WA (1999) Inhibition of human immunodeficiency virus type 1 replication in vitro in acutely and persistently infected human CD4± mononuclear cells expressing murine and humanized anti-HIV-1 tat sFv intrabodies. Hum Gene Ther 10:1453–1467

Miller AD, Miller DG, Garcia JV, Lynch CM (1993) Use of retroviral vectors for gene transfer and expression. Methods Enzymol 217:581

Mirzabekov T, Bannert N, Farzan M, Hofmann W, Kolchinsky P, Wu L, Wyatt R, Sodroski J (1999) Enhanced expression, native purification, characterization of CCR5, a principal HIV-1 coreceptor. J Biol Chem 274:28745–28750

Mirzabekov T, Kontos H, Farzan M, Marasco W, Sodroski J (2000) Paramagnetic proteoliposomes containing a pure, native and oriented seven-transmembrane segment protein CCR5. Nat Biotech 18:649–654

Moore JP (1997) Coreceptors: implications for HIV pathogenesis and therapy. Science 276:51–52

Mummidi S, Ahuja SS, Gonzalez E, Anderson SA, Santiago EN, Stephan KT, Craig FE, O'Connell P, Tryon V, Clark RA, Dolan MJ, Ahuja SK (1998) Genealogy of the CCR5 locus and chemokine system gene variants associated with altered rates of HIV-1 disease progression. Nat Med 4:768–793

Neuveut C, Jeang K-T (1996) Recombinant human immunodeficiency virus Type 1 genomes with *tat* unconstrained by overlapping reading frames reveal residues in Tat important for replication in tissue culture. J Virol 70:5572–5581

Ohage EC, Wirtz P, Barnikow J, Steipe B (1999) Intrabody construction and expression.II. A synthetic catalytic Fv fragment. J Mol Biol 291:1129–1134

Okui N, Kobayashi N, Kitamura Y (1998) Production of uninfectious human immunodeficiency virus Type 1 containing viral protein R fused to a single-chain antibody against viral integrase. J Virol 72:6960–6964

Okui N, Sakuma Ry, Kobayashi N, Yoshikura H, Kitamura T, Chiba J, Kitamura Y (2000) Packageable antiviral therapeutics against human immunodeficiency virus type 1: virion-targeted virus inactivation by incorporation of a single-chain antibody against viral integrase into progeny virions. Hum Gen Ther 11:537–546

Pelegrin M, Marin M, Oates A, Noel D, Saller R, Salmons B, Piechaczyk M (2000) Immunotherapy of a viral disease by in vivo production of therapeutic monoclonal antibodies. Hum Gene Ther 11:1407–1415

Perelson AS, Neumann AU, Markowitz M, Leonard JM, Ho DD (1996) HIV-1 dynamics in vivo: virion clearance rate, infected cell life-span, and viral generation time. Science 271:1582–1586

Pialoux G, Raffi F, Brun-Vezinet F, Meiffredy V, Flandre P, Gastaut J-A, Dellamonica P, Yeni P, Delfraissy J-F, Aboulker J-P for the Trilege (Agence Nationale de Recherches sur lei SIDA 072) Study Team (1998) A randomized trial of three maintenance regimens given after three months of induction therapy with Zidovudine, Lamivudine, and Indinavir in previously untreated HIV-1-infected patients. New Engl J Med 339:1269–1276

Platt EJ, Wehrly K, Kuhman SE, Chesebro B, Kabat D (1998) Effects of CCR5 and CD4 cell surface concentrations on infections by macrophage tropic isolates of HIV-1. J Virol 72:2855–2864

Portolano S, McLachlan SM, Rapoport B (1993) High affinity, thyroid-specific human auto antibodies displayed on the surface of filamentous phage use V genes similar to other auto antibodies. J Immunol 151:2839–2851

Poznansky MC, Foxall R, Mhashilkar A, Coker R, Jones S, Ramstedt U, Marasco W (1998) Inhibition of HIV replication and growth advantage of $CD4^+$ T-cells from HIV-infected individuals which express an intracellular antibody against HIV-1 gp 120 or tat. Hum Gene Ther 9:487–496

Poznansky MC, LaVecchio J, Olszak IT, Silva-Arietta S, Porter-Brooks J, Brody K, Adams GB, Ramstedt U, Marasco WA, Scadden DT (1999) Inhibition of HIV replication and growth advantage of $CD4^{\pm}$ T-cells and monocytes derived from $CD34^{\pm}$ cells transduced with an intracellular antibody directed against HIV-1 tat. Hum Gene Ther 10:2505–2514

Ranga U, Woffendin C, Verma S, Xu L, June CH, Bishop DK, Nabel GJ (1998) Enhanced T cell engraftment after retroviral delivery of an antiviral gene in HIV-infected individuals. Proc Natl Acad Sci USA 95:1201–1206

Reimann KA, Li JT, Voss G, Lekutis C, Tenner-Racz K, Racz P, Lin W, Montefiori DC, Lee-Parritz DE, Lu Y, Collman RG, Sodroski J, Letvin NL (1996a) An *env* gene derived from a primary human immunodeficiency virus type 1 isolate confers high in vivo replicative capacity to a chimeric simian/human immunodeficiency virus in rhesus monkeys. J Virol 70:3198–3206

Reimann KA, Li JT, Veazey R, Halloran M, Park IW, Karlsson GB, Sodroski J, Letvin NL (1996b) A chimeric simian/human immunodeficiency virus expressing a primary patient human immunodeficiency virus type 1 isolate env causes an AIDS-like disease after in vivo passage in rhesus monkeys. J Virol 70:6922–6928

Reynes J, Portales P, Segondy M, Baillat V, Andre P, Reant B, Avinens O, Couderc G, Benkirane M, Clot J, Eliaou J-F, Corbeau P (2000) $CD4^+$ T cell surface CCR5 density as a determining factor of virus load in person infected with human immunodeficiency virus type 1. J Infec Dis 181:927–932

Richardson JH, Hofmann W, Sodroski JG, Marasco WA (1998) Intrabody-mediated knockout of the high-affinity IL-2 receptor in primary human T cells using a bicistronic lentivirus vector. Gene Ther 5:635–644

Riddell SR, Elliott M, Lewinsohn DA, Gilbert MJ, Wilson L, Manley SA, Lupton SD, Overell RW, Reynolds TC, Corey L, Greenberg PD (1996) T-cell mediated rejection of gene-modified HIV-specific cytotoxic T lymphocytes in HIV-infected patients. Nat Med 10:216–223

Rock KL (1996) A new foreign policy: MHC class I molecules monitor the outside world. Immunol Today 17:131–137

Rudoll T, Phillips K, Lee S-W, Hull S, Gaspar O, Sucgang N, Gilboa E, Smith C (1996) High-efficiency retroviral vector mediated gene transfer into human peripheral blood CD4+ T lymphocytes. Gene Ther 3:695–705

Ruffing N, Sullivan N, Sharmeen L, Sodroski J, Wu J (1998) CCR5 has an expanded ligand-binding repertoire and is the primary receptor used by MCP-2 on activated T cells. Cell Immunol 189:160–168

Schroff RW, Gottlieb MS, Prince HE, Chai LL, Fahey JL (1983) Immunological studies of homosexual men with immunodeficiency and Kaposi's sarcoma. Clin Immunol Immunopathol 27:300–310

Shaheen F, Duan L, Zhu M, Bagasra O, Pomerantz R (1996) Targeting HIV-1 reverse transcriptase by intracellular expression of single chain variable fragments to inhibit early stages of the viral life cycle. J Virol 70:3392–3400

Sheets MD, Amersdorfer P, Finnern R, Sargent P, Lindqvist E, Schier R, Hemingsen G, Wong C, Gerhart JC, Marks JD (1998) Efficient construction of a large nonimmune phage antibody library: the production of high-affinity human single-chain antibodies to protein antigens. Proc Natl Acad Sci USA 83:6157–6162

Smith KY, Valdez H, Landay A, Spritzler J, Kessler HA, Connick E, Kuritzkes D, Gross B, Francis I, McCune JM, Lederman MM, the ACTG 375 Protocol Team (2000) Thymic size and lymphocyte restoration in patients with human immunodeficiency virus infection after 48 weeks of Zidovudine, Lamivudine, and Ritonavir therapy. J Infec Dis 181:141–147

Steinberger P, Andris-Widhopf J, Buhler B, Torbett BE, Barbas III CF (2000) Functional deletion of the CCR5 receptor by intracellular immunization produces cells that are refractory to CCR5-dependent HIV-1 infection and cell fusion. Proc Natl Acad Sci USA 97:805–810

Sulis E, Lusso P, Contu L (2000) Prolonged asymptomatic HIV-1 infection. N Eng J Med 342:1221–1222

Verzeletti S, Bonini C, Marktel S, Nobili N, Ciceri F, Traversari C, Bordignon C (1998) Herpes simplex virus thymidine kinase gene transfer for controlled graft-versus-host disease and graft-versus-leukemia: clinical follow-up and improved new vectors. Hum Gene Ther 9:2243–2251

Visintin M, Tse E, Axelson H, Rabbitts TH, Cattaneo A (1999) Selection of antibodies for intracellular function using a two-hybrid in vivo system. Proc Natl Acad Sci USA 96:11723–11728

Wei X, Ghosh SK, Taylor ME, Johnson VA, Emini EA, Deutsch EP, Lifson JD, Bonhoeffer S, Nowak MA, Hahn BH, Saag MS, Shaw GM (1995) Viral dynamics in human immunodeficiency virus type 1 infection. Nature 373:117–122

Woffendin C, Ranga U, Yang Z-Y, Xu L, Nabel GJ (1996) Expression of a protective gene prolongs survival of T cells in human immunodeficiency virus-infected patients. Proc Natl Acad Sci USA 93:2889–2894

Wong JK, Hezareh M, Gunthard HF, Havlir DV, Ignacio CC, Spina CA, Richman DD (1997) Recovery of replication-competent HIV despite prolonged suppression of plasma viremia. Science 278:1291–1295

Worn A, Auf der Maur A, Escher D, Honegger A, Barberis A, Pluckthun A (2000) Correlation between in vitro stability and in vivo performance of anti-GCN4 intrabodies as cytoplasmic inhibitors. J Biol Chem 275:2795–2803

Wu L, Paxton WA, Kassam N, Ruffing N, Rottman JB, Sullivan N, Choe H, Sodroski J, Newman W, Koup RA, Mackay CR (1997) CCR5 levels and expression pattern correlate with infectability by macrophage-tropic HIV-1, in vitro. J Exp Med 185:1681–1691

Yang A-G, Bai X, Huang XF, Yao C, Chen S-Y (1997) Phenotypic knockout of HIV type 1 chemokine coreceptor CCR-5 by intrakines as potential therapeutic approach for HIV-1 infection. Proc Natl Acad Sci USA 94:11567–11572

Zebedee Sl, Barbas CF III, Hom YL, et al. (1992) Human combinatorial antibody libraries to hepatitis B surface antigen. Proc Natl Acad Sci USA 89:3175–3179

Zhou P, Goldstein S, Devadas K, Tewari D, Notkins AL (1998) Cells transfected with a non-neutralizing antibody gene are resistant to HIV infection: targeting the endoplasmic reticulum and *trans*-Golgi network. J Immunol 160:1489–1496

Zuber B, Levi M, Karlsson M, Sioberg BM, Brytting M, Wahren B (1998) Immune reaction against the HIV-1 coreceptor CCR5. In: Third European Conference on AIDS Research, Munich, Germany, p 6.S1 (Abstract)

Zu Putlitz J, Skerra A, Wands JR (1999) Intracellular expression of a cloned antibody fragment interferes with hepatitis B virus surface antigen secretion. Biochem Biophys Res Comm 255:785–791

# T-Bodies as Antiviral Agents

N. Bitton[1], P. Debre[1], Z. Eshhar[2], and G. Gorochov[1]



## 1 Introduction

The immune system recognizes antigens using two main mechanisms. Whereas B cells recognize conformational or sequential epitopes on the surface of native

[1] Laboratoire d'Immunologie Cellulaire, CERVI, INSERM U543, Hopital Pitie-Salpetriere, 83 Bvd de l'Hopital, 75013 Paris, France
[2] Department of Immunology, The Weizmann Institute of Science, Rehovot 76100, Israel

proteins, T cells recognize only processed peptides displayed by antigen-presenting cells (APCs) in association with MHC molecules. The antiviral response is largely dependent on T cells, consisting of either T helper (Th) cells needed for antibody production, or cytotoxic T effector cells (CTLs).

In human and animal models, the $CD8^+$ T cell response is essential for the eradication of acute viral infections such as influenza (Ada and Jones 1986), RSV (Cannon et al. 1987) and LCMV (Byrne et al. 1984), and for control of persistent infections such as CMV (Li et al. 1994, 1995; Reusser et al. 1991; Riddell 1995; Walter et al. 1995), HSV (Posavad et al. 1997), EBV (Heslop and Rooney 1997), and HIV (Schmitz et al. 1999) (see Sect. 4). On the other hand, antibodies are important for protection against some primary (for instance by passive transfer from mother to child), and most secondary viral infections (Bachmann and Zinkernagel 1997).

Appreciation of the complementary nature of T-cell- and antibody-based immunotherapy has stimulated interest in developing new approaches that would combine the advantages of both arms of the immune response and minimize the limitations of each form of therapy. The T-body approach is based on transfecting effector cytotoxic T cells with a recombinant chimeric receptor designed for cellular targeting and subsequent effector activation. The extracellular domain of the receptor consists of an antigen-binding domain derived from a preselected antibody. The transmembrane and intracellular moieties of such chimeric receptors are derived from effector cell triggering chains. T cells bearing such receptors could recognize their specific antigen in a non-MHC-restricted manner and subsequently induce T cell activation.

Published reports of T-bodies directed against infectious targets have been so far limited to HIV-1. However, the generation of specific T cells against any antigen against which a suitable monoclonal antibody (mAb) has been raised could extend the spectrum of diseases towards which the chimeric receptor strategy can be applied. In cases in which no antibody is available, phage-display technology offers the ability to directly select antibodies with new specificities.

In this review, we discuss the rationale for the use T-bodies as antiviral agents and more specifically as HIV antivirals. Other infectious agents that could be targeted by T-bodies, as well as the technical and ethical limitations encountered by this approach, are considered.

## 2 Chimeric Receptor Design

In early models, variable antibody domains were spliced to the constant T cell receptor (TCR) domains that indirectly associate with the CD3 complex (Gross et al. 1989a,b). Chimeric antibody/TCRs contained the antibody $V_H$ and $V_L$ domains replacing the TCR $V\alpha$ and $V\beta$ domains (Gorochov et al. 1992; Gross and Eshhar 1992; Gross et al. 1989a,b; Kuwana et al. 1987). It was shown that either one of the antibody domains could be fused to each of the TCR chains (Cβ or Cα). Subsequently, a simplified design used the single chain Fv (ScFv) configuration of an antibody as the extracellular recognition unit of the chimeric receptor (CR) (ESHHAR et al. 1993). Single chain variable domains of antibodies (ScFv) are generated by joining the $V_H$ and $V_L$ regions of a mAb via a flexible linker (BIRD et al. 1988; Fig. 1). The basis for this design relies on the observation that fusion proteins utilizing signaling chains carrying a motif for tyrosine kinase activation mediate TCR signal transduction (IRVING and WEISS 1991; LETOURNEUR and KLAUSNER 1991; ORLOFF et al. 1990; ROMEO et al. 1992; ROMEO and SEED 1991). Both the FcεRIγ (ESHHAR et al. 1993; TRAN et al. 1995) and CD3ζ (ESHHAR et al. 1993; HWU et al. 1995; TRAN et al. 1995; WEIJTENS et al. 1996) chains have been successfully used as signal transducing units (Fig. 1). These two signal transduction domains were compared, and, in some cases, CD3ζ was shown to be the more efficient one (ROBERTS et al. 1998).

## 3 Combining the Advantages of Antibody- and T-Cell-Based Immunotherapy to Circumvent Viral Escape Mechanisms

The potential advantages of the T-body approach, compared to antibody- or CTL-based strategies are numerous (Fig. 2).

In AIDS patients, lymphoid tissues contain large numbers of trapped virus particles and represent the main site for ongoing HIV replication in $CD4^+$ T cells and macrophages (EMBRETSON et al. 1993). Unfortunately, antibodies do not efficiently penetrate target tissues (JAIN 1989). Antibodies are also unstable in serum and are cleared rapidly from circulation. More importantly, they are relatively inefficient at recruiting host effector functions.

A first attempt at combining humoral and cellular immunity consisted of engineering bispecific antibodies, with one arm directed against the antigen on the target cell surface and the other arm recognizing an activation molecule on the surface of endogenous killer cells. This configuration was expected to activate the effector cell in the presence of target antigen (SEGAL et al. 1988; WALDMANN et al. 1992). Several bispecific antibodies have been developed to retarget effectors to kill HIV-1-infected cells (BERG et al. 1991; CHAMOW et al. 1994; OKADA et al. 1992; TRAUNECKER et al. 1991; ZARLING et al. 1988). This process is similar to the chimeric TCR approach and does not require the transfection of T cells. On the other hand, bispecific antibodies cannot directly link antigen recognition and signal transduction (BLANK et al. 1993) and also suffer from the antibody-related limitations described above. Bispecific antibodies remain bound to the cells for only limited periods of time (48–96h), due to the proteolysis that follows their dissociation from the CTL surface (BLANK et al. 1993; BOLHUIS et al. 1992; CHAMOW et al. 1994; JAIN 1989; LANZAVECCHIA and SCHEIDEGGER 1987; MEZZANZANICA et al. 1991; RENNER et al. 1994; VAN et al.

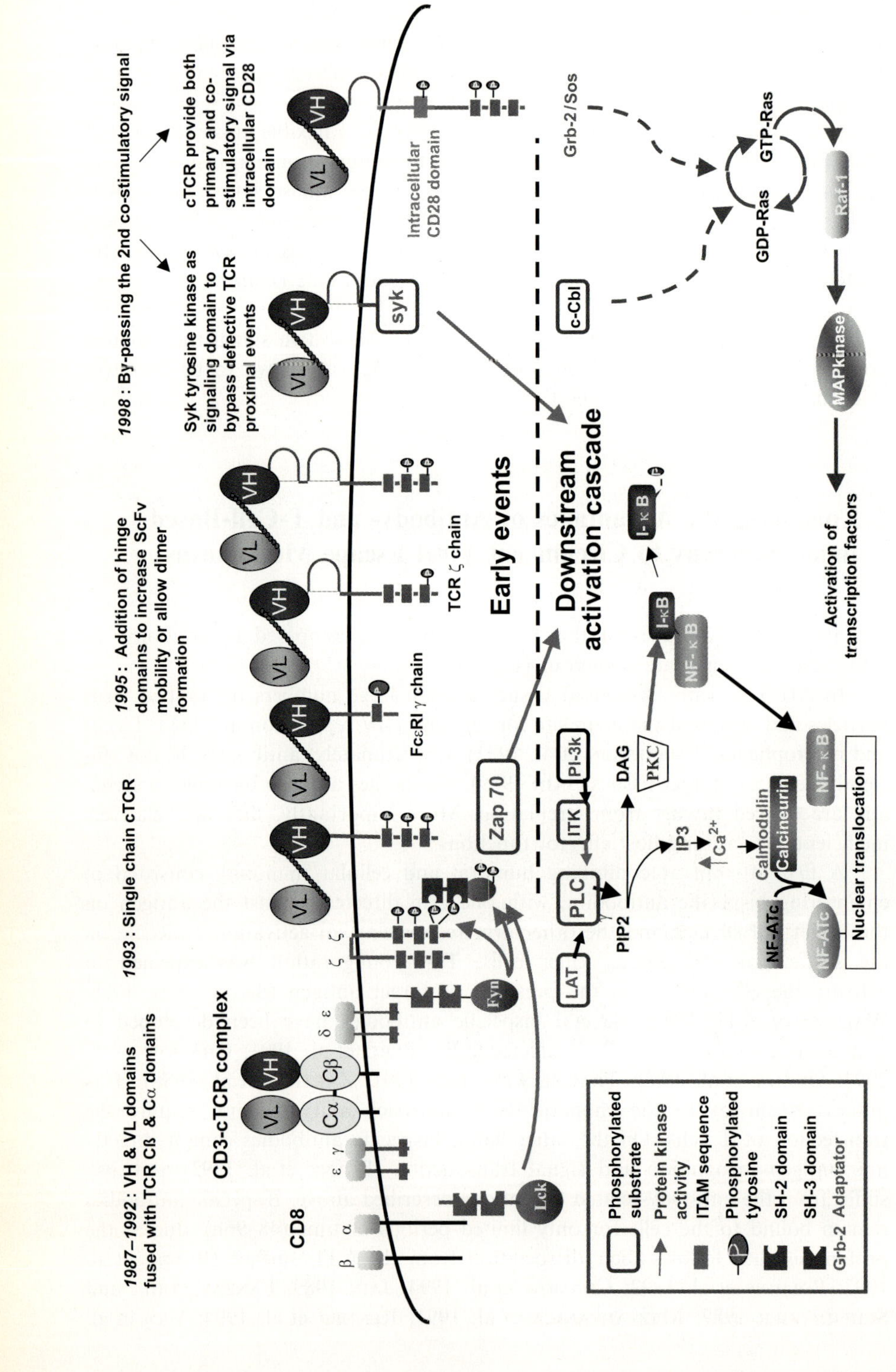

1987–1992 : VH & VL domains fused with TCR Cß & Cα domains
1993 : Single chain cTCR
1995 : Addition of hinge domains to increase ScFv mobility or allow dimer formation
1998 : By-passing the 2nd co-stimulatory signal
Syk tyrosine kinase as signaling domain to bypass defective TCR proximal events
cTCR provide both primary and co-stimulatory signal via intracellular CD28 domain
CD8
CD3-cTCR complex
VL
VH
Cα
Cβ
β
α
ε γ
δ ε
ζ ζ
FcεRI γ chain
TCR ζ chain
syk
Intracellular CD28 domain
Lck
Fyn
Zap 70
Early events
Downstream activation cascade
Grb-2/Sos
c-Cbl
GDP-Ras
GTP-Ras
Raf-1
MAPkinase
Activation of transcription factors
LAT
PLC
ITK
PI-3k
PIP2
DAG
PKC
IP3
Ca2+
I-κB
NF-κB
Calmodulin
Calcineurin
NF-ATc
Nuclear translocation
Phosphorylated substrate
Protein kinase activity
ITAM sequence
Phosphorylated tyrosine
SH-2 domain
SH-3 domain
Grb-2 Adaptator

**Fig. 1.** Chimeric T cell receptors: evolution of the family. *Sos*, Son of Sevenson (exchange factor); *PLCγ*, phospholipase C-γ1; *PI*, phosphatidyl-inositol; *IP3*, inositol tri-phosphate; *DAG*, diacyl-glycerol; *PTK*, phospho-tyrosine kinase (p59fyn, p56lck); *ITAM*, immuno-receptor tyrosine-based activation motif; *ZAP*, zeta-associated protein; *PKC*, $Ca^{2+}$-dependent protein kinase; *syk*, spleen tyrosine kinase; *ITK*, inducible T cell kinase; *LAT*, linker for activated T cells

1989). By contrast, chimeric receptors are expressed by transduced CTLs for the lifetime of the cells (Weber et al. 1996). In animal models, it was shown that $CD8^+$ memory cells survive at least 70 days in the periphery (Tanchot and Rocha 1998).

T-cell-based therapy also has several shortcomings. T cells are restricted by the MHC for antigen recognition and therefore are restricted to the individual. In addition, it is very difficult to generate large amounts of specific CTLs in vitro, and viruses frequently develop escape mechanisms from CTL responses. Interestingly, T-bodies could overcome some of these viral-defense mechanisms. For instance, some molecules such as Nef (Collins et al. 1998; Schwartz et al. 1996) or Tat (Howcroft et al. 1993; Weissman et al. 1998) may down-regulate HLA class I expression and impair recognition by CTLs, while others like the E3/19K protein of adenovirus block the antigen-processing pathway (Burgert et al. 1987). Obviously, T-body recognition, which is MHC independent, would not be affected by defective, MHC-restricted antigen presentation. Moreover, recognition by classical CTLs might be affected by antigenic variations (in case of genetically unstable

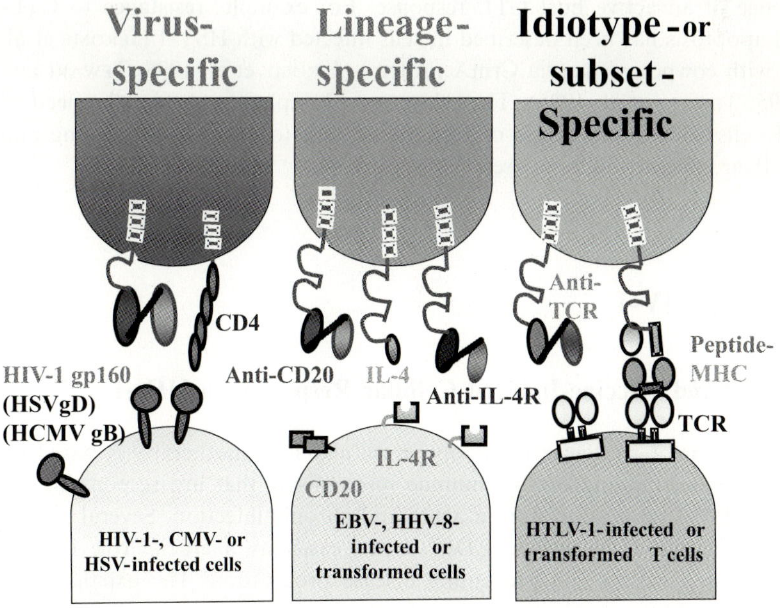


**Fig. 2.** Different targeting modes for engineered T cells. *IL-4R*, Interleukin-4 receptor

RNA viruses) that affect peptide transport, cleavage, processing (Couillin et al. 1994; Del et al. 1992; Eisenlohr et al. 1992), MHC binding (Couillin et al. 1994), or TCR recognition (Lill et al. 1992; Ploegh 1998; Price et al. 1998; Spriggs et al. 1996). Immune pressure can also result in the loss of B cell epitopes, but cells expressing chimeric receptors can be redirected against several epitopes (and, not necessarily neutralizing B cell epitopes), in order to minimize the possibilities of viral escape.

Subsequent to their specific activation, CTLs release chemokines that allow for selective transendothelial migration into sites of infection (see Sect. 6) (Baggiolini 1997, 1998; Butcher and Picker 1996; Roth et al. 1998). Viruses have developed several strategies to divert the natural inflammatory response by modulating chemokines or chemokine receptors. Several viruses encode "mimic" receptors for cytokines or chemokines which serve as decoys and compete with the chemokine receptors on endogenous effector cells (Chee et al. 1990; Neote et al. 1993; Vieira et al. 1998). Some viruses encode chemokines homologues which could act as competitive antagonists (Boshoff et al. 1997; Kledal et al. 1997; Moore et al. 1996; Nicholas et al. 1997; Penfold et al. 1999) or as agonists which subvert host immunity (by provoking a bias in Th1/Th2 response, or by recruiting host cells) (Baba et al. 1996; Dairaghi et al. 1999; Endres et al. 1999). Another "viral escape" strategy is the secretion of chemokine-binding proteins (Alcami et al. 1998; Isegawa et al. 1998; Lalani et al. 1997, 1999; Martinez et al. 1995). Altogether, such viral-encoded factors can perturb the activation and recruitment of effector cells into infected tissues.

Furthermore, additional mechanisms could contribute to viral persistence in the presence of an active host CTL response. For example, resistance to CTL-mediated apoptosis has been described in cells infected with HSV-1 (Jerome et al. 1998) or with cowpox virus (via CrmA protein) (Jerome et al. 1998; Tewari and Dixit 1995; Tewari et al. 1995). These last two examples emphasize the need to endow T cells with a new mode of recognition and to direct their homing and optimize their effector functions (see below, Sect. 6).

# 4 Targeting HIV-1

## 4.1 Natural and Vaccine-Induced Cellular Responses to HIV-1

The development of strategies for adoptive cellular immunotherapy is based on our current understanding of the immune mechanisms that are responsible for controlling HIV-1 replication during acute and chronic infection. Several studies have suggested that HIV specific $CD8^+$ CTLs can play a crucial role in controlling virus replication and preventing disease progression. The expansion of HIV-1 specific CTLs following primary infection correlates with a sharp decrease in viral replication (Borrow et al. 1994; Koup 1994), which progressively rises again with the decline in HIV-specific CTL activity (Klein and Miedema 1995). Conversely, lack of disease progression has been associated with high frequencies of CTL precursors endowed with broad specificities (Harrer et al. 1996). It was recently shown that higher frequencies of HIV-1 env-specific CTLs correlate with lower levels of plasma HIV-1 RNA- and PBMC-associated infectious virus (Musey et al. 1997). HIV-1 specific CTLs were found in individuals with repeated exposure to the virus who did not become infected (Clerici et al. 1992; Langlade et al. 1994). In vitro studies have confirmed that purified $CD8^+$ cells have the ability to inhibit the cellular replication cycle of HIV (Walker et al. 1986). Finally, during primary or chronic simian immunodefficiency virus (SIV) infection, high levels of viral replication are seen in monkeys depleted of $CD8^+$ T cells (Schmitz et al. 1999). It is therefore likely that, through the development of a vigorous CTL response soon after infection, HIV-1-specific CTLs would be able to eradicate HIV-1 infection in certain transiently infected patients and would at least be able to contain HIV-1 spread in most other patients. However, in most cases, the host response ultimately fails to control HIV replication, leading to disease progression. The progression of HIV infection despite the development of HIV-specific $CD8^+$ CTL responses suggests that cell-based immunotherapeutic strategies must overcome significant obstacles in order to be effective.

One such hurdle is the deficit in HIV-1-specific T cell help in AIDS patients (Rosenberg et al. 1997). Furthermore, under treatment with highly active anti-retroviral therapy (HAART), a significant loss of CD4+ helper memory was shown to occur (Pitcher et al. 1999). This absence of CD4 help may lead to inefficient immunological control and successful escape of the virus from its reservoirs (Gorochov 1999).

To induce specific CTLs, vaccination with live attenuated viruses has been proposed and indeed appears to be effective in animal models (Daniel et al. 1992). Nevertheless, such an approach presents too many risks to be widely applied to healthy individuals. Synthetic vaccines would probably be safer, but must contain a mixture of many different T cell epitopes in order to prevent viral escape. An appropriate epitope mix could be optimized for a given individual but the optimal composition may depend on HLA haplotype.

## 4.2 Natural and Vaccine-Induced HIV-1-Specific Humoral Responses

While HIV-specific antibodies are always found in the serum of HIV patients, their neutralizing potential appears to be limited. Vaccines have also failed to elicit neutralizing antibody responses against field isolates of HIV-1 (Hanson et al. 1993). In addition, adoptive transfer of large quantities of HIV-specific antibodies to SCID mice gave disappointing results: cell-to-cell transmission is not blocked by mAbs (Poignard et al. 1999). Nevertheless, it was recently shown that passive administration of anti-HIV-1 IgG to macaques prevented vaginal (Mascola et al. 1999) or oral (Baba et al. 1996) transmission of SIV chimeras. Thus, while antibodies are not able to clear an established viral infection, they may nevertheless play a protective role against transmission.

In vitro, a good correlation exists between efficient neutralization and high affinity of interaction between the antibody and the epitope expressed on the virus itself or on the surface of infected cells (Parren et al. 1998; Van Cott et al. 1994). On the other hand, the nature of the epitope that is targeted appears to be less important.

## 4.3 Anti-HIV-1 T-Bodies

### 4.3.1 Rational Design

The rational design of anti-HIV-1 T-bodies requires consideration of several issues, the first one being the choice of the viral epitopes to target and of the corresponding antibodies. The antibodies that are most likely to be effective as T-bodies are not necessarily those with the greatest neutralizing ability. Based on the serial triggering model proposed by Lanzavecchia (1997), it is even possible that cells expressing receptors of very high affinity would be less active against target cells than effectors expressing lower affinity receptors.

Indeed, T cells make use of receptors endowed with particularly fast dissociation rates from their cognate ligands ($t_{1/2} < 30$s) (Margulies 1997). It was therefore proposed (Bitton et al. 1998) that several non-neutralizing HIV-specific antibodies could be "recycled" as effective TCRs. We showed that an antibody that could not effectively neutralize a particular HIV variant could still be used to construct chimeric T-bodies that specifically recognize this variant (Bitton et al. 1998).

The only HIV products expressed at the surface of infected cells are MHC-bound peptides and the envelope protein of HIV-1 (gp120–gp41 complex) (Earl et al. 1991a,b; Willey et al. 1988). Although antibodies that bind particular MHC–peptide complexes have been described (Porgador et al. 1997), there are no antibodies available that recognize MHC–HIV-peptide complexes. Hence, T-bodies can only be redirected against the native envelope protein of HIV-1, as no other native HIV protein is displayed on intact cells. It should be noted that only productively infected cells express the HIV envelope, while latently infected cells do not express detectable amounts of the envelope protein. Furthermore, the HIV gp120–gp41 complex is only expressed during late stages of the HIV replication cycle (Pomerantz et al. 1990) in cells that start to actively produce virus. Strategies that would enable the targeting of latently infected cells, or cells at an earlier stage of infection, are discussed below.

T-bodies are expected to penetrate inside the virus sanctuaries, which are represented by solid organs. They could also be induced to secrete soluble antiviral cytokines at sites of viral replication. Finally, in view of the decline of CD4 T cell functions in HIV-infected patients, anti-HIV T-body-expressing cells should be engineered in a way that will maintain their function in vivo in the absence of CD4 T cell help.

### 4.3.2 Effector Cell Recipients of HIV-1-Specific Chimeric Receptors

Many different cell types have already been used successfully as CR recipients:

1. CD8 Cells. $CD8^+$ T cells were first used as recipients for chimeric receptor (cTCR) expression, both because of their killing capacity and their potential to secrete cytokines. cTCR-bearing $CD8^+$ T cells were demonstrated to specifically secrete interleukin (IL)-2 in the presence of HIV-1-infected cells (Bitton et al. 1998), to kill HIV-1-infected cells (Roberts et al. 1994) and to inhibit HIV-1 replication in vitro (Yang et al. 1997).
2. NK Cells. CD4-ζ-expressing NK cells are able to specifically kill HIV-1 infected cells and tumor cells expressing gp120 at their surface in vitro (Tran et al. 1995). The use of modified NK cells may enable generation of T-cell-independent systemic immunity to virus in HIV-infected patients.
3. Hematopoietic Progenitors. The direct introduction of cTCR genes into hematopoietic stem cells (HSCs) represents a very attractive option (Roberts et al. 1998). Modification of stem cells could circumvent the necessity to transfect and expand cells ex vivo. Gene modification of HSCs may be preferable to modification of terminally differentiated effector cells, such as T or NK cells. Indeed, multiple effector cells can be simultaneously redirected using a stem cell approach (bypassing the necessity to isolate and express mature effector cells, a process which may negatively impact on their in vivo trafficking or function). Such an approach may also provide long-lasting memory cells and, as such, could serve as a renewable source of gene-modified effector cells allowing prolonged antigen-specific immune surveillance. Current evolution of the lentiviral vectors may render transduction of HSC feasible and safe in humans (see Sect. 6.1).

## 4.4 In Vitro Experiments

Various HIV-binding domains have been used to construct cTCR. In an attempt to circumvent the viral escape mechanisms resulting from the intense mutation rate occurring during viral replication, CD4 (the high-affinity co-receptor for gp120) was chosen as an extracellular domain [coined the "universal receptor" (UR)] for the design of a first generation of anti-HIV cTCR. Such chimeric receptors were expected to interact with gp120 expressed on infected cells irrespective of the viral strain. Romeo and Seed (1991) described chimeras composed of CD4 extracellular domains (amino acids 1–369) fused to the transmembrane and intracellular domain of TCR/CD3 (ζ) or IgGFc-receptor-γ signal-transducing elements. They showed that those chimeras are capable of directing CTLs to recognize and kill cells expressing gp120.

Roberts et al. (1994) have reported the effect of such receptors as well as of chimeric receptors made from ScFv of anti-gp41 98.6 mAb (Gorny et al. 1989; Xu et al. 1991) or from the gp120-specific human mAb 447-D (Cavacini et al. 1993). Both types of cTCR are able to initiate an efficient effector T cell response against HIV-infected cells, including cytokine secretion, proliferation, and cytolytic activity upon interaction with target cells expressing surface viral antigen (ROBERTS et al. 1994).

One concern is that CD4-ζ receptors might put the UR-expressing cells at risk of being infected themselves. Indeed, T cells engineered to express high surface levels of CD4 could potentially represent a new cellular reservoir of virus in vivo. This issue has been solved recently by the development of CD4 variants that do not support infection by HIV-1, but retain their ability to bind gp120 and redirect cytolysis when expressed under the cTCR configuration (HEGE and ROBERTS 1996). Another concern was that cells expressing the CD4-based chimeric receptors could kill uninfected targets expressing MHC class II; however, the affinity of CD4 for the latter seems to be insufficient to produce an effective cross-linking of the UR that could lead to efficient T cell activation (ROBERTS et al. 1994; ROMEO and SEED 1991; TRAN et al. 1995).

For their functional activity, some receptor constructs appear to require an hinge region, such as CD8α or CD4 immunoglobulin Ig-like domain, as a spacer between the ScFv and the transmembrane unit (MORITZ and GRONER 1995). Recently, another study assessed the differences between ScFv-ζ receptors [directed against the HIV-1 gp120 protein, using an ScFv derived from the F105 antibody (BAGLEY et al. 1994)] that included alternative extracellular spacer and transmembrane domains derived from members of the immunoglobulin supergene family (PATEL et al. 1999). The authors demonstrated that modifying the extracellular protein domains of the anti-HIV cTCR significantly impacted receptor stability, substrate-binding affinity, and cTCR-mediated killing. These effects could not be simply explained by differences in the level of cell surface expression. It seems difficult to generalize those findings to all anti-HIV-1 cTCRs. By using a different gp120 recognition unit, we found that T cells expressing a cTCR without a hinge region had a lower specific activation threshold than other transfectants expressing a cTCR fitted with such spacers (Fig. 2).

## 4.5 Animal Models

As a preclinical model, HEGE et al. (1996) described the generation of T-cell-independent systemic immunity in SCID mice reconstituted with bone marrow cells transduced with CD4-ζ UR. These mice rejected the leukemic Raji cells expressing HIV-1 env. This is the first study demonstrating the efficacy of chimeric receptor hematopoietic stem cells in vivo. In addition, this study supports the role of non-T effector cells in the elimination of virally infected cells, as CD4-ζ expression was observed in vivo in myeloid and NK cells.

At present, several animal models are available for the study of cTCR as a modality against HIV-1 infection. Humanized or PBL-grafted SCID mice, and more recently the development of transgenic mice expressing human CD4 and CCR5, may be used for testing to guide the clinical development of immune-based therapy for HIV-1 (LACASSE et al. 1999).

## 4.6 Clinical Trials

The safety and efficacy of adoptive T cell therapy in humans has been established for prophylaxis against viral diseases caused by CMV (Walter et al. 1995) and EBV (Heslop et al. 1996) as well as in the treatment of hematological malignancies and melanoma (Mitchell 1989). The adoptive transfer of autologous HIV-specific CTLs isolated from the blood and expanded in vitro has been evaluated. Unfortunately, transfer of HIV-1-Nef-specific CTLs to an AIDS patient led to selection for mutant HIV variants and subsequent disease progression (Riddell et al. 1996). The transfer of HIV-1-specific lines or clones into HIV-1-infected patients has yielded only limited results so far (Lieberman et al. 1997; Riddell et al. 1996; Tan et al. 1999). However, infusion of virus-specific CTLs ($10^9$ CTLs/$m^2$) resulted in frequencies of 1%–4% of $CD8^+$ T cells in the recipients' blood, and increased HIV-specific cytotoxic activity (Brodie et al. 1999).

The data obtained from animal models, in which tumor or viral infections can be eradicated by adoptive transfer of antigen-specific T, NK or hematopoietic stem cells, together with the findings from clinical studies suggests that patients may have to receive cell dosages in excess of $10^9$ antigen-specific lymphocytes to obtain a therapeutic antiviral effect (Hege et al. 1996). Cells transfected with a chimeric receptor are particularly well suited to evaluation in a clinical trial, and can be used to obtain large populations of antigen-specific T cells within a few weeks, in contrast to the long time required for selection, characterization and expansion of native CTLs with defined specificity.

The in vivo life span of gene-modified T cells was studied in HIV-discordant syngeneic twin pairs in which peripheral blood lymphocytes from the healthy twin were retrovirally transduced with the *neo* gene only (Walker et al. 1993)and infused into the affected twin. This marker gene was detected by PCR in both blood and lymph nodes of the HIV-positive twin for at least 25 weeks post-infusion, demonstrating the possibility of adoptive T cell therapy in this disease.

A study of the adoptive transfer of syngeneic gene-modified $CD8^+$ lymphocytes in HIV-infected identical twins was started in 1995 (Walker 1996). The goal of this protocol was to redirect CTLs against HIV-infected target cells, using the CD4-ζ chimeras described above. This study was designed to determine the safety and activity of healthy UR-expressing T cells following transfer to HIV-1-infected identical twins. In an initial phase I/II study, PBMC obtained from HIV-1-seronegative donor twins were enriched for $CD8^+$ expression, activated in vitro with IL-2 and anti-CD3, and transduced with a murine retrovirus containing the CD4-ζ gene. Following dose escalation, 30 HIV-1-infected twins received up to six infusions over 1 year of either $10^{10}$ CD4-ζ-transduced or control $CD8^+$ T cells (Walker et al. 1998). In a second study, designed to test the effects of providing HIV-1-specific $CD4^+$ T cell help, 17 twins subsequently received three additional infusions of $10^{10}$ CD4-ζ-modified $CD4^+$ and $CD8^+$ T cells, at 2-week intervals (Hege et al. 1998). In this latter study, preactivated modified T cells were detected in 100% ($n = 21$) of patients receiving gene-modified $CD8^+$ T cells alone, with peak levels of $10^4$ copies/$10^6$ PBMC in 16 patients. Nevertheless, rapid clearance of modified cells was seen in nine recipients. In contrast, all 17 recipients of both gene-modified $CD4^+$ and $CD8^+$ T cells showed prolonged, high level persistence of transfected cells. Fractions of circulating gene-marked $CD4^+$ and $CD8^+$ T cells ranged from $10^3$–$>10^4$ copies/$10^6$ PBMC, increased with time (in some patients), and persisted for up to 100 days. No treatment-limiting side effects were observed. These studies demonstrate the safety of adoptive transfer of genetically engineered, HIV-1-specific T cells. Compared to gene-modified $CD8^+$ T cells alone, $CD4^+$ and $CD8^+$ T cells given together resulted in increased cell survival in the circulation and provided preliminary evidence of in vivo proliferation of the engineered cells. Moreover, this study demonstrated tissue trafficking by finding CD4-ζ T cells in rectal-mucosa-associated lymphoid tissue in two patients.

Recently, the disease course of HIV-1 infected patients receiving HAART was compared in two groups of patients administered either unmodified T cells or CD4-ζ expressing $CD4^+$ and $CD8^+$ T cells (Hege et al. 1999). The patients were infused with $10^{10}$ cells three times at 2-week intervals and were followed for 24 weeks. UR-modified T cells persisted at high levels in the blood in all patients (0.2%–0.8% of PBMC) with stable gene expression and good trafficking in rectal mucosa. The frequency of patients with viral rebound appeared to be lower in the CD4-ζ treated group. Similarly, the residual viral load was somewhat reduced in the same groups (0.5 mean log decrease in rectal HIV-1 DNA at week 24). However treatment with UR-transfected $CD4^+$ cells did not induce an increase in total CD4 counts in patients.

These promising results warrant confirmation, but it has been suggested that CD4-ζ modified T cells are associated with a decrease in tissue reservoirs of HIV-1. Therefore, CD4-ζ modified T cells might represent a useful adjuvant immunotherapy for patients with undetectable viremia on HAART (Hege et al. 2000).

## 5 Potential Targets for the Future Design of Antiviral T-Bodies

The T-body approach could be extended to other viruses that escape the natural immune response and persist in the host. Only viruses that are responsible for serious chronic disease and/or severe complications should be considered. The viral diseases described below represent good candidates, since a mAb or a ScFv directed against membrane-associated viral protein exists, while some escape mechanisms hamper the CTL response.

The ability to construct virus-specific T-bodies is dependent on the expression of viral envelope components at the surface of infected cells and on the availability of specific antibodies.

### 5.1 Herpesviridae

The herpes viruses family, including Epstein-Barr virus (EBV), human herpes virus-8 (HHV-8), human cytomegalovirus (HCMV) and herpes simplex virus (HSV) type

1 and 2, are characterized by latent infection and associated with tumors in immunosuppressed hosts. Several T-body-based approaches could be attempted for the treatment of diseases in which herpes viruses are involved:

1. AIDS-associated Kaposi's Sarcoma (AIDS-KS) is the most common malignancy associated with HIV-1 infection. The $\gamma$ human herpes virus (HHV)-8 is linked to the etiology of KS (Moore et al. 1996). Malignant KS cells express the IL4 receptor and have been successfully targeted with a recombinant cytotoxin made of IL-4 fused to exotocin (Husain et al. 1999). Similarly, CTLs could be targeted to KS via cTCRs using the extracellular portion of IL-4 or anti-IL4-receptor antibodies.

2. HSV entry is dependent on the interaction of virion glycoprotein D (gD) with one of several cellular receptors. It was found that the major neutralizing antigenic site on HSV, gD, overlaps a receptor-binding domain (Whitbeck et al. 1997). It is therefore conceivable to engineer anti-HSV chimeric receptors using the extracellular portion of the potent neutralizing DL11 antibody.

3. EBV is an ubiquitous virus that persists asymptomatically in B lymphocytes and replicates in oropharyngeal epithelial cells. B cells infected in vivo become transformed by expression of viral latency-associated transforming proteins. In immunocompromised individuals (e.g. those with HIV-1 infection, or after bone marrow transplantation), there is a high incidence of unregulated EBV-driven B cell proliferation leading to B lymphoma (Herndier et al. 1994). HIV-related malignant lymphomas usually have a poor outcome (Kaplan et al. 1997). Adoptive transfer of EBV-specific T lymphocytes has been successfully used in immunocompromised patients with EBV-associated lymphoproliferative disease. Transferred T lymphocytes can reconstitute the immune response to EBV. The gp350 glycoprotein is expressed on the surface of infected cells (Bertoni et al. 1989; Khanna et al. 1999). Anti-gp350 mAbs are available, but are not effective against chronically infected EBV cells, as gp350 is expressed only during acute infection. The only protein expressed at the cell membrane of latently EBV-infected cells is LMP-1 (latent membrane protein). LMP-1 is an analogue of a constitutively active form of CD40, a member of the tumor necrosis factor (TNF) receptor family. It may therefore be possible to redirect CTLs with anti-LMP-1 specificity in order to lyse EBV-infected cells. However, LMP1 has a very low immunogenicity (Meij et al. 1999) and antibodies that target LMP-1-expressing cells are not yet available.

Despite the lack of virus-specific antibodies, T-bodies could be easily targeted to B-cell-specific surface markers like CD19, CD20 or CD22. CD20 is expressed on the cell surface of malignant human lymphomas (Press et al. 1987), and antibodies against CD20 have been successfully used to treat such affections (Kuehnle et al. 2000; Maloney et al. 1994, 1997a,b). CD19 and CD22 immunotoxins have also been tested in animal models (Ghetie et al. 1994a,b). It is expected that the normal B cell lineage would also be affected by the treatment, but would be replenished through hematopoietic stem cell differentiation.

4. HCMV can persist latently for the lifetime of the individual; the absence of CMV-specific $CD4^+$ helper cells allows life-threatening reactivation of CMV in

AIDS patients or transplant recipients. In the natural course of infection, neutralizing antibodies are produced but they cannot halt the infectious process. HCMV uses many of the escape mechanisms described earlier (Sect. 3) to shield infected cells from both innate and specific immunity. The expression of viral proteins US2, US3 and US11 is responsible for the down-regulation of cell surface MHC I (Beersma et al. 1993; Fruh et al. 1997; Hengel et al. 1998) and provides protection against CTL-mediated clearance of infected cells. NK attack is avoided through the expression of the MHC class I homologue UL18 (Beck and Barrell 1988; Browne et al. 1990) and through the enhancement of HLA-E expression by the viral protein UL40 (Tomasec et al. 2000). Finally, the viral protein US28 leads to chemokine sequestration (Bodaghi et al. 1998; Gao and Murphy 1994). In a study of adoptive therapy with CMV-specific cytotoxic lymphocytes, conducted in 14 bone marrow transplant recipients, it was shown that CTLs persisted in vivo for up to 12 weeks (Riddell et al. 1992, 1993; Walter et al. 1995). T-bodies could be targeted against one, or several, epitopes of the CMV glycoprotein B (Boldicke et al. 1995; Ihara et al. 1998; Li et al. 1995; Ohlin et al. 1996; Schoppel et al. 1996; Speckner et al. 1999; Takekoshi et al. 1998) and are expected to remain active in the absence of CD4-specific help even in the subset of severely immunocompromised patients who suffer from CMV-related complications. However, it is unclear whether the anti-CMV antibodies currently available bind the surface of infected cells.

## 5.2 HTLV-1

Human T cell leukemia virus type 1 is the etiologic agent of adult T cell leukemia and of a chronic neurological disease known as tropical spastic paraparesis, or HTLV-1 associated myelopathy (Gessain et al. 1990; Hinuma et al. 1981; Osame et al. 1986; Poiesz et al. 1980; Yoshida et al. 1982). It is estimated that over 20 million individuals carry HTLV-1 worldwide. Although a robust immune response is elicited during invasion, infection usually persists in the host. The surface envelope glycoprotein gp46 is the primary target of neutralizing antibodies. T-bodies could be directed to gp46. In addition, HTLV-1-related lymphomas could be targeted with clone (idiotype)-specific anti-TCR chimeric receptors. In this case, the extracellular domain of the cTCR would specifically recognize the TCR of the leukemic T cells. To date, a large collection of anti-TCR β chain antibodies is available; each delineates tiny Vβ subsets in healthy controls (Aissaoui et al. 1999; Diu et al. 1993; Necker et al. 1995; Romagne et al. 1992, 1995). In lymphoma patients, the antibody that specifically recognizes the TCR of the malignant clone can be chosen as the T-body recognition element (Gorochov et al. 1995).

## 5.3 Down-Regulation of Deleterious Responses

Antiviral CTLs are also responsible for deleterious effects. For instance, they can cause brain lesions in mice infected with LCMV (Zinkernagel and Hengartner

1994) and liver necrosis of hepatitis B virus (HBV)-infected patients. Taking into account that a CTL response is usually directed against a few immunodominant epitopes, it may be possible to selectively inhibit those specificities (in cases in which they have been or could be predefined). We are currently developing a new format of cTCR in which the extracellular portion of the receptor would consist of an MHC–peptide single chain complex. Killer cells bearing such receptors would be targeted to T cells expressing the corresponding TCRs. A similar strategy could be applied to mature T cell lymphomas and to immune cells or allo-immune cells arising during graft vs host disease or autoimmune disorders. Appropriate peptide–MHC ligands would have to be selected on the cell targets from phage libraries displaying such complexes (Le Doussal et al. 2000).

## 6 Important Milestones

Before the chimeric receptor approach can be widely adopted for clinical use, several aspects of this modality must be further optimized. Mechanisms that can result in a decrease of natural specific CTL activity such as lack of CD4 T cell help, viral escape, or clonal exhaustion may also limit the in vivo use of engineered T cells and should be addressed. Finally, the safety and efficacy of gene transfer should also be optimized.

### 6.1 Gene Transfer Tissues

Genetic engineering of T lymphocytes for adoptive clinical immunotherapy calls for efficient gene transduction methods. The strategy of stable gene transfer into hematopoietic stem cells or mature effector cells must be improved. Very efficient, HIV-1 lentiviral-based systems of gene tranduction have been recently described (Miyoshi et al. 1999; Naldini et al. 1996; Poeschla et al. 1996; Reiser et al. 1996). These lentiviruses can integrate into the genome of non-dividing cells (Gallay et al. 1995, 1996, 1997). However, cellular activation remains a prerequisite for efficient T cell transduction (Unutmaz et al. 1999). Construction of lentiviral vectors pseudotyped with VSV-G envelope (glycoprotein G of the vesicular stomatitis virus) has resolved the issue of the restricted tropism of lentiviruses (Naldini et al. 1996). The VSV envelope has a broad host range, as its binding to target cells occurs via an ubiquitous phospholipid component of the cell membrane. The main remaining restriction to a wide application of HIV-based vectors in human studies is obviously safety (Otto et al. 1994). Although, there may be fewer ethical restrictions in their use for gene therapy in individuals who are already HIV-infected, one should also consider that the risk of recombination of a lentiviral vector would be higher in patients harboring an ongoing retroviral infection. In this regard, lentivectors based on SIV may be advantageous.

Another potential risk is represented by the activation of proto-cellular oncogenes by the inserted provirus. To avoid this risk, self-inactivated vectors have been developed (DOUGHERTY and TEMIN 1987; GUYER and COLLINS 1995; MILLER 1997; MIYOSHI et al. 1998; YU et al. 1986). Empirical demonstration of these novel gene therapy vectors as well as the development of large-scale production methods are goals of future research.

## 6.2 In Vivo Maintenance of Functional Properties

A critical step in the application of the T-body approach in immunotherapy is achieving functional expression of the chimeric receptor genes in human lymphocytes.

In vitro selection of transduced T cells requires co-expression of a resistance gene that could elicit an immune response in vivo. Survival of autologous $CD8^+$ HIV-specific T cell clones modified to express the hygromycin phosphotransferase gene (HyTK) is limited by the induction of a potent HyTK-specific immune response (RIDDELL et al. 1996). It should be noted that similar responses have not been reported after the infusion of specific CTLs expressing the *neo* gene (BLAESE et al. 1993; BRENNER et al. 1993; DEISSEROTH et al. 1994; KOHN et al. 1995; MERROUCHE et al. 1995; ROONEY et al. 1995; ROSENBERG et al. 1990). One of the advantages of the new generation of vectors (Sect. 6.1), allowing a high efficiency of gene transfer, would also be to by-pass the need for in vitro selection using resistance genes.

The transferred cells do not need to express a foreign gene to see their life span limited in vivo. In a recent study, Gag- and Pol-specific T cells that had been transferred to HIV-1-infected patients (TAN et al. 1999) were rapidly eliminated through apoptosis (90% of the infused cells died within 48h). The transferred CTLs displayed a mature phenotype in vivo, characterized by an up-regulation of Fas and an absence of CD28. The lack of CD28 expression could be related to the over-stimulation of the cells in vivo (before culture) or in vitro. These data suggest that HIV-specific CTLs persisted very briefly in vivo, and could therefore mediate only a transient antiviral effect. The limited survival of adoptively transferred T cells may correlate with an inadequate, endogenous, HIV-1-specific, $CD4^+$ Th response, required to sustain the transferred CTLs (KALAMS and WALKER 1998; ROSENBERG et al. 1997). Indeed, the $CD4^+$ Th deficiency itself could represent an important limiting factor in these hosts (MATLOUBIAN et al. 1994). A correlation between the decline of $CD4^+$ T cells and loss of HIV-specific $CD8^+$ T cell responses during the progression of AIDS has been reported (ROSENBERG et al. 1997). The maintenance of a $CD4^+$ Th cell response is concomitant with the vigorous CTL response seen in long-term non-progressors (ROSENBERG et al. 1997). Strategies aimed at providing help to the transferred $CD8^+$ CTLs were proposed. IL-2 or Th cells genetically modified to resist HIV have been co-infused with genetically modified effector cells (WOFFENDIN et al. 1996). Similarly, $CD8^+$ CTL expressing a chimeric receptor containing the cytoplasmic domain of IL-2 have been constructed (NELSON et al. 1994).

Indeed, direct signaling through the chimeric receptor, without appropriate co-stimulatory signals, could induce an in vivo state of anergy in the transduced cells. Therefore, the redirected effector cells would have to be pre-activated or co-stimulated to properly respond to triggering through the signaling domain of the chimeric receptor. It was demonstrated that a chimeric ScFv/γ-receptor-mediated T cell lysis of tumor cells is co-regulated by adhesion and accessory molecules (Weijtens et al. 1998). This study outlined the important role of CD2, CD3 and the CD11a/CD18 molecules in the ScFv/γ-mediated cytolysis of tumor cells. The requirement for CD11a/CD18/CD54 interactions is more apparent in the presence of relatively high levels of free antigen. Moreover, reduced expression of lymphocyte signal transduction molecules (TCR zeta chain and p56lck tyrosine kinase) in patients with chronic infectious diseases could lead to an impaired immune response (Gratama et al. 1999).

To bypass the requirement for a co-stimulatory signal, double transfectants were generated. These transfectants simultaneously express ScFv-CD28 and ScFv-CD3ζ. The increased activity of these chimeras demonstrates that antigen-specific co-stimulatory signals can also synergize with signals mediated through chimeric CD3ζ chains to secrete maximal levels of IL-2 (Alvarez and Hawkins 1996). More recently, Finney et al. (1998) improved the design of chimeric receptors in a way that allows the delivery of both primary and co-stimulatory signals by the same receptor. In such a design, the intracellular co-stimulatory signaling domain of CD28 was fused with the CD3ζ chain (Finney et al. 1998). Using a similar design, we have introduced the CD28 molecule (excluding its ligand-binding domain) in between the ScFv and the FcR γ chain and have obtained a functional receptor (Z. Eshhar, unpublished data). We also fused the ScFv, through its spacer and transmembrane domain, to the cytoplasmic protein kinase Syk; this receptor was able to by-pass defective TCR-proximal events (Fitzer et al. 1998).

In vivo, T bodies could also be anergized, or blocked through direct interaction of the chimeric receptor with soluble gp160. As previously described, cTCR-expressing T cells can be inhibited in several models by soluble antigen (Gross et al. 1989); it is known that soluble gp160 is found in the serum of HIV infected patients (Moore and Jarrett 1988; Oh et al. 1992). Roberts et al. (1994) have shown that sera from HIV infected donors does not contain sufficient levels of antibodies or free antigen to inhibit cytolytic activity of chimeric-receptor-expressing T cells. This issue was also addressed in our laboratory, where T cells redirected with the b12 anti-gp160 mAb were tested in the presence of soluble gp160. We found that 50μg/ml of free antigen is required to efficiently prevent the activation of gp160-specific T-bodies, while serum concentrations of gp160 never exceed 90ng/ml in HIV patients (Oh et al. 1992). In patients with low levels of viremia, soluble gp160 levels are well below this level, and it is therefore unlikely that soluble antigen could effectively reduce the in vivo activity of anti-HIV specific T-bodies.

## 6.3 Controlling Homing and Persistence

Although it was shown that some HIV-1-specific CTLs could migrate to lymph nodes after passive transfer and retain cytotoxic activity (Brodie et al. 1999), it can also be argued that transduced T cells are likely to change their in vivo homing properties. Indeed, retroviral transduction and propagation usually require in vitro activation in order to induce cell cycling in tissue culture.

Chemokine receptors play a key role in regulating the activation and trafficking of blood-borne cells. It was recently shown that cells expressing the chemokine receptor CCR7 lack immediate effector function, but are efficiently routed toward lymph nodes, where they stimulate dendritic cells (Sallusto et al. 1999). Conversely, upon activation through the TCR, T cells down-regulate CCR7 expression. For therapy against viral-associated lymphomas, T-bodies could be engineered to constitutively express CCR7 in order to be targeted to lymphoid tissues. In contrast to normal $CCR7^+$ memory cells, T-bodies could display immediate effector function in situ.

Other "homing" receptors could be introduced in T-bodies to control their migration to selected tissues. For instance, effectors designed to target HHV-8-associated KS or HTLV-1-related skin lymphomas should express the cutaneous lymphocyte antigen (CLA) (Picker et al. 1990). In general, expression of antigen-specific receptors is not sufficient to drive extravasation of T cells into inflamed tissues. Therefore, in addition to a new specificity, a corresponding homing address might have to be systematically transferred to T-bodies.

Another issue to consider relates to the in vivo survival of the transferred T cells. For example, the Nef protein induces an up-regulation of Fas ligand in HIV-1-infected cells (Xu et al. 1999), suggesting that HIV-specific T-bodies expressing Fas could be killed in vivo upon encounter with their targets in a Fas-dependent manner. Interestingly, infused CTL clones retrieved from the body displayed increased susceptibility to apoptosis induced by a Fas-L-expressing cell line (Zauli et al. 1999). The elucidation of cell death mechanisms may lead to the engineering of apoptosis-resistant CTL to bypass this obstacle (Tan et al. 1999). For instance, T-bodies could be engineered to over-express the FLIP inhibitory protein (Algeciras et al. 1999; French and Tschopp 1999; Irmler et al. 1997). Finally, the ability of the transduced T cells to differentiate into memory cells remains to be demonstrated.

## 6.4 Controlling the In Vivo Fate of Genetically Modified T Cells

If, as suggested above, HIV-specific CTLs are able to persist and remain functional in vivo, they could possibly mediate deleterious effects (Koenig et al. 1995). In adult mice infected with LCMV (Leist et al. 1988; Odermatt et al. 1991), LCMV-induced neurologic disease is directly mediated by virus-specific CTLs (Baenziger et al. 1986). It might therefore be necessary to control the fate and/or function of the cytotoxic $CD8^+$ T cells following transfer in vivo. Different strategies have been developed for regulating transgene expression in eukaryotic cells (Yarranton 1992). The tetracycline (Tet)-regulatable system (TRS) seems to be particularly well-suited due to the relatively low concentrations of tetracycline necessary to regulate *trans*-gene expression (Gossen et al. 1993; Gossen and Bujard 1992). The tet-suppressible expression of a cTCR in T cells has been already demonstrated in vitro (Alvarez et al. 1997).

# 7 Future Approaches

Although escape mutants were not reported during the clinical trials performed with UR-expressing cells (Walker 1996), the selection of viral mutants has been demonstrated following transfer of Nef-specific T cells (Koenig et al. 1995). The latter results emphasize the need to simultaneously target several viral epitopes, and also to employ several antiviral strategies, each relying on a different mechanism.

The effector function of the genetically engineered T-bodies should not rely exclusively on their cytotoxic activity. The chimeric-receptor-expressing cells can be induced to secrete endogenous cytokines at the target site or to serve as a platform to carry and release such transgenic cytokines at the desired site. Rosenberg et al. (1990) modified tumor-infiltrating lymphocytes (TILs) with the gene coding for TNF in an attempt to deliver high concentrations of this tumor-suppressive cytokine to the tumor site without dose-limiting systemic toxicity. A T-body approach has also been used to target lymphocytes to tumors for local toxin delivery (Chen et al. 1997), and a similar strategy could be used to deliver cytokines active against HIV infection. CTLs expressing chimeric receptors would migrate through the circulation to infected tissues and secrete ILs upon activation. In the choice of soluble factors, one possible candidate is interferon (IFN)-β, which has been shown to mediate highly potent antiviral effects in animal models (Vieillard et al. 1994, 1995).

Although new combination-drug regimens are increasingly effective (Palella et al. 1998), viral eradication has so far not been achieved (Chun et al. 1997). The tight regulation of HIV replication that is observed in successfully treated patients is nevertheless correlated with a drastic reduction of the frequency of anti-HIV CTLs (Ogg et al. 1998). The specific anti-HIV immune response should therefore be boosted, or passively transferred, in order to prevent relapse of the disease after interruption of the antiviral treatment. It was recently proposed (Cohen 1998) that latently infected cells could be compelled to express HIV antigens by injecting patients with stimulators like anti-CD3 antibodies (for T cells) or GM-CSF (for macrophages). Upon activation and proliferation, such cells would become unmasked and cleared from the system by effector-killer cells. Transduced T cells may already be used in a syngeneic setting, but their non-MHC-restricted specificity may enable their use in unrelated individuals, provided that the allogeneic response of the host is inhibited. Immunocompromised patients could meet such conditions.

# 8 Concluding Remarks

Adoptive cellular immunotherapy based on gene-modified cells is still in its infancy. The procedure remains cumbersome and potentially risky for the patient. In the case of HIV, the initial success of antiviral drug cocktails at controlling viral replication has not justified wide-scale use of ex vivo gene therapy. However, it now appears that neither classical drugs nor antibody-based therapy can achieve viral eradication. This has led to revived interest in cell-based immunotherapy, based on gene-modified cells that can be used as effectors, as well as targeting devices and vehicles for the specific delivery of soluble factors into tissues.

*Acknowledgements.* We are indebted to P.W.H.I. Parren and D.R. Burton for making our work on the anti-HIV T-bodies possible through the kind gift of b12 genes and viral isolates. We wish to thank Georges Bismuth for his critical comments on Fig. 1, and Shelley Schwartzbaum for her review of the manuscript. This work was supported by fellowships from the Ministère pour la Recherche et la Technologie (Paris, France), from the Fondation pour la Recherche Médicale (Natacha Bitton) and Université Paris VI Pitié-Salpêtrière (Z. Eshhar), by research grants from ANRS (Agence Nationale de Recherche contre le SIDA, Paris), from AFIRST (Association Franco-Israelienne pour la Recherche et la Technologie) (G. Gorochov and Z. Eshhar), and in part by the fifth framework program of the European Commission.

## References

Ada GL, Jones PD (1986) The immune response to influenza infection. Curr Top Microbiol Immunol 128:1–54

Aissaoui A, Klingel SI, Couderc J, Chateau D, Romagne F, Jambou F, Vincent A, Levasseur P, Eymard B, Maillot MC, Galanaud P, Berrih AS, Cohen KS (1999) Prevention of autoimmune attack by targeting specific T-cell receptors in a severe combined immunodeficiency mouse model of myasthenia gravis [see comments]. Ann Neurol 46:559–567

Alcami A, Symons JA, Collins PD, Williams TJ, Smith GL (1998) Blockade of chemokine activity by a soluble chemokine binding protein from vaccinia virus. J Immunol 160:624–633

Algeciras SA, Griffith TS, Lynch DH, Paya CV (1999) Cell cycle-dependent regulation of FLIP levels and susceptibility to Fas-mediated apoptosis. J Immunol 162:5205–5211

Alvarez VL, Agha MS, Hawkins RE, Russell SJ (1997) Pharmacological control of antigen responsiveness in genetically modified T lymphocytes. J Immunol 159:5889–5895

Alvarez VL, Hawkins RE (1996) Antigen-specific targeting of CD28-mediated T cell co-stimulation using chimeric single-chain antibody variable fragment-CD28 receptors. Eur J Immunol 26:2304–2309

Baba M, Imai T, Yoshida T, Yoshie O (1996) Constitutive expression of various chemokine genes in human T-cell lines infected with human T-cell leukemia virus type 1: role of the viral transactivator Tax. Int J Cancer 66:124–129

Bachmann MF, Zinkernagel RM (1997) Neutralizing antiviral B cell responses. Annu Rev Immunol 15:235–270

Baenziger J, Hengartner H, Zinkernage RM, Cole GA (1986) Induction or prevention of immunopathological disease by cloned cytotoxic T cell lines specific for lymphocytic choriomeningitis virus. Eur J Immunol 16:387–393

Baggiolini M (1998) Chemokines and leukocyte traffic. Nature 392:565–568

Baggiolini M, Dewald B, Moser B (1997) Human chemokines: an update. Annu Rev Immunol 15: 675–705

Bagley J, Dillon PJ, Rosen C, Robinson J, Sodroski J, Marasco WA (1994) Structural characterization of broadly neutralizing human monoclonal antibodies against the CD4 binding site of HIV-1 gp120. Mol Immunol 31:1149–1160

Beck S, Barrell BG (1988) Human cytomegalovirus encodes a glycoprotein homologous to MHC class I antigens. Nature 331:269–272

Beersma MF, Bijlmakers MJ, Ploegh HL (1993) Human cytomegalovirus down-regulates HLA class I expression by reducing the stability of class I H chains. J Immunol 151:4455–4464

Berg J, Lotscher E, Steimer KS, Capon DJ, Baenziger J, Jack HM, Wabl M (1991) Bispecific antibodies that mediate killing of cells infected with human immunodeficiency virus of any strain. Proc Natl Acad Sci USA 88:4723–4727

Bertoni G, Nguyen QV, Humphreys RE, Sairenji T (1989) Intracellular synthesis of Epstein-Barr virus membrane antigen gp350/220. Inhibitory effect of monensin on its expression. Intervirology 30:61–73

Bird RE, Hardman KD, Jacobson JW, Johnson S, Kaufman BM, Lee SM, Lee T, Pope SH, Riordan GS, Whitlow M (1988) Single-chain antigen-binding proteins [published erratum appears in Science 1989 Apr 28; 244(4903):409]. Science 242:423–426

Bitton N, Verrier F, Debre P, Gorochov G (1998) Characterization of T cell-expressed chimeric receptors with antibody-type specificity for the CD4 binding site of HIV-1 gp120. Eur J Immunol 28:4177–4187

Blaese RM, Culver KW, Chang L, Anderson WF, Mullen C, Nienhuis A, Carter C, Dunbar C, Leitman S, Berger M, et al. (1993) Treatment of severe combined immunodeficiency disease (SCID) due to adenosine deaminase deficiency with $CD34^+$ selected autologous peripheral blood cells transduced with a human ADA gene. Amendment to clinical research project, Project 90-C-195, January 10, 1992. Hum Gene Ther 4:521–527

Blank VC, Braakman E, Ronteltap CP, Tilly BC, Sturm E, Warnaar SO, Bolhuis RL (1993) Clustered CD3/TCR complexes do not transduce activation signals after bispecific monoclonal antibody-triggered lysis by cytotoxic T lymphocytes via CD3. J Immunol 151:2904–2914

Bodaghi B, Jones TR, Zipeto D, Vita C, Sun L, Laurent L, Arenzana SF, Virelizier JL, Michelson S (1998) Chemokine sequestration by viral chemoreceptors as a novel viral escape strategy: withdrawal of chemokines from the environment of cytomegalovirus-infected cells. J Exp Med 188:855–866

Boldicke T, Haase B, Bocher M, Lindenmaier W (1995) Human monoclonal antibodies to cytomegalovirus. Characterization and recombinant expression of a glycoprotein-B-specific antibody. Eur J Biochem 234:397–405

Bolhuis RL, Lamers CH, Goey SH, Eggermont AM, Trimbos JB, Stoter G, Lanzavecchia A di RE, Miotti S, Raspagliesi F, et al. (1992) Adoptive immunotherapy of ovarian carcinoma with bs-mAb-targeted lymphocytes: a multicenter study. Int J Cancer Suppl 7:78–81

Borrow P, Lewicki H, Hahn BH, Shaw GM, Oldstone, MB (1994) Virus-specific $CD8^+$ cytotoxic T-lymphocyte activity associated with control of viremia in primary human immunodeficiency virus type 1 infection. J Virol 68:6103–6110

Boshoff C, Endo Y, Collins PD, Takeuchi Y, Reeves JD, Schweickart VL, Siani MA, Sasaki T, Williams TJ, Gray PW, Moore PS, Chang Y, Weiss RA (1997) Angiogenic and HIV-inhibitory functions of KSHV-encoded chemokines [see comments]. Science 278:290–294

Brenner MK, Rill DR, Holladay MS, Heslop HE, Moen RC, Buschle M, Krance RA, Santana VM, Anderson WF, Ihle JN (1993) Gene marking to determine whether autologous marrow infusion restores long-term haematopoiesis in cancer patients. Lancet 342:1134–1137

Brodie SJ, Lewinsohn DA, Patterson BK, Jiyamapa D, Krieger J, Corey L, Greenberg PD, Riddell SR (1999) In vivo migration and function of transferred HIV-1-specific cytotoxic T cells. Nat Med 5: 34–41

Browne H, Smith G, Beck S, Minson T (1990) A complex between the MHC class I homologue encoded by human cytomegalovirus and beta 2 microglobulin. Nature 347:770–772

Buchacher A, Predl R, Strutzenberger K, Steinfellner W, Trkola A, Purtscher M, Gruber G, Tauer C, Steindl F, Jungbauer A, et al. (1994) Generation of human monoclonal antibodies against HIV-1 proteins; electrofusion and Epstein-Barr virus transformation for peripheral blood lymphocyte immortalization. Aids Res Hum Retroviruses 10:359–369

Burgert HG, Maryanski JL, Kvist S (1987) "E3/19K" protein of adenovirus type 2 inhibits lysis of cytolytic T lymphocytes by blocking cell-surface expression of histocompatibility class I antigens. Proc Natl Acad Sci USA 84:1356–1360

Burton DR, Pyati J, Koduri R, Sharp SJ, Thornton GB, Parren PW, Sawyer LS, Hendry RM, Dunlop N, Nara PL, et al. (1994) Efficient neutralization of primary isolates of HIV-1 by a recombinant human monoclonal antibody. Science 266:1024–1027

Butcher EC, Picker LJ (1996) Lymphocyte homing and homeostasis. Science 272:60–66

Byrne JA, Ahmed R, Oldstone MB (1984) Biology of cloned cytotoxic T lymphocytes specific for lymphocytic choriomeningitis virus. I. Generation and recognition of virus strains and H-2b mutants. J Immunol 133:433–439

Cannon MJ, Stott EJ, Taylor G, Askonas BA (1987) Clearance of persistent respiratory syncytial virus infections in immunodeficient mice following transfer of primed T cells. Immunology 62:133–138

Cavacini LA, Emes CL, Power J, Underdahl J, Goldstein R, Mayer K, Posner MR (1993) Loss of serum antibodies to a conformational epitope of HIV-1/gp120 identified by a human monoclonal antibody is associated with disease progression. J Acquir Immune Defic Syndr 6:1093–1102

Chamow SM, Zhang DZ, Tan XY, Mhatre SM, Marsters SA, Peers DH, Byrn RA, Ashkenazi A, Junghans RP (1994) A humanized, bispecific immunoadhesin-antibody that retargets $CD3^+$ effectors to kill HIV-1-infected cells. J Immunol 153:4268–4280

Chee MS, Satchwell SC, Preddie E, Weston KM, Barrell BG (1990) Human cytomegalovirus encodes three G protein-coupled receptor homologues [see comments]. Nature 344:774–777

Chen SY, Yang AG, Chen JD, Kute T, King CR, Collier J, Cong Y, Yao C, Huang XF (1997) Potent antitumour activity of a new class of tumour-specific killer cells. Nature 385:78–80

Chun TW, Stuyver L, Mizell SB, Ehler LA, Mican JA, Baseler M, Lloyd AL, Nowak MA, Fauci AS (1997) Presence of an inducible HIV-1 latent reservoir during highly active antiretroviral therapy. Proc Natl Acad Sci USA 94:13193–13197

Clerici M, Giorgi JV, Chou CC, Gudeman VK, Zack JA, Gupta P, Ho HN, Nishanian PG, Berzofsky JA, Shearer GM (1992) Cell-mediated immune response to human immunodeficiency virus (HIV) type 1 in seronegative homosexual men with recent sexual exposure to HIV-1 [see comments]. J Infect Dis 165:1012–1019

Cohen J (1998) Exploring how to get at–and eradicate–hidden HIV [news]. Science 279:1854–1855

Collins KL, Chen BK, Kalams SA, Walker BD, Baltimore D (1998) HIV-1 Nef protein protects infected primary cells against killing by cytotoxic T lymphocytes. Nature 391:397–401

Conley AJ, Gorny MK, Kessler JN, Boots LJ, Ossorio CM, Koenig S, Lineberger DW, Emini EA, Williams C, Zolla PS (1994) Neutralization of primary human immunodeficiency virus type 1 isolates by the broadly reactive anti-V3 monoclonal antibody, 447–52D. J Virol 68:6994–7000

Couillin I, Culmann PB, Gomard E, Choppin J, Levy JP, Guillet JG, Saragosti S (1994) Impaired cytotoxic T lymphocyte recognition due to genetic variations in the main immunogenic region of the human immunodeficiency virus 1 NEF protein [see comments]. J Exp Med 180:1129–1134

Dairaghi DJ, Fan RA, McMaster BE, Hanley MR, Schall TJ (1999) HHV8-encoded vMIP-I selectively engages chemokine receptor CCR8. Agonist and antagonist profiles of viral chemokines. J Biol Chem 274:21569–21574

Daniel MD, Kirchhoff F, Czajak SC, Sehgal PK, Desrosiers RC (1992) Protective effects of a live attenuated SIV vaccine with a deletion in the nef gene [see comments]. Science 258:1938–1941

Deisseroth AB, Zu Z, Claxton D, Hanania EG, Fu S, Ellerson D, Goldberg L, Thomas M, Janicek K, Anderson WF, et al. (1994) Genetic marking shows that $Ph^+$ cells present in autologous transplants of chronic myelogenous leukemia (CML) contribute to relapse after autologous bone marrow in CML. Blood 83:3068–3076

Del VM, Hengel H, Hacker H, Hartlaub U, Ruppert T, Lucin P, Koszinowski UH (1992) Cytomegalovirus prevents antigen presentation by blocking the transport of peptide-loaded major histocompatibility complex class I molecules into the medial-Golgi compartment. J Exp Med 176:729–738

Diu A, Romagne F, Genevee C, Rocher C, Bruneau JM, David A, Praz F, Hercend T (1993) Fine specificity of monoclonal antibodies directed at human T cell receptor variable regions: comparison with oligonucleotide-driven amplification for evaluation of V beta expression. Eur J Immunol 23:1422–1429

Dougherty JP, Temin HM (1987) A promoterless retroviral vector indicates that there are sequences in U3 required for 3′ RNA processing. Proc Natl Acad Sci USA 84:1197–1201

Earl PL, Koenig S, Moss B (1991a) Biological and immunological properties of human immunodeficiency virus type 1 envelope glycoprotein: analysis of proteins with truncations and deletions expressed by recombinant vaccinia viruses. J Virol 65:31–41

Earl PL, Moss B, Doms RW (1991b) Folding, interaction with GRP78-BiP, assembly, and transport of the human immunodeficiency virus type 1 envelope protein. J Virol 65:2047–2055

Eisenlohr LC, Yewdell JW, Bennink JR (1992) Flanking sequences influence the presentation of an endogenously synthesized peptide to cytotoxic T lymphocytes. J Exp Med 175:481–487

Embretson J, Zupancic M, Ribas JL, Burke A, Racz P, Tenner RK, Haase AT (1993) Massive covert infection of helper T lymphocytes and macrophages by HIV during the incubation period of AIDS [see comments]. Nature 362:359–362

Endres MJ, Garlisi CG, Xiao H, Shan L, Hedrick JA (1999) The Kaposi's sarcoma-related herpesvirus (KSHV)-encoded chemokine vMIP-I is a specific agonist for the CC chemokine receptor (CCR)8. J Exp Med 189:1993–1998

Eshhar Z, Waks T, Gross G, Schindler DG (1993) Specific activation and targeting of cytotoxic lymphocytes through chimeric single chains consisting of antibody-binding domains and the gamma or zeta subunits of the immunoglobulin and T-cell receptors. Proc Natl Acad Sci USA 90:720–724

Finney HM, Lawson AD, Bebbington CR, Weir AN (1998) Chimeric receptors providing both primary and costimulatory signaling in T cells from a single gene product. J Immunol 161:2791–2797

Fitzer AC, Schindler DG, Waks T, Eshhar Z (1998) Harnessing Syk family tyrosine kinases as signaling domains for chimeric single chain of the variable domain receptors: optimal design for T cell activation. J Immunol 160:145–154

French LE, Tschopp J (1999) Inhibition of death receptor signaling by FLICE-inhibitory protein as a mechanism for immune escape of tumors [comment]. J Exp Med 190:891–894

Fruh K, Ahn K, Peterson PA (1997) Inhibition of MHC class I antigen presentation by viral proteins. J Mol Med 75:18–27

Gallay P, Hope T, Chin D, Trono D (1997) HIV-1 infection of nondividing cells through the recognition of integrase by the importin/karyopherin pathway. Proc Natl Acad Sci USA 94:9825–9830

Gallay P, Stitt V, Mundy C, Oettinger M, Trono D (1996) Role of the karyopherin pathway in human immunodeficiency virus type 1 nuclear import. J Virol 70:1027–1032

Gallay P, Swingler S, Song J, Bushman F, Trono D (1995) HIV nuclear import is governed by the phosphotyrosine-mediated binding of matrix to the core domain of integrase. Cell 83:569–576

Gao JL, Murphy PM (1994) Human cytomegalovirus open reading frame US28 encodes a functional beta chemokine receptor. J Biol Chem 269:28539–28542

Gessain A, Saal F, Giron ML, Lasneret J, Lagaye S, Gout O, De TG, Sigaux F, Peries J (1990) Cell surface phenotype and human T lymphotropic virus type 1 antigen expression in 12 T cell lines derived from peripheral blood and cerebrospinal fluid of West Indian, Guyanese and African patients with tropical spastic paraparesis. J Gen Virol 71:333–341

Ghetie MA, Picker LJ, Richardson JA, Tucker K, Uhr JW, Vitetta ES (1994a) Anti-CD19 inhibits the growth of human B-cell tumor lines in vitro and of Daudi cells in SCID mice by inducing cell cycle arrest. Blood 83:1329–1336

Ghetie MA, Tucker K, Richardson J, Uhr JW, Vitetta ES (1994b) Eradication of minimal disease in severe combined immunodeficient mice with disseminated Daudi lymphoma using chemotherapy and an immunotoxin cocktail. Blood 84:702–707

Gorny MK, Gianakakos V, Sharpe S, Zolla PS (1989) Generation of human monoclonal antibodies to human immunodeficiency virus. Proc Natl Acad Sci USA 86:1624–1628

Gorochov G (1999) Dangerous T-cell amnesia [news; comment]. Nat Med 5:483–484

Gorochov G, Bachelez H, Cayuela JM, Legac E, Laroche L, Dubertret L, Sigaux F (1995) Expression of V beta gene segments by Sezary cells. J Invest Dermatol 105:56–61

Gorochov G, Lustgarten J, Waks T, Gross G, Eshhar Z (1992) Functional assembly of chimeric T-cell receptor chains. Int J Cancer Suppl 7:53–57

Gossen M, Bonin AL, Bujard H (1993) Control of gene activity in higher eukaryotic cells by prokaryotic regulatory elements. Trends Biochem Sci 18:471–475

Gossen M, Bujard H (1992) Tight control of gene expression in mammalian cells by tetracycline-responsive promoters. Proc Natl Acad Sci USA 89:5547–5551

Gratama JW, Zea AH, Bolhuis RL, Ochoa AC (1999) Restoration of expression of signal-transduction molecules in lymphocytes from patients with metastatic renal cell cancer after combination immunotherapy. Cancer Immunol Immunother 48:263–269

Gross G, Eshhar Z (1992) Endowing T cells with antibody specificity using chimeric T cell receptors. FASEB J 6:3370–3378

Gross G, Gorochov G, Waks T, Eshhar Z (1989a) Generation of effector T cells expressing chimeric T cell receptor with antibody type-specificity. Transplant Proc 21:127–130

Gross G, Waks T, Eshhar Z (1989b) Expression of immunoglobulin-T-cell receptor chimeric molecules as functional receptors with antibody-type specificity. Proc Natl Acad Sci USA 86:10024–10028

Guyer MS, Collins FS (1995) How is the Human Genome Project doing, and what have we learned so far? Proc Natl Acad Sci USA 92:10841–10848

Hanson DL, Horsburgh CJ, Fann SA, Havlik JA, Thompson SD (1993) Survival prognosis of HIV-infected patients. J Acquir Immune Defic Syndr 6:624–629

Harrer T, Harrer E, Kalams SA, Barbosa P, Trocha A, Johnson RP, Elbeik T, Feinberg MB, Buchbinder SP, Walker BD (1996) Cytotoxic T lymphocytes in asymptomatic long-term nonprogressing HIV-1 infection. Breadth and specificity of the response and relation to in vivo viral quasispecies in a person with prolonged infection and low viral load. J Immunol 156:2616–2623

Hege KM, Cooke KS, Finer MH, Zsebo KM, Roberts MR (1996) Systemic T cell-independent tumor immunity after transplantation of universal receptor-modified bone marrow into SCID mice. J Exp Med 184:2261–2269

Hege KM, Deeks SG, Mitsuyasu RT, Anton PA, Scadden DT, Connick E, Broad DF (1998) Co-infusion of autologous HIV-1 specific gene modified CD4 and CD8 T cells lead to improved T cell survival and tissue trafficking. 12th World AIDS Conference. Geneva. 2:42321

Hege KM, Roberts MR (1996) T-cell gene therapy. Curr Opin Biotechnol 7:629–634

Hege KM, Wagner B, Mitsuyasu R, Anton P, Scadden D, Kwok S, Lazar R, Pennathur-das R, Deeks S (1999) HIV-specific T cell gene therapy in subjects with undetectable viremia on HAART. 6th Conference on Retroviruses and Opportunistic Infections. Foundation for Retrovirolgy and Human Health, 31 Jan–4 Feb 1999, Chicago. 1:73

Hege KM, Wagner B, Mitsuyasu R, Anton P, Scadden D, Kwok S, Macken C, Deeks S (2000) HIV-specific T cell gene therapy suppresses viral local rebound in subjects on HAART. 7th Conference on Retroviruses and Opportunistic Infections. Foundation for Retrovirology and Human Health, 30 Jan–2 Feb 2000, San Francisco

Hengel H, Brune W, Koszinowski UH (1998) Immune evasion by cytomegalovirus–survival strategies of a highly adapted opportunist [see comments]. Trends Microbiol 6:190–197

Herndier BG, Kaplan LD, McGrath MS (1994) Pathogenesis of AIDS lymphomas. AIDS 8:1025–1049

Heslop HE, Ng CY, Li C, Smith CA, Loftin SK, Krance RA, Brenner MK, Rooney CM (1996) Long-term restoration of immunity against Epstein-Barr virus infection by adoptive transfer of gene-modified virus-specific T lymphocytes. Nat Med 2:551–555

Heslop HE, Rooney CM (1997) Adoptive cellular immunotherapy for EBV lymphoproliferative disease. Immunol Rev 157:217–222

Hinuma Y, Nagata K, Hanaoka M, Nakai M, Matsumoto T, Kinoshita KI, Shirakawa S, Miyoshi I (1981) Adult T-cell leukemia: antigen in an ATL cell line and detection of antibodies to the antigen in human sera. Proc Natl Acad Sci USA 78:6476–6480

Howcroft TK, Strebel K, Martin MA, Singer DS (1993) Repression of MHC class I gene promoter activity by two-exon Tat of HIV. Science 260:1320–1322

Husain SR, Kreitman RJ, Pastan I, Puri RK (1999) Interleukin-4 receptor-directed cytotoxin therapy of AIDS-associated Kaposi's sarcoma tumors in xenograft model [see comments]. Nat Med 5:817–822

Hwu P, Yang JC, Cowherd R, Treisman J, Shafer GE, Eshhar Z, Rosenberg SA (1995) In vivo antitumor activity of T cells redirected with chimeric antibody/T-cell receptor genes. Cancer Res 55:3369–3373

Ihara S, Maeda F, Takekoshi M (1998) [Neutralizing human antibody specific for human cytomegalovirus in E. coli expression system]. Nippon Rinsho 56:161–166

Irmler M, Thome M, Hahne M, Schneider P, Hofmann K, Steiner V, Bodmer JL, Schroter M, Burns K, Mattmann C, Rimoldi D, French LE, Tschopp J (1997) Inhibition of death receptor signals by cellular FLIP [see comments]. Nature 388:190–195

Irving BA, Weiss A (1991) The cytoplasmic domain of the T cell receptor zeta chain is sufficient to couple to receptor-associated signal transduction pathways. Cell 64:891–901

Isegawa Y, Ping Z, Nakano K, Sugimoto N, Yamanishi K (1998) Human herpesvirus 6 open reading frame U12 encodes a functional beta-chemokine receptor. J Virol 72:6104–6112

Jain RK (1989) Delivery of novel therapeutic agents in tumors: physiological barriers and strategies. J Natl Cancer Inst 81:570–576

Jerome KR, Tait JF, Koelle DM, Corey L (1998) Herpes simplex virus type 1 renders infected cells resistant to cytotoxic T-lymphocyte-induced apoptosis. J Virol 72:436–441

Kalams SA, Walker BD (1998) The critical need for CD4 help in maintaining effective cytotoxic T lymphocyte responses [comment]. J Exp Med 188:2199–2204

Kaplan LD, Straus DJ, Testa MA, Von RJ, Dezube BJ, Cooley TP, Herndier B, Northfelt DW, Huang J, Tulpule A, Levine AM (1997) Low-dose compared with standard-dose m-BACOD chemotherapy for non-Hodgkin's lymphoma associated with human immunodeficiency virus infection. National Institute of Allergy and Infectious Diseases AIDS Clinical Trials Group [see comments]. N Engl J Med 336:1641–1648

Khanna R, Sherritt M, Burrows SR (1999) EBV structural antigens, gp350 and gp85, as targets for ex vivo virus-specific CTL during acute infectious mononucleosis: potential use of gp350/gp85 CTL epitopes for vaccine design. J Immunol 162:3063–3069

Kledal TN, Rosenkilde MM, Coulin F, Simmons G, Johnsen AH, Alouani S, Power CA, Luttichau HR, Gerstoft J, Clapham PR, Clark LI, Wells T, Schwartz TW (1997) A broad-spectrum chemokine antagonist encoded by Kaposi's sarcoma-associated herpesvirus. Science 277:1656–1659

Klein MR, Miedema F (1995) Long-term survivors of HIV-1 infection. Trends Microbiol 3:386–391

Koenig S, Conley AJ, Brewah YA, Jones GM, Leath S, Boots LJ, Davey V, Pantaleo G, Demarest JF, Carter C, et al. (1995) Transfer of HIV-1-specific cytotoxic T lymphocytes to an AIDS patient leads to selection for mutant HIV variants and subsequent disease progression [see comments]. Nat Med 1:330–336

Kohn DB, Weinberg KI, Nolta JA, Heiss LN, Lenarsky C, Crooks GM, Hanley ME, Annett G, Brooks JS, El KA, et al. (1995) Engraftment of gene-modified umbilical cord blood cells in neonates with adenosine deaminase deficiency. Nat Med 1:1017–1023

Koup RA (1994) Virus escape from CTL recognition [comment]. J Exp Med 180:779–782

Kuehnle I, Huls MH, Liu Z, Semmelmann M, Krance RA, Brenner MK, Rooney CM, Heslop HE (2000) CD20 monoclonal antibody (rituximab) for therapy of Epstein-Barr virus lymphoma after hematopoietic stem-cell transplantationBlood 95:1502–1505

Kuwana Y, Asakura Y, Utsunomiya N, Nakanishi M, Arata Y, Itoh S, Nagase F, Kurosawa Y (1987) Expression of chimeric receptor composed of immunoglobulin-derived V regions and T-cell receptor-derived C regions. Biochem Biophys Res Commun 149:960–968

LaCasse RA, Follis KE, Trahey M, Scarborough JD, Littman DR, Nunberg JH (1999) Fusion-competent vaccines: broad neutralization of primary isolates of HIV [see comments]. Science 283: 357–362

Lalani AS, Graham K, Mossman K, Rajarathnam K, Clark LI, Kelvin D, McFadden G (1997) The purified myxoma virus gamma interferon receptor homolog M-T7 interacts with the heparin-binding domains of chemokines. J Virol 71:4356–4363

Lalani AS, Masters J, Zeng W, Barrett J, Pannu R, Everett H, Arendt CW, McFadden G (1999) Use of chemokine receptors by poxviruses. Science 286:1968–1971

Langlade DP, Ngo, Giang, Huong N, Ferchal F, Oksenhendler E (1994) Human immunodeficiency virus (HIV) nef-specific cytotoxic T lymphocytes in noninfected heterosexual contact of HIV-infected patients [see comments]. J Clin Invest 93:1293–1297

Lanzavecchia A (1997) Understanding the mechanisms of sustained signaling and T cell activation [comment]. J Exp Med 185:1717–1719

Lanzavecchia A, Scheidegger D (1987) The use of hybrid hybridomas to target human cytotoxic T lymphocytes. Eur J Immunol 17:105–111

Le Doussal JM, Piqueras B, Dogan I, Debré P, Gorochov G (2000) Phage display of peptide-major histocompatibility complex. J Immunol Methods 241:-158

Leist TP, Ruedi E, Zinkernagel RM (1988) Virus-triggered immune suppression in mice caused by virus-specific cytotoxic T cells. J Exp Med 167:1749–1754

Letourneur F, Klausner RD (1991) T-cell and basophil activation through the cytoplasmic tail of T-cell-receptor zeta family proteins. Proc Natl Acad Sci USA 88:8905–8909

Li L, Coelingh KL, Britt WJ (1995) Human cytomegalovirus neutralizing antibody-resistant phenotype is associated with reduced expression of glycoprotein H. J Virol 69:6047–6053

Li W, Crombach M, Schonian U, Yang YZ, Maisch B, Chen HZ (1994) Detection of cytomegalovirus genome by in situ hybridization in paraffin embedded endomyocardial biopsy specimens of viral myocarditis. Chin Med J (Engl) 107:888–891

Lieberman J, Skolnik PR, Parkerson GR, Fabry JA, Landry B, Bethel J, Kagan J (1997) Safety of autologous, ex vivo-expanded human immunodeficiency virus (HIV)-specific cytotoxic T-lymphocyte infusion in HIV-infected patients. Blood 90:2196–2206

Lill NL, Tevethia MJ, Hendrickson WG, Tevethia SS (1992) Cytotoxic T lymphocytes (CTL) against a transforming gene product select for transformed cells with point mutations within sequences encoding CTL recognition epitopes. J Exp Med 176:449–457

Maloney DG, Liles TM, Czerwinski DK, Waldichuk C, Rosenberg J, Grillo LA, Levy R (1994) Phase I clinical trial using escalating single-dose infusion of chimeric anti-CD20 monoclonal antibody (IDEC-C2B8) in patients with recurrent B-cell lymphoma. Blood 84:2457–2466

Maloney DG, Grillo LA, Bodkin DJ, White CA, Liles TM, Royston I, Varns C, Rosenberg J, Levy R (1997a) IDEC-C2B8: results of a phase I multiple-dose trial in patients with relapsed non-Hodgkin's lymphoma [see comments]. J Clin Oncol 15:3266–3274

Maloney DG, Grillo LA, White CA, Bodkin D, Schilder RJ, Neidhart JA, Janakiraman N, Foon KA, Liles TM, Dallaire BK, Wey K, Royston I, Davis T, Levy R (1997b) IDEC-C2B8 (Rituximab) anti-

CD20 monoclonal antibody therapy in patients with relapsed low-grade non-Hodgkin's lymphoma. Blood 90:2188–2195

Margulies DH (1997) Interactions of TCRs with MHC-peptide complexes: a quantitative basis for mechanistic models. Curr Opin Immunol 9:390–395

Martinez PL, Thompson JP, Moyer RW (1995) Mapping and investigation of the role in pathogenesis of the major unique secreted 35-kDa protein of rabbitpox virus. Virology 206:591–600

Mascola JR, Lewis MG, Stiegler G, Harris D, VanCott TC, Hayes D, Louder MK, Brown CR, Sapan CV, Frankel SS, Lu Y, Robb ML, Katinger H, Birx DL (1999) Protection of Macaques against pathogenic simian/human immunodeficiency virus 89.6PD by passive transfer of neutralizing antibodies. J Virol 73:4009–4018

Matloubian M, Concepcion RJ, Ahmed R (1994) $CD4^+$ T cells are required to sustain $CD8^+$ cytotoxic T-cell responses during chronic viral infection. J Virol 68:8056–8063

Meij P, Vervoort MB, Aarbiou J, van DP, Brink A, Bloemena E, Meijer CJ, Middeldorp JM (1999) Restricted low-level human antibody responses against Epstein-Barr virus (EBV)-ençoded latent membrane protein 1 in a subgroup of patients with EBV-associated diseases. J Infect Dis 179:1108–1115

Merrouche Y, Negrier S, Bain C, Combaret V, Mercatello A, Coronel B, Moskovtchenko JF, Tolstoshev P, Moen R, Philip T, et al. (1995) Clinical application of retroviral gene transfer in oncology: results of a French study with tumor-infiltrating lymphocytes transduced with the gene of resistance to neomycin. J Clin Oncol 13:410–418

Mezzanzanica D, Canevari S, Colnaghi MI (1991) Retargeting of human lymphocytes against human ovarian carcinoma cells by bispecific antibodies: from laboratory to clinic. Int J Clin Lab Res 21: 159–164

Miller AD (1997) Putting muscle to work for gene therapy [news; comment]. Nat Med 3:278–279

Mitchell MS (1989) Active specific immunotherapy of melanoma: clinical results and correlates. Biotechnol Ther 1:163–179

Miyoshi H, Blomer U, Takahashi M, Gage FH, Verma IM (1998) Development of a self-inactivating lentivirus vector. J Virol 72:8150–8157

Miyoshi H, Smith KA, Mosier DE, Verma IM, Torbett BE (1999) Transduction of human CD34(+) cells that mediate long-term engraftment of NOD/SCID mice by HIV vectors. Science 283: 682–686

Moore JP, Jarrett RF (1988) Sensitive ELISA for the gp120 and gp160 surface glycoproteins of HIV-1. AIDS Res Hum Retroviruses 4:369–379

Moore PS, Boshoff C, Weiss RA, Chang Y (1996) Molecular mimicry of human cytokine and cytokine response pathway genes by KSHV. Science 274:1739–1744

Moritz D, Groner B (1995) A spacer region between the single chain antibody-and the CD3 zeta-chain domain of chimeric T cell receptor components is required for efficient ligand binding and signaling activity. Gene Ther 2:539–546

Musey L, Hughes J, Schacker T, Shea T, Corey L, McElrath MJ (1997) Cytotoxic-T-cell responses, viral load, and disease progression in early human immunodeficiency virus type 1 infection [see comments]. N Engl J Med 337:1267–1274

Naldini L, Blomer U, Gallay P, Ory D, Mulligan R, Gage FH, Verma IM, Trono D (1996) In vivo gene delivery and stable transduction of nondividing cells by a lentiviral vector [see comments]. Science 272:263–267

Necker A, Michalaki H, Raynal MC, Rocher C, Diu A, Romagne F (1995) Monoclonal antibodies to human T-cell receptor variable beta regions. Ann NY Acad Sci 756:106–109

Nelson BH, Lord JD, Greenberg PD (1994) Cytoplasmic domains of the interleukin-2 receptor beta and gamma chains mediate the signal for T-cell proliferation. Nature 369:333–336

Neote K, DiGregorio D, Mak JY, Horuk R, Schall TJ (1993) Molecular cloning, functional expression, and signaling characteristics of a C-C chemokine receptor. Cell 72:415–425

Nicholas J, Ruvolo V, Zong J, Ciufo D, Guo HG, Reitz MS, Hayward GS (1997) A single 13-kilobase divergent locus in the Kaposi sarcoma-associated herpesvirus (human herpesvirus 8) genome contains nine open reading frames that are homologous to or related to cellular proteins. J Virol 71:1963–1974

Odermatt B, Eppler M, Leist TP, Hengartner H, Zinkernagel RM (1991) Virus-triggered acquired immunodeficiency by cytotoxic T-cell-dependent destruction of antigen-presenting cells and lymph follicle structure. Proc Natl Acad Sci USA 88:8252–8256

Ogg GS, Jin X, Bonhoeffer S, Dunbar PR, Nowak MA, Monard S, Segal JP, Cao Y, Rowland JS, Cerundolo V, Hurley A, Markowitz M Ho DD, Nixon DF, McMichael AJ (1998) Quantitation of HIV-1-specific cytotoxic T lymphocytes and plasma load of viral RNA. Science 279:2103–2106

Oh SK, Cruikshank WW, Raina J, Blanchard GC, Adler WH, Walker J, Kornfeld H (1992) Identification of HIV-1 envelope glycoprotein in the serum of AIDS and ARC patients. J Acquir Immune Defic Syndr 5:251–256

Ohlin M, Owman H, Mach M, Borrebaeck CA (1996) Light chain shuffling of a high affinity antibody results in a drift in epitope recognition. Mol Immunol 33:47–56

Okada H, Momota H, Azuma T, Hattori T, Okada N (1992) Specific cytolysis of HIV-infected cells by lymphocytes armed with bifunctional antibodies. Immunol Lett 31:247–252

Orloff DG Ra CS, Frank SJ, Klausner RD, Kinet JP (1990) Family of disulphide-linked dimers containing the zeta and eta chains of the T-cell receptor and the gamma chain of Fc receptors. Nature 347:189–191

Osame M, Izumo S, Igata A, Matsumoto M, Matsumoto T, Sonoda S, Tara M, Shibata Y (1986) Blood transfusion and HTLV-I associated myelopathy [letter]. Lancet 2:104–105

Otto E, Jones TA, Vanin EF, Stambaugh K, Mueller SN, Anderson WF, McGarrity GJ (1994) Characterization of a replication-competent retrovirus resulting from recombination of packaging and vector sequences. Hum Gene Ther 5:567–575

Palella FJ, Delaney KM, Moorman AC, Loveless MO, Fuhrer J, Satten GA, Aschman DJ, Holmberg SD (1998) Declining morbidity and mortality among patients with advanced human immunodeficiency virus infection. N Engl J Med 338:853–860

Parren PW, Mondor I, Naniche D, Ditzel HJ, Klasse PJ, Burton DR, Sattentau QJ (1998) Neutralization of human immunodeficiency virus type 1 by antibody to gp120 is determined primarily by occupancy of sites on the virion irrespective of epitope specificity. J Virol 72:3512–3519

Patel SD, Moskalenko M, Smith D, Maske B, Finer MH, McArthur JG (1999) Impact of chimeric immune receptor extracellular protein domains on T cell function. Gene Ther 6:412–419

Penfold ME, Dairaghi DJ, Duke GM, Saederup N, Mocarski ES, Kemble GW, Schall TJ (1999) Cytomegalovirus encodes a potent alpha chemokine. Proc Natl Acad Sci USA 96:9839–9844

Picker LJ, Michie SA, Rott LS, Butcher EC (1990) A unique phenotype of skin-associated lymphocytes in humans. Preferential expression of the HECA-452 epitope by benign and malignant T cells at cutaneous sites. Am J Pathol 136:1053–1068

Pitcher CJ, Quittner C, Peterson DM, Connors M, Koup RA, Maino VC, Picker LJ (1999) HIV-1-specific $CD4^+$ T cells are detectable in most individuals with active HIV-1 infection, but decline with prolonged viral suppression [see comments]. Nat Med 5:518–525

Ploegh HL (1998) Viral strategies of immune evasion. Science 280:248–253

Poeschla E, Corbeau P, Wong SF (1996) Development of HIV vectors for anti-HIV gene therapy. Proc Natl Acad Sci USA 93:11395–11399

Poiesz BJ, Ruscetti FW, Gazdar AF, Bunn PA, Minna JD, Gallo RC (1980) Detection and isolation of type C retrovirus particles from fresh and cultured lymphocytes of a patient with cutaneous T-cell lymphoma. Proc Natl Acad Sci USA 77:7415–7419

Poignard P, Sabbe R, Picchio GR, Wang M, Gulizia RJ, Katinger H, Parren PW, Mosier DE, Burton DR (1999) Neutralizing antibodies have limited effects on the control of established HIV-1 infection in vivo. Immunity 10:431–438

Pomerantz RJ, Trono D, Feinberg MB, Baltimore D (1990) Cells nonproductively infected with HIV-1 exhibit an aberrant pattern of viral RNA expression: a molecular model for latency. Cell 61:1271–1276

Porgador A, Yewdell JW, Deng Y, Bennink JR, Germain RN (1997) Localization, quantitation, and in situ detection of specific peptide-MHC class I complexes using a monoclonal antibody. Immunity 6:715–726

Posavad CM, Koelle DM, Shaughnessy MF, Corey L (1997) Severe genital herpes infections in HIV-infected individuals with impaired herpes simplex virus-specific $CD8^+$ cytotoxic T lymphocyte responses. Proc Natl Acad Sci USA 94:10289–10294

Press OW, Appelbaum F, Ledbetter JA, Martin PJ, Zarling J, Kidd P, Thomas ED (1987) Monoclonal antibody 1F5 (anti-CD20) serotherapy of human B cell lymphomas. Blood 69:584–591

Price DA, Meier UC, Klenerman P, Purbhoo MA, Phillips RE, Sewell AK (1998) The influence of antigenic variation on cytotoxic T lymphocyte responses in HIV-1 infection. J Mol Med 76:699–708

Reiser J, Harmison G, Kluepfel SS, Brady RO, Karlsson S, Schubert M (1996) Transduction of nondividing cells using pseudotyped defective high-titer HIV type 1 particles. Proc Natl Acad Sci USA 93:15266–15271

Renner C, Jung W, Sahin U, Denfeld R, Pohl C, Trumper L, Hartmann F, Diehl V, van LR, Pfreundschuh M (1994) Cure of xenografted human tumors by bispecific monoclonal antibodies and human T cells. Science 264:833–835

Reusser P, Riddell SR, Meyers JD, Greenberg PD (1991) Cytotoxic T-lymphocyte response to cytomegalovirus after human allogeneic bone marrow transplantation: pattern of recovery and correlation with cytomegalovirus infection and disease. Blood 78:1373–1380

Riddell SR (1995) Pathogenesis of cytomegalovirus pneumonia in immunocompromised hosts. Semin Respir Infect 10:199–208

Riddell SR, Elliott M, Lewinsohn DA, Gilbert MJ, Wilson L, Manley SA, Lupton SD, Overell RW, Reynolds TC, Corey L, Greenberg PD (1996) T-cell mediated rejection of gene-modified HIV-specific cytotoxic T lymphocytes in HIV-infected patients [see comments]. Nat Med 2:216–223

Riddell SR, Gilbert MJ, Greenberg PD (1993) $CD8^+$ cytotoxic T cell therapy of cytomegalovirus and HIV infection. Curr Opin Immunol 5:484–491

Riddell SR, Watanabe KS, Goodrich JM Li CR, Agha ME, Greenberg PD (1992) Restoration of viral immunity in immunodeficient humans by the adoptive transfer of T cell clones [see comments]. Science 257:238–241

Roberts MR, Cooke KS, Tran AC, Smith KA, Lin WY, Wang M, Dull TJ, Farson D, Zsebo KM, Finer MH (1998) Antigen-specific cytolysis by neutrophils and NK cells expressing chimeric immune receptors bearing zeta or gamma signaling domains. J Immunol 161:375–384

Roberts MR, Qin L, Zhang D, Smith DH, Tran AC, Dull TJ, Groopman JE, Capon DJ, Byrn RA, Finer MH (1994) Targeting of human immunodeficiency virus-infected cells by $CD8^+$ T lymphocytes armed with universal T-cell receptors. Blood 84:2878–2889

Romagne F, Besnardeau L, Malissen B (1992) A versatile method to produce antibodies to human T cell receptor V beta segments: frequency determination of human V beta $2^+$ T cells that react with toxic-shock syndrome toxin-1. Eur J Immunol 22:2749–2752

Romagne F, Kanagawa O, David AJ, Peyrat MA, Bonneville M, Necker A (1995) TCRBV23 specificity of two monoclonal antibodies revealed by a panel of human V beta chains expressed in mouse cells. J Immunol Methods 186:313–322

Romeo C, Amiot M, Seed B (1992) Sequence requirements for induction of cytolysis by the T cell antigen/Fc receptor zeta chain. Cell 68:889–897

Romeo C, Seed B (1991) Cellular immunity to HIV activated by CD4 fused to T cell or Fc receptor polypeptides. Cell 64:1037–1046

Rooney CM, Smith CA Ng CY, Loftin S, Li C, Krance RA, Brenner MK, Heslop HE (1995) Use of gene-modified virus-specific T lymphocytes to control Epstein-Barr-virus-related lymphoproliferation. Lancet 345:9–13

Rosenberg ES, Billingsley JM, Caliendo AM, Boswell SL, Sax PE, Kalams SA, Walker BD (1997) Vigorous HIV-1-specific $CD4^+$ T cell responses associated with control of viremia [see comments]. Science 278:1447–1450

Rosenberg SA, Aebersold P, Cornetta K, Kasid A, Morgan RA, Moen R, Karson EM, Lotze MT, Yang JC, Topalian SL, et al. (1990) Gene transfer into humans–immunotherapy of patients with advanced melanoma, using tumor-infiltrating lymphocytes modified by retroviral gene transduction [see comments]. N Engl J Med 323:570–578

Roth SJ, Diacovo TG, Brenner MB, Rosat JP, Buccola J, Morita CT, Springer TA (1998) Transendothelial chemotaxis of human alpha/beta and gamma/delta T lymphocytes to chemokines. Eur J Immunol 28:104–113

Sallusto F, Palermo B, Lenig D, Miettinen M, Matikainen S, Julkunen I, Forster R, Burgstahler R, Lipp M, Lanzavecchia A (1999) Distinct patterns and kinetics of chemokine production regulate dendritic cell function. Eur J Immunol 29:1617–1625

Schmitz JE, Kuroda MJ, Santra S, Sasseville VG, Simon MA, Lifton MA, Racz P, Tenner RK, Dalesandro M, Scallon BJ, Ghrayeb J, Forman MA, Montefiori DC, Rieber EP, Letvin NL, Reimann KA (1999) Control of viremia in simian immunodeficiency virus infection by $CD8^+$ lymphocytes. Science 283:857–860

Schoppel K, Hassfurther E, Britt W, Ohlin M, Borrebaeck CA, Mach M (1996) Antibodies specific for the antigenic domain 1 of glycoprotein B (gpUL55) of human cytomegalovirus bind to different substructures. Virology 216:133–145

Schwartz O, Marechal V, Le GS, Lemonnier F, Heard JM (1996) Endocytosis of major histocompatibility complex class I molecules is induced by the HIV-1 Nef protein. Nat Med 2:338–342

Segal DM, Qian JH, Garrido MA, Perez P, Winkler DF, Wunderlich JR, Snider DP, Valdayo MJ, Titus JA (1988) Targeting of cytotoxic cells against tumors with heterocrosslinked, bispecific antibodies. Princess Takamatsu Symp 19:323–331

Speckner A, Glykofrydes D, Ohlin M, Mach M (1999) Antigenic domain 1 of human cytomegalovirus glycoprotein B induces a multitude of different antibodies which, when combined, results in incomplete virus neutralization. J Gen Virol:2183–2191

Spriggs MK, Armitage RJ, Comeau MR, Strockbine L, Farrah T, Macduff B, Ulrich D, Alderson MR, Mullberg J, Cohen JI (1996) The extracellular domain of the Epstein-Barr virus BZLF2 protein binds the HLA-DR beta chain and inhibits antigen presentation. J Virol 70:5557–5563

Takekoshi M, Maeda F, Tachibana H, Inoko H, Kato S, Takakura I, Kenjyo T, Hiraga S, Ogawa Y, Horiki T, Ihara S (1998) Human monoclonal anti-HCMV neutralizing antibody from phage display libraries. J Virol Methods 74:89–98

Tan R, Xu X, Ogg GS, Hansasuta P, Dong T, Rostron T, Luzzi G, Conlon CP, Screaton GR, McMichael AJ, Rowland JS (1999) Rapid death of adoptively transferred T cells in acquired immunodeficiency syndrome. Blood 93:1506–1510

Tanchot C, Rocha B (1998) The organization of mature T-cell pools. Immunol Today 19:575–579

Tewari M, Dixit VM (1995) Fas-and tumor necrosis factor-induced apoptosis is inhibited by the poxvirus crmA gene product. J Biol Chem 270:3255–3260

Tewari M, Telford WG, Miller RA, Dixit VM (1995) CrmA, a poxvirus-encoded serpin, inhibits cytotoxic T-lymphocyte-mediated apoptosis. J Biol Chem 270:22705–22708

Tomasec P, Braud VM, Rickards C, Powell MB, McSharry BP, Gadola S, Cerundolo V, Borysiewicz LK, McMichael AJ, Wilkinson GW. Surface expression of HLA-E, an inhibitor of natural killer cells, enhanced by human cytomegalovirus gpUL40. Science 287:1031

Tran AC, Zhang D, Byrn R, Roberts MR (1995) Chimeric zeta-receptors direct human natural killer (NK) effector function to permit killing of NK-resistant tumor cells and HIV-infected T lymphocytes. J Immunol 155:1000–1009

Traunecker A, Lanzavecchia A, Karjalainen K (1991) Bispecific single chain molecules (Janusins) target cytotoxic lymphocytes on HIV infected cells. EMBO J 10:3655–3659

Unutmaz D, KewalRamani VN, Marmon S, Littman DR (1999) Cytokine signals are sufficient for HIV-1 infection of resting human T lymphocytes. J Exp Med 189:1735–1746

Van DJ, Tsuruo T, Segal DM, Bolhuis RL, Colognola R, van de, Griend RJ, Fleuren GJ, Warnaar SO (1989) Bispecific antibodies reactive with the multidrug-resistance-related glycoprotein and CD3 induce lysis of multidrug-resistant tumor cells. Int J Cancer 44:738–743

VanCott TC, Bethke FR, Polonis VR, Gorny MK, Zolla PS, Redfield RR, Birx DL (1994) Dissociation rate of antibody-gp120 binding interactions is predictive of V3-mediated neutralization of HIV-1. J Immunol 153:449–459

Vieillard V, Lauret E, Maguer V, Jacomet C, Rozenbaum W, Gazzolo L, De ME (1995) Autocrine interferon-beta synthesis for gene therapy of HIV infection: increased resistance to HIV-1 in lymphocytes from healthy and HIV-infected individuals. AIDS 9:1221–1228

Vieillard V, Lauret E, Rousseau V, De ME (1994) Blocking of retroviral infection at a step prior to reverse transcription in cells transformed to constitutively express interferon beta. Proc Natl Acad Sci USA 91:2689–2693

Vieira J, Schall TJ, Corey L, Geballe AP (1998) Functional analysis of the human cytomegalovirus US28 gene by insertion mutagenesis with the green fluorescent protein gene. J Virol 72:8158–8165

Waldmann TA, Pastan IH, Gansow OA, Junghans RP (1992) The multichain interleukin-2 receptor: a target for immunotherapy. Ann Intern Med 116:148–160

Walker CM, Moody DJ, Stites DP, Levy JA (1986) $CD8^+$ lymphocytes can control HIV infection in vitro by suppressing virus replication. Science 234:1563–1566

Walker R, Blaese RM, Carter CS, Chang L, Klein H, Lane HC, Leitman SF, Mullen CA, Larson M (1993) A study of the safety and survival of the adoptive transfer of genetically marked syngeneic lymphocytes in HIV-infected identical twins. Hum Gene Ther 4:659–680

Walker RE (1996) A phase I/II pilot study of the safety of the adoptive transfer of syngeneic gene-modified cytotoxic T lymphocytes in HIV-infected identical twins. Hum Gene Ther 7:367–400

Walker RE, Bechtel CM, Natarajan V, Blaese RM, Hege KM, Roberts MR, Lane HC (1998) In vivo persistence of genetically modified, HIV-1 specific syngeneic lymphocytes in HIV-1 discordant identical twins. 12th World AIDS Conference. June 28–July 3, Marathon Multimedia, Geneva. 2:42322

Walter EA, Greenberg PD, Gilbert MJ, Finch RJ, Watanabe KS, Thomas ED, Riddell SR (1995) Reconstitution of cellular immunity against cytomegalovirus in recipients of allogeneic bone marrow by transfer of T-cell clones from the donor [see comments]. N Engl J Med 333:1038–1044

Weber J, Fenyo EM, Beddows S, Kaleebu P, Bjorndal A (1996) Neutralization serotypes of human immunodeficiency virus type 1 field isolates are not predicted by genetic subtype. The WHO network for HIV isolation and characterization. J Virol 70:7827–7832

Weijtens ME, Willemsen RA, Valerio D, Stam K, Bolhuis RL (1996) Single chain Ig/gamma gene-redirected human T lymphocytes produce cytokines, specifically lyse tumor cells, recycle lytic capacity. J Immunol 157:836–843

Weijtens ME, Willemsen RA, van KB, Bolhuis RL (1998) Chimeric ScFv/gamma receptor-mediated T-cell lysis of tumor cells is coregulated by adhesion and accessory molecules. Int J Cancer 77:181–187

Weissman JD, Brown JA, Howcroft TK, Hwang J, Chawla A, Roche PA, Schiltz L, Nakatani Y, Singer DS (1998) HIV-1 tat binds TAFII250 and represses TAFII250-dependent transcription of major histocompatibility class I genes. Proc Natl Acad Sci USA 95:11601–11606

Whitbeck JC, Peng C, Lou H, Xu R, Willis SH, Ponce de, Leon M, Peng T, Nicola AV, Montgomery RI, Warner MS, Soulika AM, Spruce LA, Moore WT, Lambris JD, Spear PG, Cohen GH, Eisenberg RJ (1997) Glycoprotein D of herpes simplex virus (HSV) binds directly to HVEM, a member of the tumor necrosis factor receptor superfamily and a mediator of HSV entry. J Virol 71:6083–6093

White SM, Renda M, Nam NY, Klimatcheva E, Zhu Y, Fisk J, Halterman M, Rimel BJ, Federoff H, Pandya S, Rosenblatt JD, Planelles V (1999) Lentivirus vectors using human and simian immunodeficiency virus elements. J Virol 73:2832–2840

Willey RL, Bonifacino JS, Potts BJ, Martin MA, Klausner RD (1988) Biosynthesis, cleavage, degradation of the human immunodeficiency virus 1 envelope glycoprotein gp160. Proc Natl Acad Sci USA 85:9580–9584

Woffendin C, Ranga U, Yang Z, Xu L, Nabel GJ (1996) Expression of a protective gene-prolongs survival of T cells in human immunodeficiency virus-infected patients. Proc Natl Acad Sci USA 93:2889–2894

Xu JY, Gorny MK, Palker T, Karwowska S, Zolla PS (1991) Epitope mapping of two immunodominant domains of gp41, the transmembrane protein of human immunodeficiency virus type 1, using ten human monoclonal antibodies. J Virol 65:4832–4838

Xu XN, Laffert B, Screaton GR, Kraft M, Wolf D, Kolanus W, Mongkolsapay J, McMichael AJ, Baur AS (1999) Induction of Fas ligand expression by HIV involves the interaction of Nef with the T cell receptor zeta chain. J Exp Med 189:1489–1496

Yang OO, Tran AC, Kalams SA, Johnson RP, Roberts MR, Walker BD (1997) Lysis of HIV-1-infected cells and inhibition of viral replication by universal receptor T cells. Proc Natl Acad Sci USA 94:11478–11483

Yarranton GT (1992) Inducible vectors for expression in mammalian cells. Curr Opin Biotechnol 3: 506–511

Yoshida M, Miyoshi I, Hinuma Y (1982) A retrovirus from human leukemia cell lines: its isolation, characterization, and implication in human adult T-cell leukemia (ATL) Princess Takamatsu Symp 12:285–294

Yu SF, von RT, Kantoff PW, Garber C, Seiberg M, Ruther U, Anderson WF, Wagner EF, Gilboa E (1986) Self-inactivating retroviral vectors designed for transfer of whole genes into mammalian cells. Proc Natl Acad Sci USA 83:3194–3198

Zarling JM, Moran PA, Grosmaire LS, McClure J, Shriver K, Ledbetter JA (1988) Lysis of cells infected with HIV-1 by human lymphocytes targeted with monoclonal antibody heteroconjugates. J Immunol 140:2609–2613

Zauli G, Gibellini D, Secchiero P, Dutartre H, Olive D, Capitani S, Collette Y (1999) Human immunodeficiency virus type 1 Nef protein sensitizes CD4(+) T lymphoid cells to apoptosis via functional upregulation of the CD95/CD95 ligand pathway. Blood 93:1000–1010

Zinkernagel RM, Hengartner H (1994) T-cell-mediated immunopathology versus direct cytolysis by virus: implications for HIV and AIDS. Immunol Today 15:262–268

# Subject Index

# Current Topics in Microbiology and Immunology

## Volumes published since 1989 (and still available)

Vol. 217: **Jessberger, Rolf; Lieber, Michael R. (Eds.):** Molecular Analysis of DNA Rearrangements in the Immune System. 1996. 43 figs. IX, 224 pp. ISBN 3-540-61037-5

Vol. 218: **Berns, Kenneth I.; Giraud, Catherine (Eds.):** Adeno-Associated Virus (AAV) Vectors in Gene Therapy. 1996. 38 figs. IX,173 pp. ISBN 3-540-61076-6

Vol. 219: **Gross, Uwe (Ed.):** Toxoplasma gondii. 1996. 31 figs. XI, 274 pp. ISBN 3-540-61300-5

Vol. 220: **Rauscher, Frank J. III; Vogt, Peter K. (Eds.):** Chromosomal Translocations and Oncogenic Transcription Factors. 1997. 28 figs. XI, 166 pp. ISBN 3-540-61402-8

Vol. 221: **Kastan, Michael B. (Ed.):** Genetic Instability and Tumorigenesis. 1997. 12 figs.VII, 180 pp. ISBN 3-540-61518-0

Vol. 222: **Olding, Lars B. (Ed.):** Reproductive Immunology. 1997. 17 figs. XII, 219 pp. ISBN 3-540-61888-0

Vol. 223: **Tracy, S.; Chapman, N. M.; Mahy, B. W. J. (Eds.):** The Coxsackie B Viruses. 1997. 37 figs. VIII, 336 pp. ISBN 3-540-62390-6

Vol. 224: **Potter, Michael; Melchers, Fritz (Eds.):** C-Myc in B-Cell Neoplasia. 1997. 94 figs. XII, 291 pp. ISBN 3-540-62892-4

Vol. 225: **Vogt, Peter K.; Mahan, Michael J. (Eds.):** Bacterial Infection: Close Encounters at the Host Pathogen Interface. 1998. 15 figs. IX, 169 pp. ISBN 3-540-63260-3

Vol. 226: **Koprowski, Hilary; Weiner, David B. (Eds.):** DNA Vaccination/Genetic Vaccination. 1998. 31 figs. XVIII, 198 pp. ISBN 3-540-63392-8

Vol. 227: **Vogt, Peter K.; Reed, Steven I. (Eds.):** Cyclin Dependent Kinase (CDK) Inhibitors. 1998. 15 figs. XII, 169 pp. ISBN 3-540-63429-0

Vol. 228: **Pawson, Anthony I. (Ed.):** Protein Modules in Signal Transduction. 1998. 42 figs. IX, 368 pp. ISBN 3-540-63396-0

Vol. 229: **Kelsoe, Garnett; Flajnik, Martin (Eds.):** Somatic Diversification of Immune Responses. 1998. 38 figs. IX, 221 pp. ISBN 3-540-63608-0

Vol. 230: **Kärre, Klas; Colonna, Marco (Eds.):** Specificity, Function, and Development of NK Cells. 1998. 22 figs. IX, 248 pp. ISBN 3-540-63941-1

Vol. 231: **Holzmann, Bernhard; Wagner, Hermann (Eds.):** Leukocyte Integrins in the Immune System and Malignant Disease. 1998. 40 figs. XIII, 189 pp. ISBN 3-540-63609-9

Vol. 232: **Whitton, J. Lindsay (Ed.):** Antigen Presentation. 1998. 11 figs. IX, 244 pp. ISBN 3-540-63813-X

Vol. 233/I: **Tyler, Kenneth L.; Oldstone, Michael B. A. (Eds.):** Reoviruses I. 1998. 29 figs. XVIII, 223 pp. ISBN 3-540-63946-2

Vol. 233/II: **Tyler, Kenneth L.; Oldstone, Michael B. A. (Eds.):** Reoviruses II. 1998. 45 figs. XVI, 187 pp. ISBN 3-540-63947-0

Vol. 234: **Frankel, Arthur E. (Ed.):** Clinical Applications of Immunotoxins. 1999. 16 figs. IX, 122 pp. ISBN 3-540-64097-5

Vol. 235: **Klenk, Hans-Dieter (Ed.):** Marburg and Ebola Viruses. 1999. 34 figs. XI, 225 pp. ISBN 3-540-64729-5

Vol. 236: **Kraehenbuhl, Jean-Pierre; Neutra, Marian R. (Eds.):** Defense of Mucosal Surfaces: Pathogenesis, Immunity and Vaccines. 1999. 30 figs. IX, 296 pp. ISBN 3-540-64730-9

Vol. 237: **Claesson-Welsh, Lena (Ed.):** Vascular Growth Factors and Angiogenesis. 1999. 36 figs. X, 189 pp. ISBN 3-540-64731-7

Vol. 238: **Coffman, Robert L.; Romagnani, Sergio (Eds.):** Redirection of Th1 and Th2 Responses. 1999. 6 figs. IX, 148 pp. ISBN 3-540-65048-2

Vol. 239: **Vogt, Peter K.; Jackson, Andrew O. (Eds.):** Satellites and Defective Viral RNAs. 1999. 39 figs. XVI, 179 pp. ISBN 3-540-65049-0

Vol. 240: **Hammond, John; McGarvey, Peter; Yusibov, Vidadi (Eds.):** Plant Biotechnology. 1999. 12 figs. XII, 196 pp. ISBN 3-540-65104-7

Vol. 241: **Westblom, Tore U.; Czinn, Steven J.; Nedrud, John G. (Eds.):** Gastroduodenal Disease and Helicobacter pylori. 1999. 35 figs. XI, 313 pp. ISBN 3-540-65084-9

Vol. 242: **Hagedorn, Curt H.; Rice, Charles M. (Eds.):** The Hepatitis C Viruses. 2000. 47 figs. IX, 379 pp. ISBN 3-540-65358-9

Vol. 243: **Famulok, Michael; Winnacker, Ernst-L.; Wong, Chi-Huey (Eds.):** Combinatorial Chemistry in Biology. 1999. 48 figs. IX, 189 pp. ISBN 3-540-65704-5

Vol. 244: **Daëron, Marc; Vivier, Eric (Eds.):** Immunoreceptor Tyrosine-Based Inhibition Motifs. 1999. 20 figs. VIII, 179 pp. ISBN 3-540-65789-4

Vol. 245/I: **Justement, Louis B.; Siminovitch, Katherine A. (Eds.):** Signal Transduction and the Coordination of B Lymphocyte Development and Function I. 2000. 22 figs. XVI, 274 pp. ISBN 3-540-66002-X

Vol. 245/II: **Justement, Louis B.; Siminovitch, Katherine A. (Eds.):** Signal Transduction on the Coordination of B Lymphocyte Development and Function II. 2000. 13 figs. XV, 172 pp. ISBN 3-540-66003-8

Vol. 246: **Melchers, Fritz; Potter, Michael (Eds.):** Mechanisms of B Cell Neoplasia 1998. 1999. 111 figs. XXIX, 415 pp. ISBN 3-540-65759-2

Vol. 247: **Wagner, Hermann (Ed.):** Immunobiology of Bacterial CpG-DNA. 2000. 34 figs. IX, 246 pp. ISBN 3-540-66400-9

Vol. 248: **du Pasquier, Louis; Litman, Gary W. (Eds.):** Origin and Evolution of the Vertebrate Immune System. 2000. 81 figs. IX, 324 pp. ISBN 3-540-66414-9

Vol. 249: **Jones, Peter A.; Vogt, Peter K. (Eds.):** DNA Methylation and Cancer. 2000. 16 figs. IX, 169 pp. ISBN 3-540-66608-7

Vol. 250: **Aktories, Klaus; Wilkins, Tracy, D. (Eds.):** Clostridium difficile. 2000. 20 figs. IX, 143 pp. ISBN 3-540-67291-5

Vol. 251: **Melchers, Fritz (Ed.):** Lymphoid Organogenesis. 2000. 62 figs. XII, 215 pp. ISBN 3-540-67569-8

Vol. 252: **Potter, Michael; Melchers, Fritz (Eds.):** B1 Lymphocytes in B Cell Neoplasia. 2000. XIII, 326 pp. ISBN 3-540-67567-1

Vol. 253: **Gosztonyi, Georg (Ed.):** The Mechanisms of Neuronal Damage in Virus Infections of the Nervous System. 2001. approx. XVI, 270 pp. ISBN 3-540-67617-1

Vol. 254: **Privalsky, Martin L. (Ed.):** Transcriptional Corepressors. 2001. 25 figs. XIV, 190 pp. ISBN 3-540-67569-8

Vol. 255: **Hirai, Kanji (Ed.):** Marek's Disease. 2001. 22 figs. XII, 294 pp. ISBN 3-540-67798-4

Vol. 256: **Schmaljohn, connie S.; Nichol, Stuart T. (Eds.):** Hantaviruses . 2001, 24 figs. XI, 196 pp. ISBN 3-540-41045-7

Vol. 257: **van der Goot, Gisou (Ed.):** Pore-Forming Toxins, 2001. 19 figs. IX, 166 pp. ISBN 3-540-41386-3

Vol. 258: **Takada, Kenzo (Ed.):** Epstein-Barr Virus and Human Cancer. 2001. 38 figs. IX, 233 pp. ISBN 3-540-41506-8

Vol. 259: **Hauber, J., Vogt, P. K. (Ed.):** Nuclear Export of Viral RNAs. 2001. 19 figs. IX, 131 pp. ISBN 3-540-41278-6